INTERNATIONAL ASTRONOMICAL UNION
UNION ASTRONOMIQUE INTERNATIONALE

SYMPOSIUM No. 36
HELD IN LUNTEREN, THE NETHERLANDS, 24–27 JUNE, 1969

ULTRAVIOLET STELLAR SPECTRA AND RELATED GROUND-BASED OBSERVATIONS

EDITED BY

L. HOUZIAUX

Département d'Astrophysique, Faculté des Sciences, Mons, Belgium

AND

H. E. BUTLER

Royal Observatory, Edinburgh, United Kingdom

D. REIDEL PUBLISHING COMPANY

DORDRECHT-HOLLAND

1970

Published on behalf of
the International Astronomical Union
by
D. Reidel Publishing Company, Dordrecht, Holland

Library of Congres Catalog Card Number 79–115887
ISBN-13: 978-94-010-3295-7   e-ISBN-13: 978-94-010-3293-3
DOI:10.1007/978-94-010-3293-3

To the memory of

ARMIN JOSEPH DEUTSCH

(1918–1969)

whose interest in stellar spectra of all wavelengths led to the

organizing of this symposium

INTRODUCTION

Two years ago, just before the Prague meeting of the International Astronomical Union, Armin Deutsch made the bold suggestion that the space spectroscopists hold a joint symposium with the ground-based observers. At that time the rocket observations of stellar spectra seemed too meagre to make such a meeting worthwhile, but we proceeded in the hope that there would be significant new results available by 1969.

IAU Commissions 29 and 44, on Stellar Spectra and Observations from Outside the Terrestrial Atmosphere respectively, agreed to sponsor the symposium so that the organization was given to the Joint Working Group of these commissions. Consequently, the Organizing Committee, which met first in Prague, consisted of A. Deutsch, M. W. Feast, L. Houziaux, V. G. Kurt, N. G. Roman, J. Sahade, A. B. Underhill, and R. Wilson, with myself as Chairman. Later COSPAR was invited to join in sponsoring the symposium and T. Chubb was added as their representative.

We were specially pleased when C. de Jager invited us to the Netherlands and offered the services of the Utrecht Observatory for the local organization. He suggested we hold the meeting at the new Lunteren Conference Centre located in a wooded area some 35 km east of Utrecht. The modern facilities of the Centre and the hospitality of its staff contributed much to the enjoyment of our four days there.

The main purpose of the symposium was to bring together the space astronomers working in the far-ultraviolet and ground-based astronomers studying related problems. For too long space astronomers have been set apart by the special techniques necessary to obtain their data with rockets and satellites. As scientists we really ought to centre our attention on the objects we are studying rather than the methods appropriate for a particular wavelength range. In fact as the emphasis in space astronomy turns away from techniques to the celestial objects being analysed, it might be appropriate to disband Commission 44 and let its members join commissions on stellar spectra, the Sun, or the interstellar medium.

In order to keep the symposium a manageable size, the range of topics was rather strictly limited. The Lyman continuum absorption by interstellar hydrogen provides a fundamental short-wavelength cutoff for most stellar spectroscopy so that X-rays were not directly on the programme. The Sun was given only minimal treatment because its ultraviolet spectrum has been discussed at many previous meetings and most of the new results concern resolution of the disk or measurements shortward of 912 Å, both observations very peculiar to the Sun. Discussion of instrumentation also was discouraged except for the very important problem of the absolute calibration of fluxes in both the ultraviolet and visual, so essential to relating the observations to theoretical stellar models. As a preview to the major session on stellar energy distributions it seemed appropriate to include reports on interstellar extinction and the nature of the grains. The new ultraviolet data on the extinction curve are essential for deducing the true energy distribution of a star as well as providing additional constraints on models of grains. Similarly, following the reports of stellar spectra a discussion of interstellar lines was included, to cover both the Lyman-α line that already is a prominent feature in the moderate-dispersion spectra of O and early B stars and the weaker lines that should normally appear with slightly higher dispersion. Finally, many ultraviolet space observations are influenced by the background emissions from stars and gas in the Galaxy, so that a session was devoted to this topic.

Among the most outstanding reports of the symposium were those by the representatives of the Smithsonian Astrophysical Observatory and the University of Wisconsin on the first results from the Orbiting Astronomical Observatory. Everyone knows the years of effort and many disappointments that preceded these spectacular observations. We are specially grateful to the astronomers of these institutions for taking time from their busy observing runs to prepare the material for presentation at Lunteren. At the same time we were just as excited to learn of the initial results of ultraviolet stellar observations by several other countries, including France, U.S.S.R. and the United Kingdom, which will complement the United States data. A Japanese astronomer already has published rocket stellar observations and we expect three or four more countries to have balloon or rocket results in the near future.

In several ways this conference may be a milestone in ultraviolet space astronomy. Firstly, as just mentioned, many countries are now contributing to the observations. Secondly, there has been a sudden increase in the rate data as obtained. As one reviewer put it, "In the next half hour the papers you hear will increase the available data by a factor of ten!" Finally, the space astronomers now have sufficient reliable ultraviolet measurements that we can talk usefully with the ground-based observers and the theoreticians must take account of both parts of the spectrum in constructing their models.

Several individuals and organizations should be mentioned for their contributions to this symposium. The IAU Executive Committee provided funds to assist several astronomers to attend. The Ministry of Education and Sciences of the Dutch Government and the Leids Kerkhoven-Bosscha Fonds provided further financial assistance for the conference. We are particularly grateful to the many members of the staff of the Utrecht Observatory who, under the direction of H. Lamers and J. B. Vogel, coordinated all the local arrangements so well. Most of all though, it was particularly gratifying to the Organizing Committee that so many astronomers were willing to come from all parts of the world to discuss their common interest in stars and the interstellar medium.

DONALD C. MORTON

1969 August 13,
Camp I at 5600 meters,
Noshaq Mountain, Afghanistan

TABLE OF CONTENTS

PART III / INTERSTELLAR ABSORPTION AND EMISSION

A. *Absorption Lines*

LIST OF PARTICIPANTS

C. Arpigny, Institut d'Astrophysique, Cointe-Sclessin, Belgium

C. A. Barth, Laboratory for Atmospheric and Space Physics, Boulder, Colo., U.S.A.

R. C. Bless, Washburn Observatory, Madison, Wis., U.S.A.

J. van Boeckel, European Space Research and Technology Center, Noordwijk, The Netherlands

T. den Boggende, Space Research Laboratory, Utrecht, The Netherlands

A. Boksenberg, University College, London, United Kingdom

G. Boldt, Max Planck-Institut für Physik und Astrophysik, Munich, Germany

R. M. Bonnet, Laboratoire de Physique stellaire et planétaire, Verrières-le-Buisson, France

J. Borgman, Kapteyn Astronomical Laboratory, Groningen, The Netherlands

B. Brinkman, Space Research Laboratory, Utrecht, The Netherlands

M. Burger, Astronomical Institute, Utrecht, The Netherlands

W. M. Burton, Culham Laboratory, Abingdon, United Kingdom

H. E. Butler, Royal Observatory, Edinburgh, United Kingdom

J. W. Campbell, Royal Observatory, Edinburgh, United Kingdom

A. M. Cantu, Osservatorio Astrofisico di Arcetri, Firenze, Italy

G. R. Carruthers, Naval Research Laboratory, Washington D.C., U.S.A.

T. A. Chubb, Naval Research Laboratory, Washington D.C., U.S.A.

D. D. Clark, European Space Research and Technology Center, Noordwijk, The Netherlands

J. Collet, European Space Research Organisation, Neuilly-sur-Seine, France

P. S. Conti, Lick Observatory, Santa Cruz, Calif., U.S.A.

G. Courtès, Laboratoire d'Astronomie spatiale, Marseille, France

R. J. Davis, Smithsonian Astrophysical Observatory, Cambridge, Mass., U.S.A.

J. Davis, Chatterton Astronomy Department, School of Physics, Sidney, Australia

J. P. Delaboudinière, Laboratoire de Physique stellaire et planétaire, Verrières-le-Buisson, France

A. J. Deutsch, Mount Wilson and Palomar Observatories, Pasadena, Calif., U.S.A.

N. A. Dimov, Crimean Astrophysical Observatory, Nauchny, Crimea, U.S.S.R.

R. J. van Duinen, Kapteyn Astronomical Laboratory, Groningen, The Netherlands

J. Emming, Space Research Laboratory, Utrecht, The Netherlands

F. Engström, Stockholm Observatory, Saltsjobaden, Sweden

R. Faraggiana, Osservatorio, Trieste, Italy

M. W. Feast, Radcliffe Observatory, Pretoria, South Africa

L. de Feiter, Space Research Laboratory, Utrecht, The Netherlands

K. Fredga, Astronomical Institute, Utrecht, The Netherlands

M. Fulchignoni, Laboratorio di Astrofisica, Frascati, Italy

O. Gingerich, Smithsonian Astrophysical Observatory, Cambridge, Mass., U.S.A.

L. Gratton, Laboratorio di Astrofisica, Frascati, Italy

J. M. Greenberg, Sterrewacht te Leiden, Leiden, The Netherlands

H. G. Groth, Universitäts-Sternwarte, Munich, Germany

M. Hack, Osservatorio Trieste, Italy

L. Hansen, University Observatory, Copenhagen, Denmark

W. Haupt, Astronomisches Institut, Ruhr-Universität, Bochum, Germany

D. Hayes, Rensselaer Polytechnic Institute, Troy, N.Y., U.S.A.

A. G. Hearn, Observatoire de Nice, France

J. R. Heintze, Astronomical Institute, Utrecht, The Netherlands

J. Hekela, Observatory Ondrejov, Czechoslovakia

K. G. Henize, NASA Manned Space Flight Center, Houston, Tex., U.S.A.

G. H. Herbig, Lick Observatory, Santa Cruz, Calif., U.S.A.

C. Heynekamp, Space Research Laboratory, Utrecht, The Netherlands

A. M. Hieronimus, Centre National d'Études Spatiales, Bretigny-sur-Orge, France

L. Houziaux, Département d'Astrophysique, Faculté des Sciences, Mons, Belgium

H. C. van de Hulst, Sterrewacht te Leiden, Leiden, The Netherlands

C. M. Humphries, Royal Observatory, Edinburgh, United Kingdom

J. Hutchings, Dominion Astrophysical Observatory, Victoria, British Columbia, Canada

C. de Jager, Astronomical Institute, Utrecht, The Netherlands

E. B. Jenkins, Princeton University Observatory, Princeton, N.J., U.S.A.

C. Jordan, Culham Laboratory, Abingdon, United Kingdom

H. E. Jorgensen, University Observatory, Copenhagen, Denmark

B. Kovatschev, Bulgarian Academy of Sciences, Department of Astronomy, Sofia, Bulgaria

L. V. Kuhi, Berkeley Astronomy Department, Berkeley, Calif., U.S.A.

V. G. Kurt, Sternberg Astronomical Institute, Moscow, U.S.S.R.

H. J. Lamers, Astronomical Institute, Utrecht, The Netherlands

M. Lehmann, Observatoire de Genève, Sauverny, Switzerland

P. Lemaire, Laboratoire de Physique stellaire et planétaire, Verrières-le-Buisson, France

M. C. Lortet, Institut d'Astrophysique, Paris, France

D. Malaise, Institut d'Astrophysique, Cointe-Ougrée, Belgium

L. Meredith, Goddard Space Flight Center, Greenbelt, Md., U.S.A.

D. C. Morton, Princeton University Observatory, Princeton, N.J., U.S.A.

E. A. Müller, Observatoire de Genève, Sauverny, Switzerland

C. Navach, Observatoire de Genève, Sauverny, Switzerland

J. B. Oke, California Institute of Technology, Pasadena, Calif., U.S.A.

J. H. Oort, Sterrewacht te Leiden, Leiden, The Netherlands

N. Panagia, Laboratorio di Astrofisica, Frascati, Italy

E. Peytremann, Observatoire de Genève, Sauverny, Switzerland

S. R. Pottasch, Kapteyn Astronomical Laboratory, Groningen, The Netherlands

F. Praderie, Observatoire de Meudon, Meudon, France

V. K. Prokofiev, Crimean Astrophysical Observatory, Nauchny, Crimea, U.S.S.R.

J. Provost, Observatoire de Nice, France

N. K. Reay, Culham Laboratory, Abingdon, United Kingdom

J. B. Rogerson Jr, Princeton University Observatory, Princeton, N.J., U.S.A.

N. G. Roman, NASA Headquarters, Washington D.C., U.S.A.

R. J. Rutten, Astronomical Institute, Utrecht, The Netherlands

D. Samain, Laboratoire de Physique stellaire et planétaire, Verrières-le-Buisson, France

M. Schneider, Observatoire de Nice, France

A. Severny, Crimean Astrophysical Observatory, p/o Nauchny, Crimea, U.S.S.R.

A. M. Smith, Goddard Space Flight Center, Greenbelt, Md., U.S.A.

A. Snijders, Astronomical Institute, Utrecht, The Netherlands

L. Spitzer Jr, Princeton University Observatory, Princeton, N.J., U.S.A.

P. Solomon, Columbia University, Department of Astronomy, New York, N.Y., U.S.A.

T. P. Stecher, Goddard Space Flight Center, Greenbelt, Md., U.S.A.

G. C. Sudbury, Royal Observatory, Edinburgh, United Kingdom

R. A. Sunyaev, Institute of Applied Mathematics, Moscow, U.S.S.R.

J. P. Swings, Institut d'Astrophysique, Cointe-Ougrée, Belgium

G. L. Tagliaferri, Observatorio Astrofisico di Arcetri, Firenze, Italy

V. M. Tiyt, Institute of Physics and Astronomy, Tartu, Estonia, U.S.S.R.

A. B. Underhill, Astronomical Institute, Utrecht, The Netherlands

J. C. Vial, Laboratoire de Physique stellaire et planétaire, Verrières-le-Buisson, France

R. Viotti, Laboratorio di Astrofisica, Frascati, Italy

H. Visser, Astronomical Institute, Utrecht, The Netherlands

M. Viton, Laboratoire d'Astronomie spatiale, Marseille, France

G. Walker, University of British Columbia, Vancouver, B.C., Canada

R. M. West, University Observatory, Copenhagen, Denmark

N. C. Wickramasinghe, Institute of Theoretical Astronomy, University of Cambridge, Cambridge, United Kingdom

R. Wilson, Culham Laboratory, Abingdon, Berkshire, United Kingdom

H. van Woerden, Kapteyn Astronomical Laboratory, Groningen, The Netherlands

PART I

STELLAR FLUXES

A. ABSOLUTE CALIBRATION

ABSOLUTE INTENSITY CALIBRATION METHODS
IN THE VACUUM UV REGION

G. BOLDT[†]

Max-Planck-Institut für Physik und Astrophysik,
Institut für extraterrestrische Physik,
8046 Garching b. München, W. Germany

Abstract. As a summary of the principal results presented at the ESRO symposium on Calibration Methods in the Vacuum Ultra Violet (Munich, 1968) a description is given of three different absolute intensity calibration methods. These are the branching ratio method, the synchrotron radiation method and the black-body radiation method, and they define the present state of the art.

1. Introduction

Since space technology opened the possibility of observing stars and stellar systems in the ultraviolet, the question of intensity calibration in that region has become of special interest to astronomers. As a result of this interest, the members of the ESRO Working Group on 'Stars and Stellar Systems' under the chairmanship of Prof. Gratton agreed at the end of 1967 to organize a symposium on calibration methods in the UV in order to find out the present state of the art. This meeting was held in Munich in May 1968 and gave a good idea about what is going on in this field at different laboratories [1].

When considering intensity calibration problems one has to discuss two questions, namely how to produce an intensity standard and how to calibrate an astronomical instrument with the help of the intensity standard. The first of these two points is the more fundamental problem and in accordance with its aim, the symposium was more or less related to this point.

An intensity standard may be realized by a source or by a detector, but in any case it has to fulfil the following conditions:

(1) All parameters which influence the intensity of the source, or as the case may be the efficiency of the detector, must be known.

(2) The dependence of the intensity or of the efficiency on these parameters has to be known.

(3) It must be possible to adjust and to check all these parameters in a simple way, at any time, with tolerances that guarantee an acceptable precision in intensity or efficiency.

Starting from these conditions the study of all so-called intensity standards leads to the result that development based on sources should be preferred to one based on detectors. Therefore the following description of some intensity standards takes only sources into consideration.

[†] Deceased January 18, 1970.

Houziaux and Butler (eds.), Ultraviolet Stellar Spectra and Ground-Based Observations, 5–11.

2. Production of Intensity Standards

The present state of the art in the field of intensity standards is defined by three different methods which are all based on sources. These are:

(1) The branching ratio method (the first description of the method was given by Griffin and McWhirter from the Culham group).

(2) The synchrotron radiation method (here the work which has been done at DESY in Hamburg in cooperation with the Heidelberg group by Haensel, Labs, Lembke and Pitz has especially to be mentioned).

(3) The black-body radiation method (which has been developed by Boldt in Munich).

A. THE BRANCHING RATIO METHOD

Figure 1 illustrates the principle of this method and, as an example of its application, the calibration of a UV monochromator carried out by McWhirter and co-workers in Culham.

In principle one needs two spectral lines coming from the same upper level. One of these lines must lie in the visible or near UV of the spectrum, where its intensity can be measured by comparison with an approved intensity standard (tungsten ribbon lamp or carbon arc). The other line must lie in the vacuum UV. If the ratio between the transition probabilities of these two lines is known and if, furthermore, the plasma emitting these lines is optically thin for both lines so that proportionality between intensity and transition probability is known, then the intensity of the UV line can be calculated.

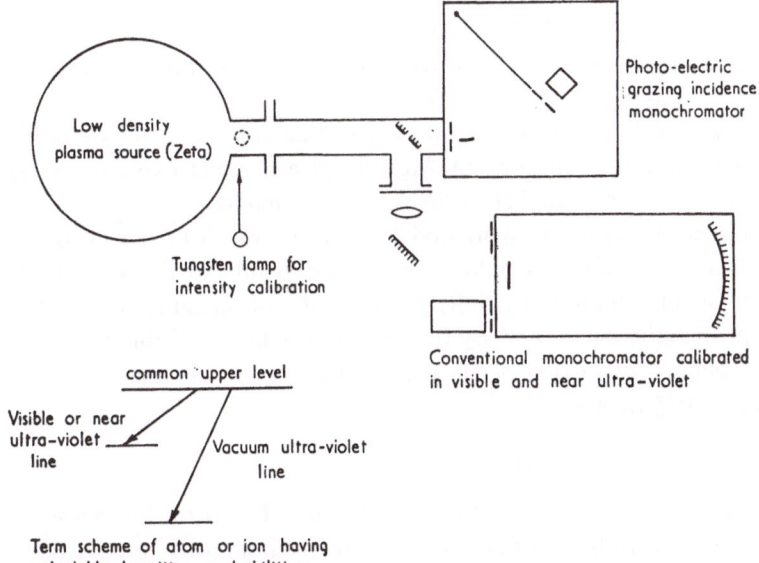

Fig. 1.   The principle of the branching ratio method and an example
of its application.

The instrumental equipment for the application of this method uses ZETA as the source, with a grazing incidence monochromator for the vacuum UV and a conventional monochromator for the visible and near UV. Light coming from the plasma is split into two beams by a mirror system, so that the UV line and the visible line can be measured simultaneously. After this the plasma source is replaced by a tungsten ribbon lamp which calibrates the monochromator for the visible light.

TABLE I

List of lines used in the branching ratios calibration

Transition	λ (Å)	w (upper)	A (sec^{-1})
H 1–3	1025.72	18	5.575×10^7
2–3	6562.80	18	4.410×10^7
He II 1–4	243.03	32	1.278×10^{7a}
2–4	1215.1	32	8.419×10^{6a}
3–4	4685.68	32	8.986×10^{6a}
He II 1–5	237.33	50	4.125×10^{6a}
3–5	3203.15	50	2.201×10^{6a}
He II 1–3	256.32	18	5.575×10^{7a}
2–3	1640.4	18	4.410×10^{7a}
C IV 2s–3p	312.43	6	4.56×10^9
3s–3p	5801.51	4	3.19×10^7
N V 2s–3p	209.28	6	1.20×10^{10}
3s–3p($P_{3/2}$)	4603.83	4	4.15×10^7
O VI 2s–3p	150.10	6	2.59×10^{10}
3s–3p($P_{3/2}$)	3811.35	4	5.10×10^7

[a] Hydrogen values. These should be multiplied by $Z^4 = 16$ for He II.

Table I gives a list of lines which have been used for this calibration method by McWhirter and co-workers.

It is obvious that ZETA can be replaced by other sources, e.g. by a hollow cathode lamp (this has already been done by McWhirter and other experimenters). Furthermore, we can expect that the list of lines will be extended.

In any case the branching ratio method is an elegant one for transfering the approved methods of absolute intensity calibration in the visible to the vacuum UV region.

The error of this method is mainly given by the uncertainty in our knowledge of transition probability ratios and by the uncertainty in the visible intensity standard. The impression is that further development of this method should finally lead to an accuracy of $\pm 10\%$ to $\pm 20\%$.

B. THE SYNCHROTRON RADIATION METHOD

Our knowledge about the synchrotron radiation is based on the Schwinger theory, which describes the radiation being emitted by a circulating relativistic electron. A number of investigations have shown that this theory can be applied to synchrotron radiation with very good accuracy. Consequently the relative intensity distribution mainly depends on the radius of the synchrotron and on the end energy of the electron and can therefore be calculated.

Figure 2 shows the result of such calculations applied to the DESY machine and taking into consideration special observational conditions. As one can see, there is only a very weak dependence on the end energy of the electron. This is also true for shorter wavelengths outside this diagram. The reason for this is the very high energy of the synchrotron, by which the maximum intensity is shifted to very short wavelengths. The maxima of the curves shown in the diagram lie between 0.3 and 30 Å.

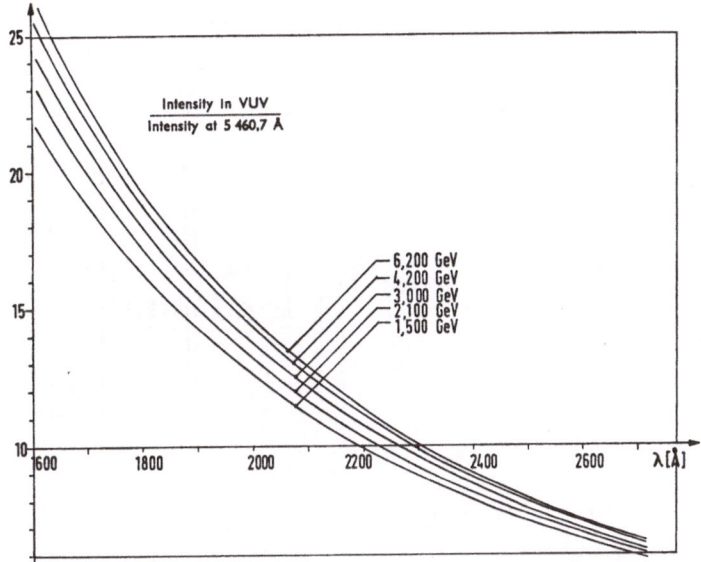

Fig. 2.   Relative intensity distribution of the DESY synchrotron radiation (calculated for special observational conditions).

Hence exact knowledge of the energy is not critical if one restricts the observation of the synchrotron radiation to the vacuum UV. Consequently the synchrotron is a very excellent source for relative intensity calibrations in the range between the visible and vacuum UV.

In order to get absolute intensity values of the synchrotron radiation one has to measure the number of electrons in the synchrotron, which is proportional to the absolute intensity of the radiation. The accuracy of such measurements is not better than ±20 to ±25%. This means that the error in the absolute intensity of the synchrotron radiation will also be at least ±20 to ±25%.

Better results can be obtained by absolute calibration of the synchrotron radiation in the visible with the help of an approved intensity standard source (e.g. tungsten ribbon lamp). Because of its very good quality as a relative intensity standard, the synchrotron can then transfer the absolute calibration from the visible to the vacuum UV. This procedure has been carried out at DESY.

The accuracy which can be achieved in this way for absolute vacuum UV intensities can be estimated to be about $\pm 10\%$.

C. THE BLACK-BODY RADIATION METHOD

The source which can produce black-body radiation is a wall-stabilized cascade arc such as is shown in Figure 3 (see reference for further details). The plasma in the arc-channel having a pressure of 1 atm is thermal and consists mainly of nitrogen and carbon. These two elements emit a certain number of optically thick lines in the

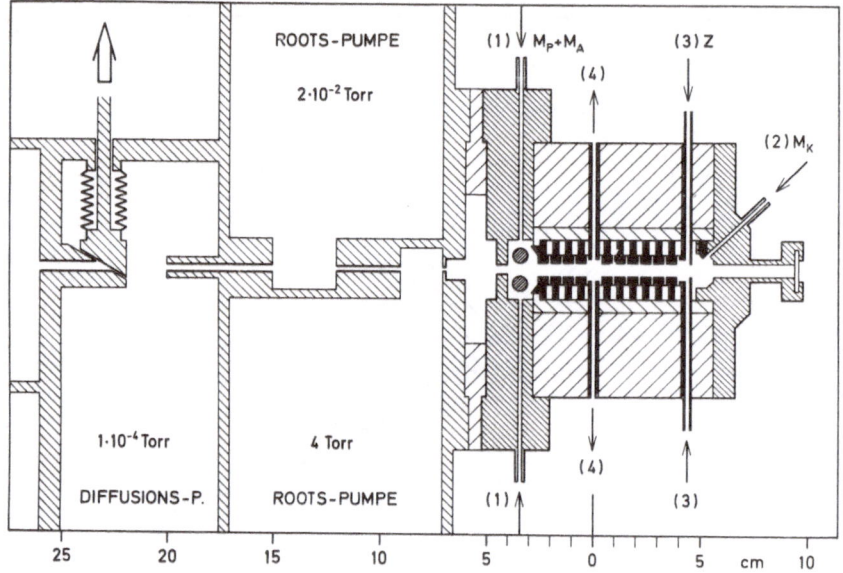

Fig. 3. Wall-stabilized cascade arc for producing black-body radiation in the far UV.

region from 1100 to 2500 Å. These lines can leave the plasma without disturbance by absorption and self-absorption via an argon window, which is realized by an argon plasma flow from the end of the arc-channel to the gas outlet in the middle part of the channel.

Since the absolute intensity in the centre of the optically thick lines is given by the Planck function, one has only to measure the plasma temperature in order to get the absolute intensities. This has been done using four different spectroscopic methods. The result was 12540 K with an accuracy of ± 1 to $\pm 2\%$. Consequently the black body intensities are known with an accuracy of ± 10 to $\pm 20\%$.

The arc plasma is connected to the vacuum system by a three stage differential pumping system.

Figure 4 shows the Planck function for 12540 K, which is the temperature of the arc plasma. On this curve the positions of the observed optically thick lines are indicated. The interpolation between these points provides the absolute intensity scale

G. BOLDT

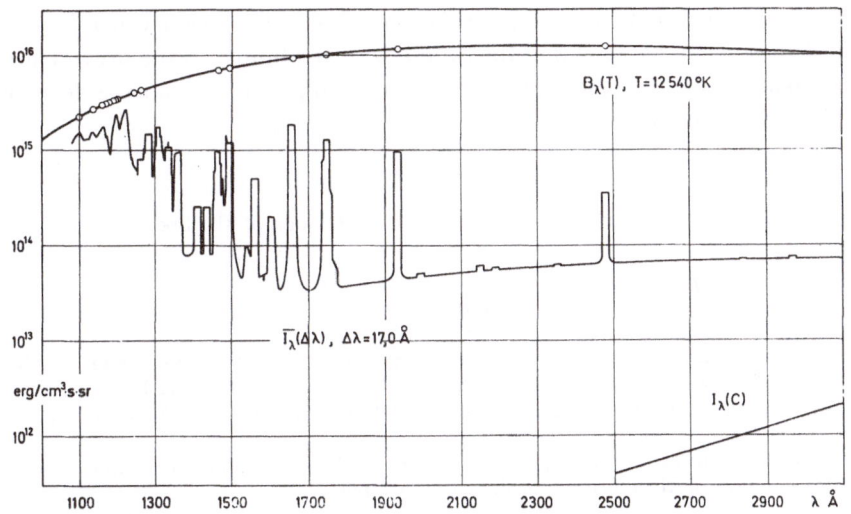

Fig. 4. Planck function of 12 540 K, absolute intensity distribution in the arc spectrum and absolute intensity distribution in the spectrum of the carbon arc.

at any point of the spectrum between 1100 and 2500 Å. This enables us to evaluate the absolute intensity distribution of the whole arc spectrum between 1100 and 2500 Å. The curve denoted by $I_\lambda(\Delta\lambda)$ shows the result of this operation, where a spectral resolution of 17.0 Å has been chosen. It stands to reason that this can be done also for other resolutions.

Attention should be drawn to the continuum between 1700 and 3100 Å. The intensity of this continuum has been measured between 1700 and 2500 Å, put into a potential law, and then extrapolated up to 3100 Å. Since no remarkable jump of the continuum intensity is to be expected in this region, this extrapolation is acceptable.

Thus one has finally obtained the absolute intensity distribution in the whole arc spectrum between 1100 and 3100 Å without using any other intensity standard.

The curve in the lower part of Figure 4, denoted by $I_\lambda(C)$, gives the absolute intensity distribution of the carbon arc, which is a well-known intensity standard in the visible and near UV. The comparison of the radiation of both sources in the overlapping region between 2500 and 3100 Å is consistent within limits of 1% at 2500 Å and 5% at 3100 Å. From this result one may gather that the estimated accuracy of the absolute intensity in the arc spectrum (± 10 to $\pm 20\%$) is realistic or perhaps even a little too pessimistic.

3. Conclusion

The fundamental problem in the field of absolute intensity calibration is the development of primary standards. The present state of the art in this field is defined by three different methods, which are all based on sources and which should be able to guarantee an accuracy of the order of $\pm 10\%$.

The next aim should be to compare these methods with each other in order to find out if they really agree within the estimated error limits. In the present situation these comparison experiments are of much more importance than the development of other intensity standards.

Reference

[1] 'Calibration Methods in the Ultraviolet and X-Ray Regions of the Spectrum', ESRO SP-33, December 1968.

Discussion

Malaise: In order to use the monochromatic flux as a calibrated source, you should be able to put the two curves (B_λ and actual output) of the figure on the same intensity scale. To do so, you have to measure the efficiency of the spectrograph at least for one wavelength, can you indicate how this was done?

Boldt: All curves shown in the diagram are related to the source-intensity-output. They do not give any information on the efficiency of the monochromator.

AN INVESTIGATION OF THE PROPERTIES OF VACUUM-ULTRAVIOLET RADIATION DETECTORS

VALDUR M. TIYT

Academy of Sciences of Estonian S.S.R., U.S.S.R.

Abstract. At the Institute of Physics and Astronomy of the Academy of Sciences of the Estonian S.S.R. there is a laboratory for the research on the properties of detectors for vacuum ultraviolet radiation. The laboratory has also the necessary measuring equipment, including a beam double vacuum monochromator for the wavelengths 1000–2700 Å with a balancing-out system of registration attached to it. The last-named device automatically eliminates effects on the results of measurement caused by the distribution of energy in the spectrum of the light source as well as by absorption in the apparatus.

Earlier the absolute efficiency of detectors was determined by means of a thermocouple but at present it is calculated from the data obtained from the photoionisation cross sections of gases.

The present report throws light on some results of the investigation of the properties of radiation detectors under different operating conditions.

Discussion

Campbell:

(1) In your use of sodium salicylate to transfer your absolute quantum efficiency measurements to other wavelengths, what evidence do you have of the uniformity of the quantum efficiency with wavelength?

(2) Have you measured the quantum efficiency of the sodium salicylate at wavelength shorter than 2537 Å?

Tiyt:

(1) We have not directly investigated the constancy with wavelength of the quantum efficiency of sodium salicylate. But we suppose that if we have a fresh layer of sodium salicylate the quantum efficiency is constant. In course of calibration of the two-beam monochromator we compare fresh and old sodium salicylate layers in order to detect if there is some variation with time.

(2) I have calibrated the screen by wavelength 2537 Å with a thermocouple.

Wilson: Have you made a direct comparison between your ionization chamber and thermocouple? You can do this by observing a source in the pass band of the former, using the latter as an energy detector.

Tiyt: No, I have not made a direct comparison between ionization chamber and thermocouple.

Houziaux and Butler (eds.), Ultraviolet Stellar Spectra and Ground-Based Observations, 12.
All Rights Reserved. Copyright © *1970 by the IAU.*

RECENT ABSOLUTE CALIBRATION WORK
AT PALOMAR MOUNTAIN

J. B. OKE and R. SCHILD

*Mount Wilson and Palomar Observatories, Carnegie Institution of Washington,
and California Institute of Technology, U.S.A.*

Abstract. A four inch hermetically sealed reflecting telescope has been attached to the prime-focus scanner to provide an observing system for absolute calibration work. Light sources, including a platinum and a copper furnace, are observed at a distance of approximately 1200 feet. Since the telescope focal length is only 13 inches these sources look like stars. Thus bright stars such as α Lyr are compared directly with the light sources under identical instrumental conditions.

It has been found that horizontal extinction can be abnormally large; therefore it is measured several times each night. The absolute calibration is being derived at present from 3500 Å to 11 000 Å and will be extended to 3200 Å as soon as possible. The calibration lies approximately midway between that measured recently by Hayes and that adopted some years ago by Oke. A very preliminary value for the absolute flux from α Lyr at 5556 Å is 3.5×10^{-20} ergs sec^{-1} cm^{-2} Hz^{-1} or 3.4×10^{-9} ergs sec^{-1} cm^{-2} Å^{-1}.

1. Introduction

The largest inherent uncertainty in the measurement of absolute fluxes and spectral energy distributions of astronomical objects is the absolute calibration of the primary standard star, α Lyr. A summary of the early work, which was largely photographic, has been given by Code (1960). More recent photoelectric absolute calibration work has been reported by Bahner (1963), Wolff *et al.* (1968), Haritonov (1963) and Willstrop (1965). These photoelectric studies make use of calibrated ribbon filament lamps and they disagree with each other in some cases by as much as 10%.

The present work was undertaken to try to improve the absolute calibration of α Lyr. To achieve the highest possible accuracy the following decisions were made: (1) The most stable possible photoelectric system, telescope, and spectrometer were built and used. (2) The equipment was installed at an excellent photometric site, namely Palomar Mountain. (3) Absolute black-body light sources were used directly, eliminating the need for calibrated strip lamps which must be calibrated against the black body sources.

2. Light Sources

Two light sources have been used. These are black bodies generated by either a platinum or copper furnace. In both cases the observed radiation comes from the base of a sight tube which is immersed in a crucible of metal. Care is taken to insure that radiation is received only from the bottom of the sight tube and not from the cooler walls. All observations are made during the stable temperature plateaux which occur while the platinum or copper is in the process of melting or freezing. Apart from small corrections for emissivity, etc., the radiation is that of a black body at the melting point of the metal being used in the crucible. The platinum is heated by means of a 10

Houziaux and Butler (eds.), Ultraviolet Stellar Spectra and Ground-Based Observations, 13–17.
All Rights Reserved. Copyright © 1970 by the IAU.

kilowatt RF induction generator which heats the platinum directly. The copper is heated by means of a simple electric furnace. Both the platinum and copper black body sources follow closely the designs of the National Bureau of Standards (Wensel *et al.*, 1931; Lee, 1969).

The melting point of copper is very close to that of gold which provides the fundamental standard at high temperatures. The melting point is taken to be 1357.9 K which is on the new 1969 practical temperature scale which has been defined to agree as closely as possible with the thermodynamic temperature scale. Because the copper melting point is low, this source could only be observed from 6300 Å to 11 000 Å. The uncertainty in the temperature, which is much less than 1 K, introduces uncertainties in the absolute fluxes in this wavelength range of only a few tenths of 1% i.e., negligible compared with most of the other errors.

The temperature of melting platinum on the new 1969 practical-temperature scale is quoted as 2045 K. However, the uncertainty in this number is approximately 2°. The platinum black body is sufficiently hot that it can be observed from 3400 Å to 11 000 Å and therefore is also useful in the blue and violet.

3. Observing Equipment

Since both the furnaces and the star α Lyr have to be observed with the same equipment, it is necessary to have a suitable telescope. The telescope used is a 4 inch diameter $f/3.3$ Newtonian reflector. The telescope has quartz windows and is hermetically sealed so that the mirrors do not become dusty or age appreciably. Since the focal length is only 13 inches, the black bodies, which are approximately 1200 feet from the telescope, look like star images. The light from the telescope is analyzed using the prime-focus scanner from the 200-inch telescope, operated as a monochromator. The monochromatic output light is measured by photomultiplier tubes and high-speed pulse counting circuitry.

Since the spectrum scanner is a single-pass grating instrument great care must be taken to eliminate overlapping orders and scattered light. Various sets of glass filters are used for this purpose. Tests show that with the chosen filter combinations contamination is approximately 1% or less and suitable corrections can be made.

The telescope and spectrometer are mounted in the 18-inch dome at Palomar, using the 18-inch Schmidt telescope as a polar-axis mount. Observations can readily be made of both the light sources and bright stars such as α Lyr.

4. Atmospheric Extinction

The star or stars being observed are measured several times during the night at various zenith distances. These measures can be used to determine the atmospheric extinction at one wavelength relative to some standard wavelength with an accuracy of approximately 1%. The absolute extinction at say 5556 Å, is somewhat more difficult to determine.

A much more serious problem is the horizontal extinction between the light source and the telescope. If the atmosphere is homogeneous the horizontal extinction is $\frac{1}{20}$ of the vertical extinction and is never more than a few per cent. Unfortunately there can sometimes be haze near the ground which increases the horizontal extinction by large amounts but has practically no effect on vertical extinction. During each night when observations were being obtained an attempt was made to measure horizontal extinction by observing a quartz iodide lamp mounted first near the light source, and then on a tower half way between the telescope and light source. This technique has been most useful in detecting changes in the extinction, but also shows that the extinction on very good nights is not abnormal. Results given below were obtained only on the clearest possible nights.

5. Observations

On any given night observations were made either with a blue sensitive photomultiplier tube from 3500 Å to 6100 Å or with a red tube from 5800 Å to 11 000 Å. The band passes were 25 Å and 100 Å respectively for the blue and red observations. The furnace being used was run almost all night and observations were made on a great

TABLE I

Observational data

Date (1969)	Wavelength range	Exit slit	Furnace
April 30–May 1	5840 Å–11000 Å	100 Å	Copper
May 1–2	5840 Å–11000 Å	100 Å	Platinum
May 23–24	3500 Å– 6100 Å	25 Å	Platinum
May 24–25	3500 Å– 6100 Å	25 Å	Platinum

many plateaux. All wavelengths being used could normally be measured at least once during a melt-plus-freeze cycle. Interspersed among the furnace measurements were observations of α Lyr and the quartz iodide lamp being used to measure extinction. The results given here are based on observations outlined in Table I. Observations had been made on many nights prior to those listed, to improve techniques.

6. Results

As indicated above the copper furnace provides an excellent radiation source for wavelengths above 6300 Å. The uncertainty in the platinum point is at least 2 K and there are further uncertainties caused by the geometry of the furnace. The actual radiating temperature of the platinum furnace can be estimated in two ways: (1) The absolute fluxes from the copper and platinum furnaces can be measured and compared at some specific wavelength which in our case is 6370 Å. This provides essentially a photo-electric pyrometer measurement. (2) The relative colors of the two sources can be compared over the wavelength range from 6370 Å to 11 000 Å. The former measurement is the more sensitive of the two. Both (1) and (2) indicate that the radiating temperature at the platinum point is 2038 K with an uncertainty of approximately 3 K.

This low temperature is partially due to the fact that the sight tube does not provide a true thermodynamic enclosure because (a) there is an aperture to the outside world and (b) the upper part of the sight tube is at a lower temperature than the platinum point. De Vos (1954) has calculated the effects of using a cylindrical sight tube open at one end. Using a depth to radius value of 15, the effective emissivity is 0.995 for diffuse reflection. This leads to a temperature correction of nearly 3 K. In fact, the geometry

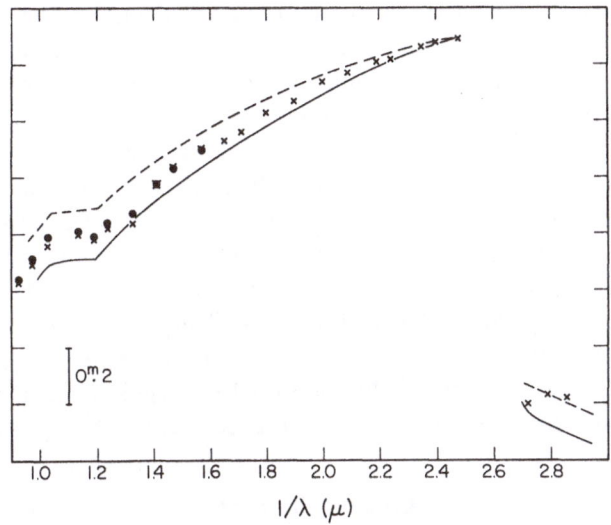

Fig. 1. The absolute spectral energy distribution of α Lyr. The ordinate is in magnitudes or −2.5 log f_ν + const. Results from the copper furnace are shown by solid dots and those from the platinum furnace by crosses. The calibration of Hayes is shown by the solid curve, that adopted by Oke in 1964 by the broken curve.

used in the present furnace is somewhat different from that for which the calculations were made, and the thorium oxide sight tube may now be strictly a diffuse reflector. But, since the corrections appear to be of approximately the right size, the experimentally derived radiating temperature of 2038 K at the platinum point is adopted.

Preliminary results are shown in Figure 1. Here the absolute flux in terms of magnitudes ($-2.5 \log f_\nu$) where f_ν is the flux in ergs sec^{-1} cm^{-2} Hz^{-1} is plotted against $1/\lambda$ where λ is the wavelength in microns. The results from the copper and platinum furnaces are made to coincide at $1/\lambda = 1.54$. For comparison the calibration determined by Wolff *et al.* (1968) and that adopted by Oke (1964) are also shown. These have been normalized at $1/\lambda = 2.48$.

Our calibration is between that used by Oke and that of Hayes. It should here be emphasized that these results are still preliminary, particularly below the Balmer jump, since (a) further observations will be made and (b) small corrections, totalling approximately 1 to 2% are not yet included. These results, particularly as they bear on the melting point of platinum and the Balmer jump, are being checked through observation of another standard source in the form of a tungsten ribbon lamp which has been calibrated at the United States National Bureau of Standards.

It is possible to make a preliminary estimate of the accuracy of the data. As noted above, the uncertainty in the melting point of copper produces no significant error from 6370 Å to 11 000 Å. Below 6370 Å, where the platinum calibration must be used, an error of 3 K in the adopted radiating temperature at the melting point introduces an error of 1.5% at 3500 Å relative to 6370 Å and 1% at 4167 Å relative to 6370 Å. The measurements themselves have an intrinsic inaccuracy, when finally averaged, of less than 1% and the atmospheric extinction for α Lyr is uncertain by approximately 1% above 4000 Å and 2% below this wavelength. The horizontal extinction on the nights actually used for calibration should be accurate to better than 1%. Combining the above errors suggests that the relative fluxes from 4036 Å to 11 000 Å are not in error by more than 3%, below 4036 Å the uncertainty is approximately 4%.

The fluxes for α Lyr shown in Figure 1 are all relative to some wavelength. To obtain a true absolute flux it is necessary to specify the absolute flux at some particular wavelength. The furnace sources are ideal for this since they are in fact very nearly black bodies. Using the copper furnace results we find for α Lyr at a wavelength of 5556 Å ($1/\lambda = 1.80$) a preliminary flux of 3.5×10^{-20} ergs sec^{-1} cm^{-2} Hz^{-1} or 3.4×10^{-9} ergs sec^{-1} cm^{-2} Å^{-1}. The uncertainty is approximately 3%. This differs markedly from the corresponding value of 3.8×10^{-9} ergs sec^{-1} cm^{-2} Å^{-1} adopted by Code (1960) and the value of 3.7×10^{-9} ergs sec^{-1} cm^{-2} Å^{-1} given by Willstrop (1965).

Acknowledgement

This work was supported by the Office of Naval Research through Contract No. N00014-67-A-0094-0005.

References

Bahner, K.: 1963, *Astrophys. J.* **138**, 1314.
Code, A. D.: 1960, in *Stars and Stellar Systems*, Vol. VI, Stellar Atmospheres (ed. by J. L. Greenstein). University of Chicago Press, Chicago, Chap. 2, p. 50.
De Vos, J. C.: 1954, *Physica* **20**, 669.
Haritonov, A. V.: 1963, *Astron. Zh.* **40**, 339 (*Soviet Astron.* **7**, 258).
Lee, R. D.: 1969, preprint.
Oke, J. B.: 1964, *Astrophys. J.* **140**, 689.
Wensel, H. T., Roeser, W. F., Barbrow, L. E., and Caldwell, F. R.: 1931, *N.B.S. J. Res.* **6**, 1103.
Willstrop, R. V.: 1965, *Mem. Roy. Astron. Soc.* **69**, 83.
Wolff, S. C., Kuhi, L. V., and Hayes, D.: 1968, *Astrophys. J.* **152**, 871.

Discussion

Morton: It is important to realize how important this absolute visual calibration is for interpretation of the ultraviolet data. We must know the energy corresponding to a particular apparent visual magnitude to obtain the observed ratio of ultraviolet to visual fluxes for comparing the observations with the model atmospheres.

GENERAL DISCUSSION

Wilson: How did you deal with the peculiar geometry of the synchrotron emission? Were you able to fill your photometer or was it necessary to build up the calibration by a spatial scan?

Bless: We do not collimate the radiation but rather carry out a point-by-point calibration over the aperture of the detector system.

We calibrate small filter photometers as a whole in the synchrotron radiation beam on a point-by-point basis since the radiation bundle is not collimated. These photometers are to be flown in Aerobees to check the OAO calibration.

Carruthers: The early thermocouple calibrations of sodium salicylate indicated a 15% lower response at 584 Å relative to 1216 Å. We recently have also confirmed this by comparison of measurements using a windowless, parallel-plate standard ionization chamber, with argon (100% photoionization at 584 Å) and with methyl iodide (87% at 1216 Å), with measurements of the same lines using a sodium-salicylate coated photomultiplier.

Davis: The Smithsonian Astrophysical Observatory has used both sodium salicylate and thermocouples in deriving our Celescope-OAO calibration standards. The quantum efficiency of sodium salicylate depends upon deposition techniques, and should not be relied upon as a standard. Our comparison with the Wisconsin calibration gives agreement to within 3% at all wavelengths between 1000 and 3000 Å. We use standard lamps to maintain the calibration of our thermocouples and other standard detectors.

Bless: I would like to mention work being done by Tim Fairchild of our group in Wisconsin. We use as a standard source synchrotron radiation from a storage ring rather than from a synchrotron which in our case simply injects a batch of electrons into the ring at 50 MeV. The ring then accelerates the electrons to any energy up to 245 MeV. The advantages of a storage ring over a synchrotron are the long lifetimes (hours) of the particles in the ring thus providing a stable source of UV radiation; most importantly, the number of electrons radiating synchrotron radiation can be accurately determined. This source therefore gives an energy distribution of known absolute intensity, rather than just a relative distribution. The number of electrons is found by observing the decay of the last several (e.g. 30) electrons in the ring by noting the incremental decrease in intensity as each electron leaves the beam. This process can be followed to the last electron. It is then a simple matter to calculate the number of electrons (typically several hundred) circulating in the ring when the calibration was made.

So far, this source has been used to calibrate photomultipliers and interference filter photometers with half-widths of 200–300 Å in the 2000–3000 Å region. The sys-

Houziaux and Butler (eds.), Ultraviolet Stellar Spectra and Ground-Based Observations, 18–19.
All Rights Reserved. Copyright © *1970 by the IAU.*

tem to be calibrated is mounted in a tank attached tangentially to the ring and the sensitivity mapped out point-by-point. Since the beam is highly polarized, calibrations are carried out with the detector system at various orientations about the longitudinal axis.

By the end of this summer it will be possible to extend calibrations down to 1000 Å.

Clark: What vacuum do you have in the storage ring?

Bless: 10^{-9} mm.

Henize: Boldt concentrated entirely on 'source-type' calibration methods whereas the practical calibration of flight instruments still seems to depend largely on 'detector-type' calibration. Can someone discuss the relative accuracies of these two methods?

Boldt: My report was related to intensity-*standards*, i.e. to instruments which are stable and reproducible with respect to their intensity or efficiency. I don't know any detector which is stable in its efficiency for more than a few months. Consequently I had to restrict myself to sources.

Bonnet: I wonder whether such big ground based calibration sources can be used as onboard calibration standards.

Bless: We use on OAO $\frac{1}{2}$ inch diameter Cerenkov sources as calibration checks. These are mounted in the photometer filter wheels and check the performance of the detector and electronics. They have performed very well.

B. INTERSTELLAR EXTINCTION

INTERSTELLAR EXTINCTION

(Introductory lecture)

H. C. VAN DE HULST

Sterrewacht Leiden, Leiden, The Netherlands

In the introductory talk as presented at the symposium I expressed the expectation that it could be thoroughly superseded by the papers and discussions remarks presented at the symposium.

It has.

Discussion

Greenberg: I would like to take this opportunity to reemphasize certain key observations in the UV which I believe will help to distinguish between absorption (metallic) and dielectric grains.

(1) The prediction based on the optical properties of elongated dielectric grains that the *extinction* in the UV relative to the visible is less for stars seen perpendicular to magnetic fields than for those seen along magnetic fields (roughly this corresponds respectively to stars with high and low degrees of polarization).

(2) The polarization by dielectric particles decreases more slowly in the UV than the polarization by absorbing particles.

Wickramasinghe: The condition for a grain to remain cold (i.e. at ~ 3 K) is that Q_{abs} (vis) $\leqslant Q_{abs}$ (0.1 cm).

It turns out that this condition is satisfied provided that the conductivity of silicates lies between 6×10^{11} and 5×10^{10} sec^{-1}.

The data on silicates do not seem to be incompatible with this requirement.

Houziaux and Butler (eds.), Ultraviolet Stellar Spectra and Ground-Based Observations, 23.
All Rights Reserved. Copyright © 1970 by the IAU.

ULTRAVIOLET INTERSTELLAR EXTINCTION
FROM A COMPARISON OF ε PERSEI AND ζ PERSEI

THEODORE P. STECHER

NASA, Goddard Space Flight Center, Greenbelt, Md., U.S.A.

Abstract. Hall's and Stebbins and Whitford's extinction pair has been used to determine interstellar extinction in the ultraviolet. The extinction shows a feature suggestive of graphite. Previous observations of Stecher (1965) are confirmed. The results are presented in the figures.

Hall (1937) and Stebbins and Whitford (1943, 1945) used this pair of stars (ε and ζ Per) in their pioneering determination of interstellar extinction. In the current study they were selected for the same reasons, i.e., they are a bright pair of similar spectral type with about a magnitude of extinction between them and they are close together in the sky. The difference in energy distribution due to spectral type (ε is a B0.5 V and ζ is a B1 Ib) is small in the visible. Larger differences in the energy distribution due to spectral type are expected in the ultraviolet and in particular, the supergiant is expected

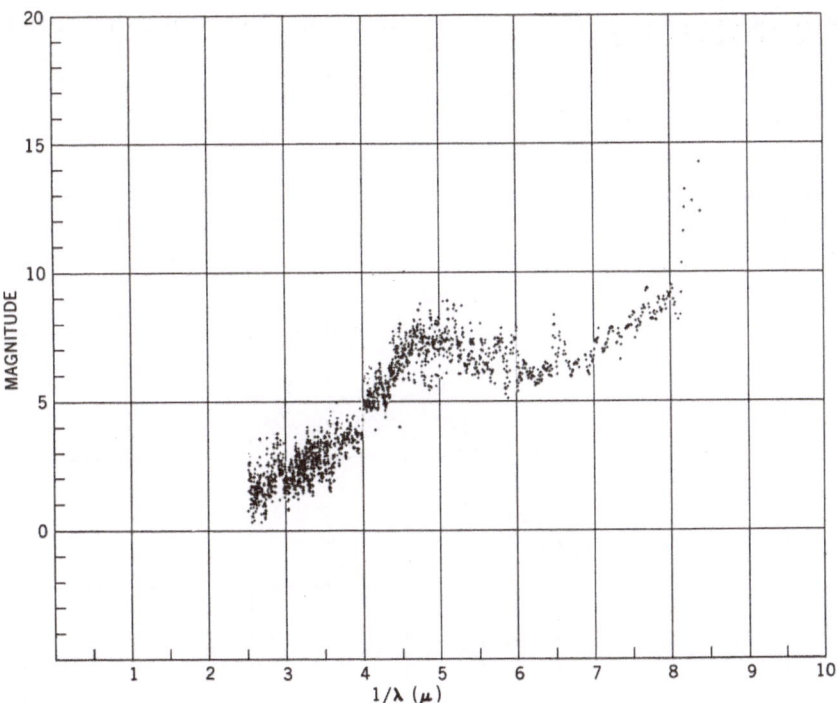

Fig. 1. Interstellar extinction in magnitudes as a function of inverse wavelength determined from ζ and ε Per scanned from short to long wavelengths. The curve is normalized to B−V=1 mag. and V~0.

Houziaux and Butler (eds.), Ultraviolet Stellar Spectra and Ground-Based Observations, 24–27.

to have a circumstellar shell. From previous measurements of stellar energy distributions (Stecher, 1969a), the spectral type difference was found to be small compared with effects of interstellar extinction which also becomes larger in the ultraviolet.

The observational material consists of two spectrophotometric scans of each star obtained with a 32 cm telescope which was mounted in an Aerobee rocket. The spectral resolution was 10 Å. There were three exit slits in the scanner with a photomultiplier for each covering the spectral ranges from 1100 Å to 2400 Å, 1600 Å to 3200 Å and 2400 Å to 4000 Å. The two stars were observed sequentially near the peak of the rocket flight. The absolute pointing error contributed less than 10 Å uncertainty in the wavelength. A description of the instrument appears elsewhere (Stecher, 1969b).

The data from each of the detectors for each scan of ε Per was averaged over 2 Å intervals and divided by that of ζ Per for the same interval. The magnitude was then formed and multiplied by 3.57. This is from Johnson *et al.* (1966) to normalize the observations to $B-V=1$ since the visual magnitudes are nearly equal, $V \sim 0$, and no further normalization was included. Figure 1 presents the difference in magnitudes, 3.57 (ζ Per $-$ ε Per) with the scan from short to long wavelengths. Figure 2 presents the difference in magnitudes 3.57 (ζ Per $-\varepsilon$ Per) with the scan going from long wavelengths to short wavelengths. The spectrometer was sensitive to the diffuse Lyman α line of atomic hydrogen in the geocorona. This sensitivity extends as a

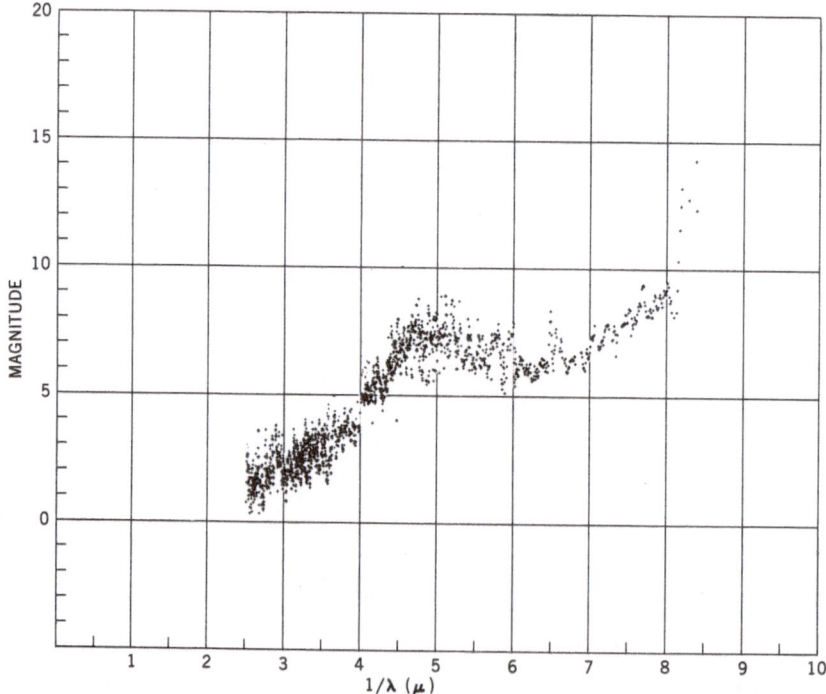

Fig. 1. Interstellar extinction in magnitudes as a function of inverse wavelength determined from ζ and ε Per scanned from short to long wavelengths. The curve is normalized to $B-V=1$ mag. and $V\sim0$.

decreasing function for about 100 Å on either side of Lyman $\dot{\alpha}$. If removed, the interstellar Lyman α line would give very large extinction for about 30 Å.

The large feature in the extinction curve at $4.7\mu^{-1}$ corresponds to the feature that is expected if the particles are graphite. In graphite, the feature is due to the transition to the conduction band of the π electron (Stecher and Donn, 1965; Wickramasinghe and Guillaume, 1965). While this feature could arise from other processes it would appear to be quite a remarkable coincidence if this were the case. At the shorter wavelengths graphite is insufficient to account for all the extinction. A molecular gas such as H_2^+ (Stecher and Williams, 1969) in the proper state of vibrational relaxation could make up the difference and should be considered as well as coatings. In either case, variation would probably occur at the shorter wavelengths. Some features that probably result from atomic transitions in a circumstellar shell are present. For instance, the carbon IV resonance transition at 1549 Å is quite clearly present.

Acknowledgement

I would like to express my appreciation to C.R. Summers for assistance in reducing the data.

References

Hall, J. S.: 1937, *Astrophys. J.* **85**, 150.
Johnson, H. L., Mitchell, R. I., Iriate, B., and Wisniewski, W. Z.: 1966, *Comm. of the Lunar and Planetary Laboratory* **4**, 99.
Stebbins, J. and Whitford, A. E.: 1943, *Astrophys. J.* **98**, 20.
Stebbins, J. and Whitford, A. E.: 1945, *Astrophys. J.* **102**, 318.
Stecher, T. P.: 1965, *Astrophys. J.* **142**, 1683.
Stecher, T. P.: 1969a, *Astrophys. J.* (in press).
Stecher, T. P.: 1969b, *Astron. J.* **74**, 98.
Stecher, T. P. and Donn, B.: 1965, *Astrophys. J.* **142**, 1681.
Stecher, T. P. and Williams, D. A.: 1969, *Astrophys. Letters* **4**, 99.
Wickramasinghe, N. C. and Guillaume, C.: 1965, *Nature* **207**, 366.

Discussion

Wickramasinghe: I merely wish to comment that this spectral feature at 2200 Å was predicted for graphite independently by Stecher and Donn, and by myself several years ago. This feature is due to a solid-state property of graphite involving a transition of π-electrons to the conduction band and it is very impressive that both the central wavelength and widths computed for this case coincide so closely with these observations. I feel this is a good indication of a graphite component of the grains.

Stecher: Yes, the signature of graphite is the bump we have here. There are of course other possibilities.

Underhill: The dip in your extinction curve for the ultraviolet falls in a region where we expect the lines in supergiants to be much stronger and entirely different in character from those of main sequence stars. What check have you that this effect is not more serious than you estimate?

Stecher: With this 10 Å resolution, one would expect to see the lines in ζ Per if the lines were that strong. This is apparently not the case.

Wilson: With regard to your suggestion that the far ultraviolet extinction may be produced by H_2^+, have you (a) estimated the density required and (b) considered whether you can maintain this density bearing in mind that the extinction process in this case destroys the carrier.

Stecher: D. A. William and I have looked into H_2^+ and can put an absolute upper limit on the density of 3×10^{-4} cm $^{-3}$. Investigation of production methods give an appreciable fraction of this. We expect to publish this in the near future.

Conti: Do the observations then represent a linear relation from 1 μ through to the UV with a hump in the UV at 2200 Å?

Stecher: The extinction is less than λ^{-1} below the hump but λ^{-1} gives a good first approximation.

OBSERVATIONS OF INTERSTELLAR EXTINCTION
IN THE ULTRAVIOLET WITH THE OAO SATELLITE

R. C. BLESS and B. D. SAVAGE

Space Astronomy Laboratory, Washburn Observatory,
University of Wisconsin, Madison, Wis., U.S.A.

Abstract. We present a preliminary analysis of a number of spectrophotometric scans of reddened and unreddened early-type stars obtained with the OAO-A2 satellite. The principal findings are: (1) all the extinction curves show a 'bump' at $1/\lambda \approx 4.5$; (2) most extinction curves show a minimum at $1/\lambda \approx 5.5$–6.0; (3) all the extinction curves show a rapid rise in the far ultraviolet; and (4) there appear to be real differences in the far ultraviolet extinction. A brief comparison of our observed extinction curves is made with theoretical particle models.

In this paper we give a preliminary analysis of spectrophotometric scans of a number of early-type stars made over the spectral region 1050–3300 Å from the OAO-A2 satellite. These scans were made by two objective grating instruments with apertures of 15×20 cm, one sensitive over the interval 1050–1800 Å with ~ 15 Å resolution, the second sensitive from 1800–3800 Å with ~ 20 Å resolution. The gratings, moving in discrete steps about equal to the resolution, sweep the spectrum past a slot behind which is located an Ascop 541 F photomultiplier for the shorter wavelength instrument and an EM1 6256B photomultiplier for the longer wavelength instrument. With the 8-sec per step integration time used for these scans, 5^{m} early-type stars can be scanned in about 15 min for each of the two wavelength intervals. Both analog and digital output signals are recorded.

Figure 1 shows representative scans over the 1050–1800 Å interval of σ Ori and ζ Oph, both O9.5 V stars, but the latter $0^{m}.28$ redder in $(B-V)$ than the former. Note the strong resonance lines of Si IV and C IV and the blend of stellar N V, stellar Si III, and mostly interstellar Lyman α. The large number of spectral features makes wavelength identifications and comparisons of two spectra straightforward.

The question of the identity of the ultraviolet spectra of two stars of the same visual color and spectral type is of obvious importance for extinction determinations. Figure 2 gives the results of comparisons of unreddened stars of types B1 II and B3 V. The arrows in the figure indicate the flux ratios for the pairs as obtained from the visual magnitudes. The ultraviolet spectra of these pairs of stars of the same type are remarkably similar. In addition, Figure 2 compares a B1 V and a B1 II star which also have nearly identical continua. Very similar flux ratio plots have been made from several other unreddened pairs having the same spectral types and luminosity classes III and V, and also for early-type unreddened pairs in the 1800–3300 Å region. We may conclude that, unless this admittedly small sample of stars is atypical, two early-type stars with the same visual spectral type and of luminosity classes in the range II to V have essentially the same continuous spectrum in the ultraviolet; differences in the continua of two stars of the same spectral type but different colors may therefore be attributed entirely to extinction by the interstellar medium.

Houziaux and Butler (eds.), Ultraviolet Stellar Spectra and Ground-Based Observations, 28–35.

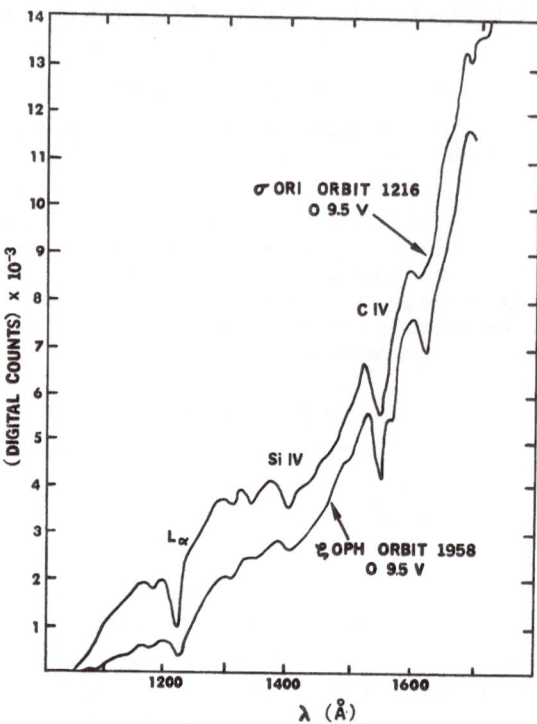

Fig. 1. Spectral scans of σ Ori ($B-V = -.24$) and ζ Oph ($B-V = +.02$) made with the short wavelength scanner on OAO-A2. Digital counts for an 8-sec integration time are plotted vs. wavelength (in Å). Background counts have been subtracted out.

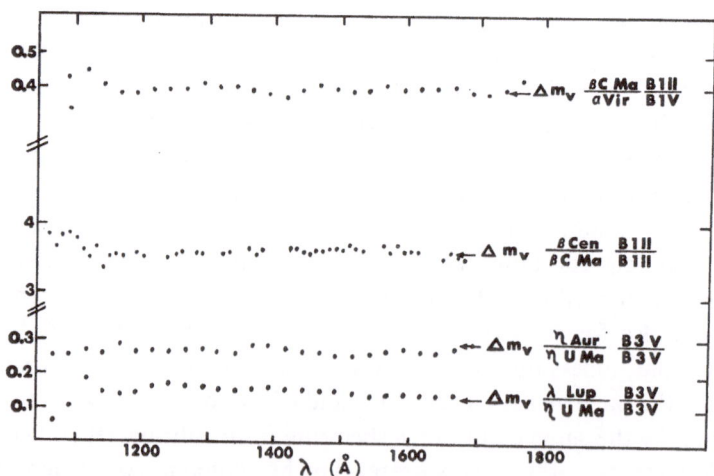

Fig. 2. Flux ratios for a number of pairs of unreddened stars as obtained from the OAO short wavelength scanner data. The Δm_v arrows indicate the flux ratios predicted from the difference in the visual magnitudes for each pair.

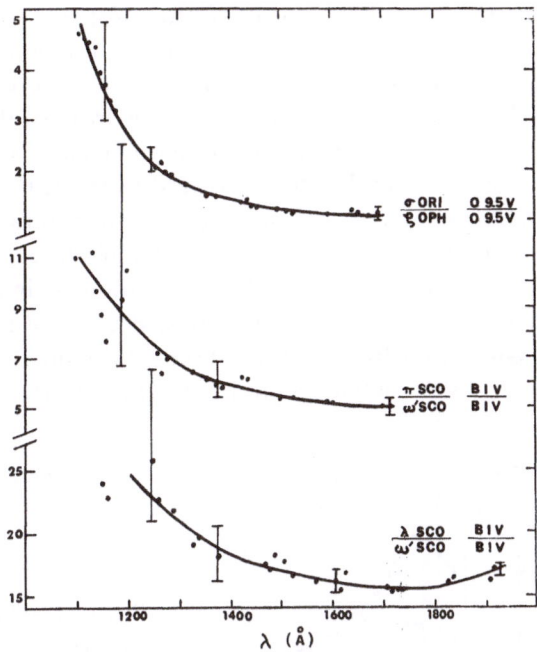

Fig. 3.   Flux ratios for three pairs of stars, one unreddened, the other reddened as obtained from the short wavelength scanner data. See Table I for data concerning each star. Error bars indicate generous error estimates. Most of error is due to uncertainty of scattered light background.

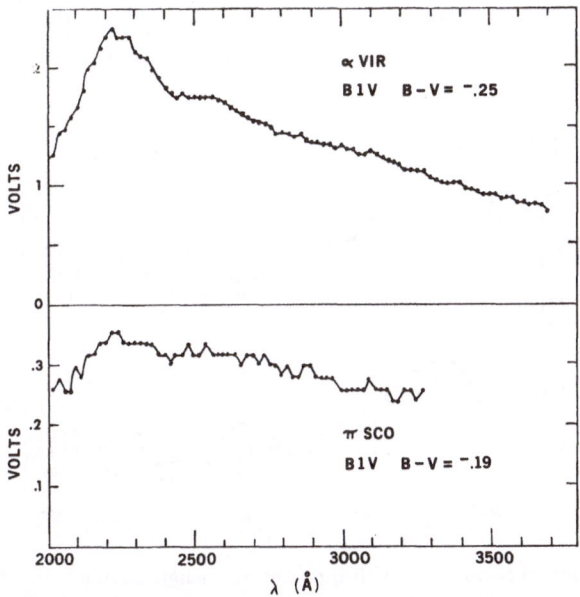

Fig. 4.   Spectral scans of α Vir ($B-V=-.25$) and π Sco ($B-V=-.19$) made with the long wavelength scanner on OAO-A2. Analog data (volts) are plotted vs. wavelength (in Å).

Point-by-point comparisons of three pairs of stars observed with the short wave-length instrument and with different reddening are given in Figure 3. Flux ratios were computed at points which avoid the strong absorption lines. The error bars shown are generous estimates of the uncertainties caused primarily by possible error in the background count determination. This background, which is chiefly scattered light, is found by observations of objects at wavelengths shortward of the LiF cutoff of the Fabry optics. The background is assumed constant in the 1050–1800 Å range.

Figure 4 gives two sample scans of a reddened and unreddened star with the longer wavelength instrument. We have plotted analog data although digital data gives better intensity resolution. The spectral resolution is 20 Å. Because of the lower resolution and the absence of many strong lines for early B stars it is more difficult to establish a wavelength scale in the 1800–3300 Å spectral range than in the shorter wavelength

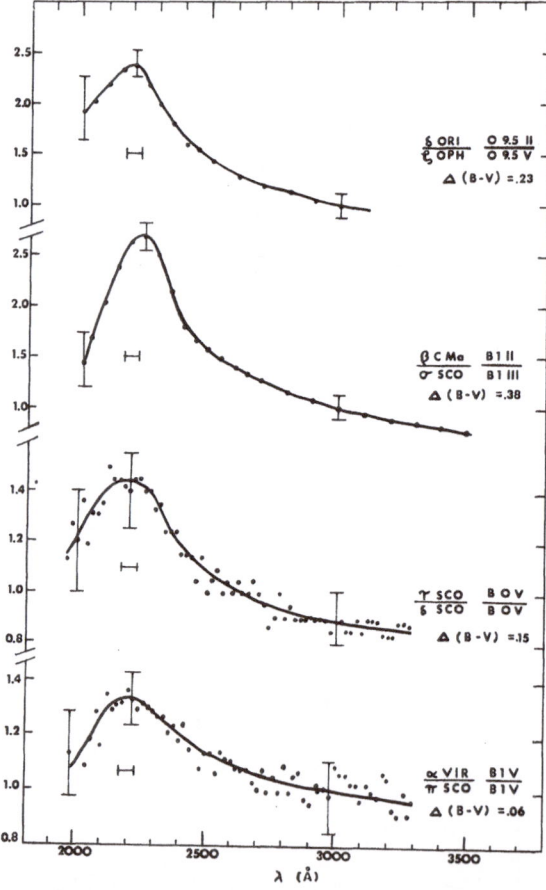

Fig. 5.   Flux ratios for stars observed with the long wavelength scanner. In addition to error bars on the ordinate (resulting mainly from an uncertainty in scattered light background) a wavelength determination error bar of ± 30 Å is indicated. All these curves except the τ Sco–δ Sco curve are normalized to a flux ratio of 1.0 at 3000 Å.

range. However the MgII doublet near 2800 Å is usually visible in early B stars and makes the wavelength assignment possible. We have not included a correction for the background radiation in these data for the longer wavelength scanner; it should be possible to do so in the future. Note the considerable difference in the appearance of the two spectra in Figure 4 with a $(B-V)$ difference of only $0^m.06$.

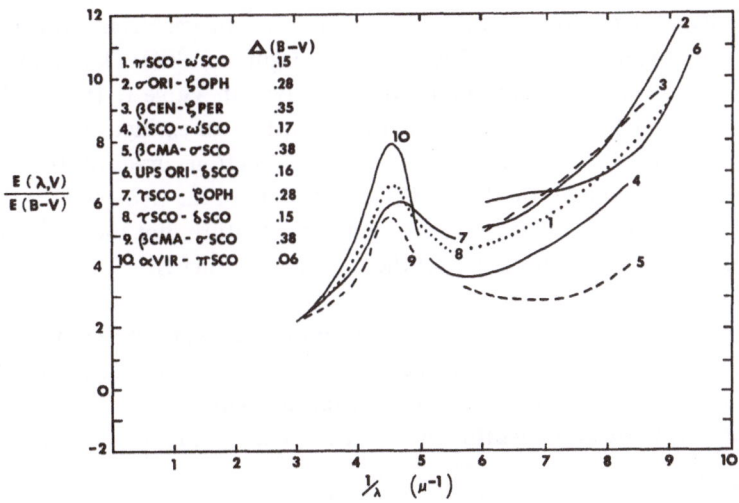

Fig. 6.   Preliminary reddening curves for the indicated pairs. For more data concerning the stars see Table I. The curves obtained from the long wavelength scanner data are normalized to $E(3333\,\text{Å}-V)/E(B-V) = 2.2$. Error bars are not shown but can be easily estimated by referring to Figures 3 and 5.

TABLE I

Data on stars used in reddening analysis

HD	Name	Spectral type	V	$B-V$	Region scanned (Å)	Orbit of observation
24398	ζ Per	B1 Ib	2.83	+.13	1050–1800	892
36485–86	δ Ori	O9.5 II	2.20	−.21	1050–3500	955, 1247
36512	υ Ori	B0 V	4.61	−.27	1050–1800	1215
37468	σ Ori	O9.5 V	3.76	−.24	1050–1800	963, 1238
44743	β CMa	B1 II	1.98	−.23	1050–3500	971
116658	α Vir	B1 V	0.97	−.24	1050–3500	795, 815
122451	β Cen	B1 II	0.61	−.23	1050–3500	1802
143018	π Sco	B1 V	2.90	−.19	1050–3500	1621
143275	δ Sco	B0 V	2.32	−.11	1050–3500	1628
144470	ω¹ Sco	B1 V	3.97	−.04	1050–1800	1632
147165	σ Sco	B1 III	2.89	+.14	1050–3500	1595, 1717
149438	τ Sco	B0 V	2.82	−.26	2000–3500	1609
149757	ζ Oph	O9.5 V	2.57	+.02	1050–3500	1704, 1955
158926	λ¹ Sco	B1 V	1.62	−.20	1050–1800	2219

Flux ratio curves are given in Figure 5 for the longer wavelength scanner. Except for the τ Sco–δ Sco pair the flux ratios have been arbitrarily normalized to 1.0 at 3000 Å. The marked 'bump' centered at about 2200 Å seems to be a characteristic feature of the interstellar extinction.

We have compiled several extinction curves from comparison of pairs of stars observed with the two OAO scanners. The reddened stars are in Orion, Scorpius, Ophiuchus, and Perseus. These curves are given in Figure 6 where the ratio of $(\lambda - V)$ excess to $(B-V)$ excess for the longer wavelength scanner comparisons is adjusted to 2.2 at $1/\lambda = 3$, the mean value found by Boggess and Borgman (1964). Data concerning the stars used for the extinction determinations are given in Table I. It is evident that a slight error in $(B-V)$ can cause a considerable shift of an extinction curve along the ordinate. For example, the difference between the extinction curves for the comparison pairs 1 and 4 is likely due to the 10 to 20% uncertainty in the $\Delta(B-V)$ for these pairs. Thus the shape of the curves is the more significant feature, not the vertical position.

Several comments can be made about these extinction curves:

(1) All show the 'bump' at $1/\lambda \approx 4.5$.

(2) Most show a minimum at $1/\lambda \approx 5.5$–6.0 but curves 5 and possibly 6 deviate significantly from the shape of the others, curve 5 having a minimum at $1/\lambda \approx 7.0$.

(3) All show a rapid rise in extinction in the far ultraviolet.

(4) Apparently there are real differences in the far UV extinction.*

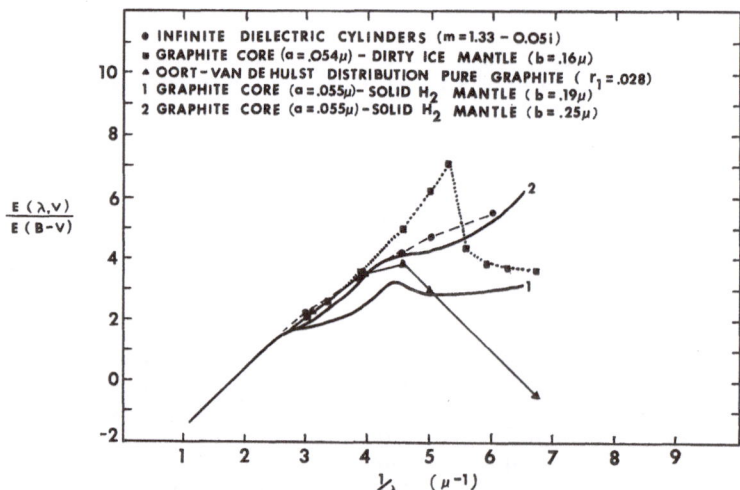

Fig. 7. Shown are a number of theoretical reddening curves: ● infinite dielectric cylinders of Greenberg and Shah (1969); ■ graphite-core and dirty ice mantle of Greenberg and Shah (1969); ▲ Oort–Van de Hulst distribution of pure graphite from Krishna Swamy and O'Dell (1967); and 1,2 graphite core-H₂ mantle from Wickramasinghe and Nandy (1968). m = dielectric constant, a = core radius, b = mantle radius.

* For OAO results near the Trapezium see on p. 107 the comment by Bless in the discussion following Carruthers' paper.

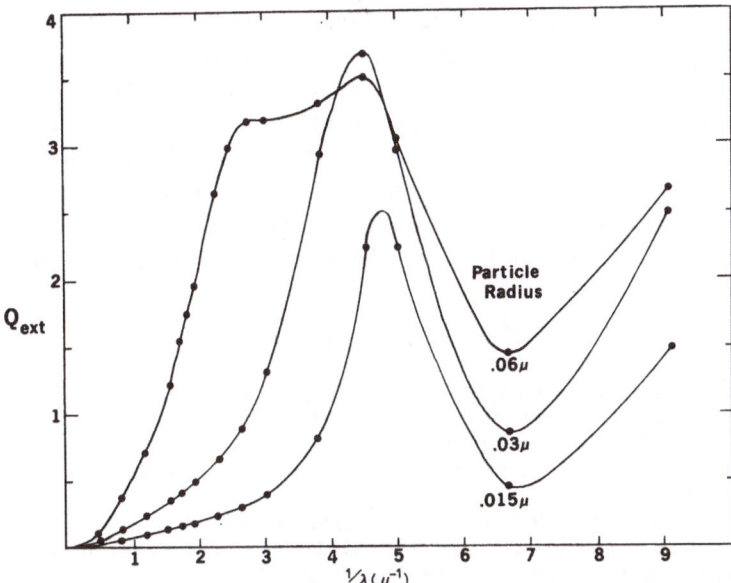

Fig. 8. $Q_{\text{extinction}}$ for various sizes of pure graphite particles from Grevesse-Guillaume and Wickramasinghe (1966). Note that the peak in extinction for small graphite particles occurs at $1/\lambda \sim 4.5$.

The extinction curve adopted by Bless *et al.* (1968) from various filter photometry measurements shows the same general trend as those given here, but both the 'bump' and the minima found in the OAO results were smoothed over.

Finally, we would like to compare these results with predictions of a few of the models of interstellar particles recently published and shown in Figure 7. The infinite dielectric cylinder model calculated by Greenberg and Shah (1969) shows no 'bump' at 2200 Å. The graphite core with a dirty ice mantle considered by Greenberg and Shah (1969) has an extinction 'bump' at shorter wavelengths to a degree inconsistent with these observations. The extinction by an Oort–Van de Hulst distribution of sizes of pure graphite particles (Krishna Swamy and O'Dell, 1967) gives a peak at the correct wavelength but does not give the far ultraviolet extinction increase. One must be careful in interpreting the numerous extinction calculations for pure graphite, however, because some results are based on old inaccurate models of the electrical properties of graphite (Krishna Swamy and Wickramasinghe, 1966) while others are based on laboratory measurements of these electrical properties (Grevesse-Guillaume and Wickramasinghe, 1966). Several extinction curves from the paper of Grevesse-Guillaume and Wickramasinghe (1966) are shown in Figure 8. One can see that pure graphite particles of small size display the 'bump' observed in the OAO data. The particle model with a rather large graphite core coated with a H_2 mantle (Wickramasinghe and Nandy, 1968) retains the 2200 Å 'bump' characteristic of graphite if the mantle is thin (see Figure 7) and also shows the far UV extinction increase. The 'bump' is not as pronounced as the observations indicate, however.

We must emphasize that agreement or disagreement between the observations and any of the particle models given above is relevant only for that particular model; changes in the parameters of a particle type can produce marked changes in its extinction characteristics. In all probability interstellar particles have a variety of sizes, shapes, and compositions whose parameters may be modified by local conditions.

A considerable amount of OAO scanner data remains to be analyzed as well as all of the filter photometry of reddened stars. This large body of extinction data may, hopefully, set sharper limits to the properties of interstellar grains than has heretofore been possible.

References

Bless, R. C., Code, A. D., and Houck, T. E.: 1968, *Astrophys. J.* **153**, 561.
Boggess, A. and Borgman, J.: 1964, *Astrophys. J.* **140**, 1636.
Greenberg, J. M. and Shah, G.: 1969, *Physica* **41**, 92.
Grevesse-Guillaume, C. and Wickramasinghe, N. C.: 1966, *Monthly Notices Roy. Astron. Soc.* **132**, 471.
Krishna Swamy, K. S. and O'Dell, C. R.: 1967, *Astrophys. J.* **147**, 937.
Krishna Swamy, K. S. and Wickramasinghe, N. C.: 1966, *Monthly Notices Roy. Astron. Soc.* **132**, 193.
Wickramasinghe, N. C. and Nandy, K.: 1968, *Nature* **219**, 1347.

Discussion

Wickramasinghe: You pointed out that pure graphite has an absorption feature at 2200 Å. It turns out that the band continues to persist in highly impure graphite as well. It has been observed in soots as well as in charcoal.

Bless: We don't insist on pure graphite, but simply point out that it is one of the particle types which could account for the bump.

Wilson: With regard to the theoretical extinction curves you showed for the two composite particles, each with a graphite core but one with an ice mantle and the other with solid hydrogen, I noted that the feature near 2200 Å was much more pronounced in one than in the other. Since the feature is supposed to be caused by the graphite, which in each case is surrounded by a dielectric mantle, I find it rather surprising. Can you offer a physical explanation?

Bless: Any thick mantle on a graphite core will smooth out the 2200 Å feature. If graphite is responsible for the bump then at least some of the particles must have thin mantles or no mantle at all.

Van de Hulst: Do your observations agree with those by Stecher?

Bless: Yes, the main features agree well.

ON DIELECTRIC MODELS OF INTERSTELLAR GRAINS*

J. MAYO GREENBERG†

Leiden Observatory, Leiden, The Netherlands

and

R. STOECKLY

Rensselaer Polytechnic Institute

Abstract. Silicate, ice, and silicate core-ice mantle particles are considered with a view to describing or predicting ultraviolet features in both extinction and polarization which depend on the dielectric nature of the particles.

Introduction

With the discovery that silicates [1] probably exist in circumstellar envelopes we have another candidate for the materials constituting the particles which produce interstellar extinction and polarization. However, on the basis of present theories for the formation of the silicates in cool stellar atmospheres and by using reasonable estimates of mass loss per star and the total number of such stars it does not seem possible to account for more than a small fraction of the total interstellar extinction by grains which are exclusively formed by stars.

The full range of possibilities seems to be that the interstellar space contains a mixture of particles in the following categories: (a) ices, (b) silicates, (c) graphite (or soot), (d) silicate cores with ice mantles, (e) graphite cores with ice mantles.

With the already present and future availability of detailed observations in the ultraviolet as far as 1000 Å it will be useful to anticipate criteria for appraising the suitability of the various grain models. Furthermore the theoretical basis on which the model calculations are made may be tested. Such factors as the degree of smoothness of the particles and the degree of their internal homogeneity begin to become important as the wavelength of the probing radiation decreases. These effects are under investigation theoretically and with the aid of the Rensselaer Polytechnic Institute microwave scattering laboratory. In this paper we will primarily limit ourselves to a discussion of theoretical results based on calculations for smooth particles made up of homogeneous materials.

In view of the optical properties of the material we have mentioned we divide the particles into those which are principally dielectric (ices, silicates) and those which are principally metallic (graphite) and look for possible differences in the ultraviolet with the assumption that the visible extinction and polarization are equally well matched. In the next section we review some of the basic model parameters for dielectric and

* Work supported in part by National Aeronautics and Space Administration Grant No. NsG113 and National Science Foundation Grant No. GP7553.
† On leave from Rensselaer Polytechnic Institute 1968–69.

compound particles. In the following section we examine these models for differences from metallic models in the ultraviolet. In Section 4 we discuss briefly some possible polarization features which may be associated with the diffuse 4430 Å line. Finally in Section 5 we give an approximate theoretical calculation which indicates how some model parameters may be modified by including a roughness effect.

2. Model Parameters for Dielectric Interstellar Grains

Using a constant index of refraction of $m = 1.33 - 0.05i$ to represent ice it has been shown [2] that we achieve an excellent match (see Figure 1) of theoretical to observed extinction from the infrared to the ultraviolet with a distribution of cylindrical (perfectly aligned) particle sizes given by $n(a) = 49 \exp[-5(a/0.2)^3] + \exp[-5(a/0.6)^3]$. The wavelength dependence of polarization is almost equally well matched. We expect the polarization to be even further improved as a consequence of two factors which reduce the polarization at long wavelength without modifying it significantly at intermediate wavelengths: (a) The larger particles are less well oriented than the small ones [3], (b) If we use a proper magnetic orientation distribution the particle sizes needed to produce the extinction will be slightly reduced (see next section) and the smaller particles produce less polarization at the long wavelengths.

In view of the uncertainties [4] associated with the existence of interstellar ice and also because of the strong suggestion of silicates in interstellar space, it is strongly indicated that the above model should be considered in an alternative manner. Noting that silicate materials have an index of refraction of the form $m \approx 1.66 - im''$

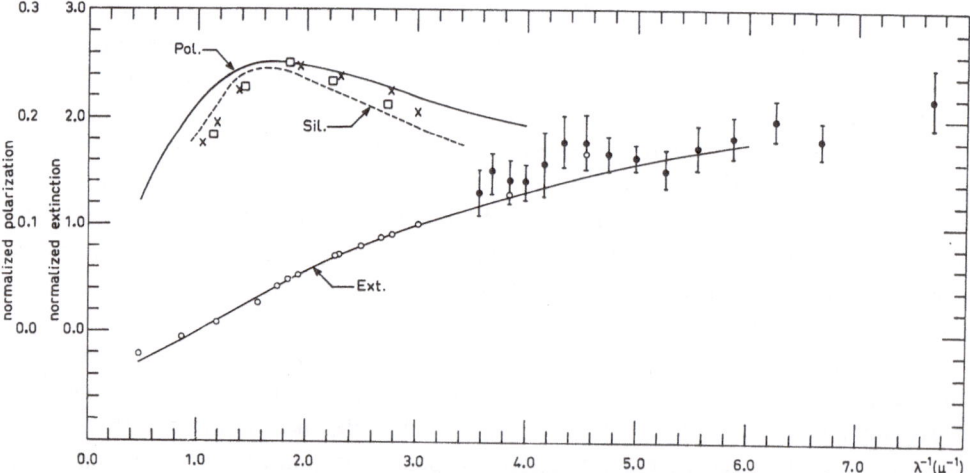

Fig. 1. Extinction and polarization by dielectric grains. Lower curve and upper solid curve (labeled × and Pol. respectively) are extinction and polarization by a size distribution $n(a) = 49 \exp[-5(a/0.2)^3] + \exp[-5(a/0.6)^3]$ of cylinders with $m = 1.33 - 0.05i$. Extinction observations are from [7]. Polarization observations (crosses and squares) are as averaged in [5]. Dashed curve is polarization by a size distribution $n = \exp[-5(a/0.1)^3]$ of cylinders with $m = 1.66$. The theoretical polarization curves are normalized to a difference of 1 magnitude of extinction between 10000 Å and 3330 Å.

we may readily see that if ice-like particles ($m' = 1.33$, $m'' = -0.05$) of an appropriate size give a match to the extinction curve, then silicate particles of one half this size will serve equally well since the dimensionless parameter $(m' - 1)a/\lambda$ will be the same. We then arrive at the possibility that the small particles ($\exp[-5(a/0.2)^3]$ for ices) can be replaced by silicate particles with a size distribution of the form $n(a) = \exp[-5(a/0.1)^3]$. The 'large' particles may then be particles with silicate cores and ice mantles. The mantle size distribution would have a smaller cut-off size than the one given by $\exp[-5(a/0.6)^3)]$ (cut-off radius $\approx 0.6\mu$) for the pure ices because such core-mantle particles act as effectively larger particles – at least in the intermediate ($1 \leqslant \lambda^{-1} \leqslant 3$) region. A mantle distribution with a cut-off below about 0.45μ is entirely consistent with the grain growth mechanism by ice accretion [5].

3. Ultraviolet Features

A. DEPENDENCE OF ULTRAVIOLET EXTINCTION ON POLARIZATION

It was suggested a number of years ago [6] that the wavelength dependence of extinction toward the short wavelengths depends on the degree and kind of orientation of the particles. Furthermore it was noted that this variability of ultraviolet (relative to visible) extinction would be greater for dielectric particles than for metallic or other types of particles. The original calculations were approximate ones for prolate spheroids whose ratio of length to width is 2:1. The essential idea of the interpretation is that: (1) an elongated dielectric particle seen sideways acts for long wavelengths according to its small dimension even though its asymptotic extinction is directly proportional to its long dimension; (2) an elongated particle seen end-on acts for long wavelengths according to its long dimension while its asymptotic ($\lambda \to 0$) extinction is independent of its long dimension, i.e. it acts larger for large wavelengths and smaller for short wavelengths than its orthogonally oriented counterpart. In Table I we present for illustration the wavelength dependence of extinction for ice cylinders of 0.1μ radius which are in the cases labeled $\psi = 0°$ and $\psi = 90°$ spinning in perfect Davis and Greenstein orientation about axes perpendicular to and along the line of sight respectively. As expected the particles seen exclusively sideways, ($\psi = 90°$) and for which zero polarization is expected, act as if they are smaller than those seen partly sidewise and end-on ($\psi = 0°$) for which the polarization would be a maximum. We therefore expect that – other parameters being equal – the extinction in the ultraviolet will be higher for regions with low polarization than for regions with high polarization.

B. WAVELENGTH DEPENDENCE OF POLARIZATION

As we see in Figure 1 the polarization decreases slowly for dielectric particles in the ultraviolet. Metallic particles which give an equally good representation of the polarization in the intermediate region give significantly less polarization in the ultraviolet.

C. THE EXTINCTION 'HUMP' AT ABOUT 2200 Å

The extinction hump at about 2200 Å [7] is now apparently well confirmed [8] although the possibility of a selection effect can not be excluded until the same feature is demonstrated for stars other than hot bright nearby young stars.

The fact that the ices have an absorption edge at about this wavelength leads to a *drop* [2, 5] in the extinction at this wavelength with a subsequent rather flat continuation rather than the rise which is observed. This is true for the graphite core-ice mantle particles as well as for the pure ices. The absorptivity (m'') of the silicates is not well known in this region, but if it turns out that it remains fairly small beyond the ab-

TABLE I

Extinction by ice cylinders of 0.1μ radius spinning about axes parallel ($\psi=90°$) and perpendicular ($\psi=0°$) to the line of sight

$\lambda(\mu)$	$\lambda^{-1}(\mu^{-1})$	Δm	
		$\psi=0°$	$\psi=90°$
0.14	7.15	1.497	2.221
0.16	6.25	2.202	3.362
0.18	5.55	2.457	3.559
0.20	5.00	2.259	2.917
0.22	4.54	2.110	2.392
0.24	4.17	1.894	1.947
0.26	3.84	1.615	1.589
0.28	3.57	1.384	1.351
0.30	3.33	1.211	1.178
0.32	3.12	1.086	1.035
0.34	2.94	0.9956	0.9085
0.36	2.77	0.9193	0.7834
0.38	2.63	0.8615	0.6770
0.42	2.38	0.7778	0.5181
0.45	2.27	0.7271	0.4367
0.48	2.08	0.6816	0.3805
0.50	2.00	0.6502	0.3509
0.60	1.67	0.4876	0.2637
0.70	1.43	0.3139	0.2026
0.80	1.25	0.1991	0.1542
1.0	1.00	0.0931	0.0897
2.5	0.4	0.0037	0.0043
5.0	0.2	0.0050	0.0059

sorption edge (<0.05) then the model proposed in the previous section would satisfy all the necessary requirements. The effect of the absorption edge of the silicates could easily be masked by surface roughness effects which incidentally become more important in the ultraviolet. The small silicate particles would then provide the continuing increase in the extinction on which would be superposed the extinction by the silicate core-ice mantle particles the ice mantles of which produce the hump. Detailed calculations on this model are in progress [9].

4. Diffuse Lines

Following the procedure described in [5], detailed calculations have been made of the shape of both the extinction and polarization in the region around an absorption band centered at 4430 Å in either small silicate or ice particles. In both cases (the silicates being smaller than the ices) the absorption band becomes highly asymmetric, being even more so in the silicates than the ices. We also find that the degree of polarization follows an identical trend with the absorption, i.e. a higher polarization at the longer wavelengths and a smaller polarization at the shorter wavelengths than one would get with no absorption band present. We expect a similar result for the ultraviolet hump at about 2200 Å.

5. Surface Roughness of Small Particles

To illustrate the probable effect of surface roughness we show a theoretical calculation based on a ray approximation [10].

For a smooth sphere with an index of refraction near unity the total cross section is given by

$$C = 8\pi a^2 \int_0^\infty x \sin^2 \eta \, dx \tag{1}$$

where

$$\eta = \frac{\varrho}{2}(1 - x^2)^{1/2}$$

$$\varrho = \frac{4\pi a}{\lambda}(m' - 1).$$

The quantity η is simply the phase shift of a ray passing a distance xa from the optic axis. We picture the surface roughness on a distribution of particles as introducing a statistical fluctuation on this phase shift. For simplicity we preserve the cylindrical symmetry of the problem. Furthermore we let the phase fluctuation be independent of x. The total phase is then given by $\eta' = \frac{1}{2}\varrho(1-x^2)^{1/2} + \varepsilon\varrho$ where the average value of ε is set equal to zero. Substituting η' into Equation 1 and averaging we obtain

$$\langle Q_r - Q_s \rangle = 2 \langle \sin^2 \varepsilon\varrho \rangle (2 - Q_s), \tag{2}$$

where Q_r and Q_s are the extinction efficiencies, $C/\pi a^2$, for the rough and smooth spheres respectively.

Since the quantity $(2 - Q_s)$ is positive [11] for $\varrho \lesssim 2$ and is negative in the range $2 \lesssim \varrho \lesssim 6$ we see that the rough sphere is first (longer wavelengths) more efficient in extinction and then, in the region of the major resonance, less efficient than the equivalent smooth sphere. This result is very much like that obtained by having introduced a small imaginary part in the index of refraction.

If the particle material has an intrinsically variable absorptivity the effect of surface roughness would be to smooth over this variability because its effect is larger when the *actual* absorptivity is smaller. This could be important, as noted previously, in masking the onset of absorption edges in the ultraviolet.

References

[1] Knacke, R. F., Gaustad, J. E., Gillett, F. C., and Stein, W. A.: 1969, *Astrophys. J.* **155**, L189.
[2] Greenberg, J. M., and Shah, G. A.: 1969, *Physica* **41**, 92.
[3] Greenberg, J. M.: 1969, *Physica* **41**, 67.
[4] Knacke, R. F., Cudaback, D. D. and Gaustad, J. E.: 1969, *Astrophys. J.* **158**, 151.
 Johnson, H. L.: 1968, *Astrophys. J.* **154**, L125.
[5] Greenberg, J. M.: 1968, in *Stars and Stellar Systems*, Vol. VII (ed. by B. M. Middlehurst and L. H. Aller), University of Chicago Press, Chicago and London, Chapter 6, p. 221.
[6] Greenberg, J. M.: 1960, *Lowell Obs. Bull.* **4**, 285.
 Greenberg, J. M.: 1960, *J. Appl. Phys.* **31**, 82.
 Wilson, R.: 1960, *Monthly Notices Roy. Astron. Soc.* **120**, 51.
[7] Stecher, T. P.: 1965, *Astrophys. J.* **142**, 1683.
[8] Other papers at this meeting.
[9] Stoeckly, R. and Greenberg, J. M.: unpublished.
[10] Van de Hulst, H. C.: 1946, *Rech. Astron. Obs. Utrecht XI*, Part 1.
 Greenberg, J. M.: 1960, *J. Appl. Phys.* **31**, 82.
[11] Van de Hulst, H. C.: 1957, *Light Scattering by Small Particles*, Wiley, New York (section 11.2).

Discussion

Wickramasinghe: Would you not expect the position of the absorption peak in relatively large icy particles ($r \sim 10^{-5}$ cm) to be rather strongly size dependent?

Greenberg: With the size distribution I have used I do not anticipate a large size effect.

Bless: Could you elaborate on your remark concerning the relation between visual polarization and UV extinction?

Greenberg: For dielectric grains the extinction curves depend very significantly on orientation. As a consequence of the fact that elongated particles seen end-on act in the longer wavelength region like larger particles seen sideways, the extinction in the UV produced perpendicular to magnetic fields drops more (or rises less steeply) in the UV than the extinction produced along magnetic fields.

Van de Hulst: The polarization in the UV by conducting particles will drop more rapidly than that produced by dielectric particles. A comparison between the polarization in the 2500–3000 Å region with that in the visual will be an important observational criterion between metallic and dielectric particles.

Wickramasinghe: We may be rather naïve to think in terms of a single type of grain in reflection nebulae or even in the general interstellar medium. It could be that we have a mixture of graphite, silicates, ices and whatever else might form.

Bless: OAO filter photometry has been obtained for several reflection nebulae, including the Merope nebula, and also for the background at several hundred points in the sky. This may give some scattering data helpful in determining the nature of interstellar particles.

EXTINCTION CURVES FOR GRAPHITE-SILICATE GRAIN MIXTURES

N. C. WICKRAMASINGHE

Institute of Theoretical Astronomy, University of Cambridge, Cambridge, England

Abstract. Mixtures of graphite particles ejected from carbon stars and silicates from oxygen-rich giants are capable of producing excellent fits to the observed interstellar extinction curve. The fits extend over the entire wavelength range of the observations, and include a hump at ~ 2200 Å in the ultraviolet due to the graphite components of the mixtures. The agreement with the observed extinction curve remains good if the small silicate particles acquire mantles of either ice or solid H_2. If no mantles are present around the silicate grains comparable mass densities of graphite and silicates are indicated; if mantles of either ice or solid H_2 are present the mass density of silicates may be an order of magnitude below that of graphite in the interstellar medium.

1. Introduction

The recent discovery of a broad emission feature at $\sim 10\mu$ in the spectra of several oxygen-rich giants has led to the belief that silicate particles are produced in these stars and are ejected into the interstellar medium (Woolf and Ney, 1969; Knacke *et al.*, 1969; Ney and Allen, 1969; Stein and Gillett, 1969; Gilman, 1969). The carbon stars, on the other hand, are expected to produce graphite particles which are also ejected into the interstellar medium (Hoyle and Wickramasinghe, 1962; Wickramasinghe, 1967). Interstellar grains may thus be regarded as a mixture of at least two distinct, highly refractory components – graphite and silicates. The astrophysical consequences of such a grain mixture have already been discussed in an earlier paper (Hoyle and Wickramasinghe, 1969). We present here in more detail a discussion of the optical properties of such mixtures with particular reference to the interstellar extinction curve in the rocket ultraviolet.

2. Graphite Particles

The interstellar extinction curve is now known over a wide waveband $\sim 0.3 \leqslant \lambda^{-1} \leqslant \leqslant 9\mu^{-1}$, and the requirement of matching this entire curve might be expected to impose stringent constraints on theoretical models of the grains. The original rocket ultraviolet observation of Stecher (1965) besides showing a generally rising extinction curve for $\lambda < 3000$ Å also indicated a rather conspicuous hump at ~ 2200 Å. This hump which has recently been confirmed independently by Stecher (1969) and by Bless and Savage (1969) coincides with a similar hump present in the calculated extinction curves for graphite particles (Wickramasinghe and Guillaume, 1965; Stecher and Donn, 1965). The graphite feature arises due to the transition of π-electrons to the conduction band of graphite. The astronomical observations could thus be interpreted as implying the presence of a graphite component to the grains causing the extinction of starlight.

Houziaux and Butler (eds.), Ultraviolet Stellar Spectra and Ground-Based Observations, 42–49.

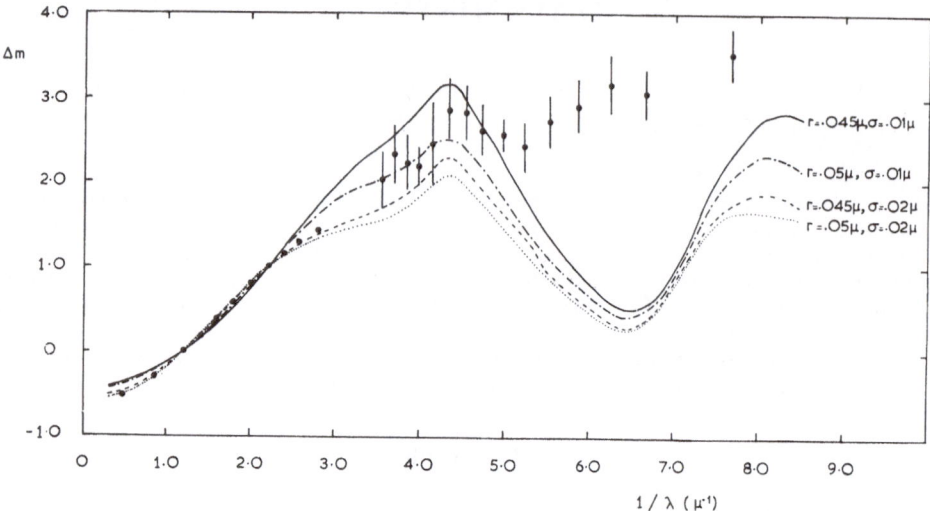

Fig. 1. Normalised extinction curves for size distributions of graphite particles defined by Equation (1) with $\bar{r}$, σ values characterising the several curves denoting the mean radii and dispersions. Normalisation is to $\Delta m = 0$ at $\lambda^{-1} = 1.22\mu^{-1}$, $\Delta m = 1$ at $\lambda^{-1} = 2.22\mu^{-1}$. The observational points are those of Johnson (1965) for Cygnus in the infrared, Nandy (1964) for Cygnus in the optical and of Stecher (1965), and Bless *et al.* (1968) in the ultraviolet. The observations are also normalised in the same manner as the theoretical curves.

We have calculated the extinction resulting from a size-distribution of graphite particles defined by

$$n(a) \propto \exp\left[-(a - \bar{r})^2/(2\sigma^2)\right] \tag{1}$$

for various values of $\bar{r}$ and σ. $n(a)\mathrm{d}a$ represents the number of grains per cm³ with radii between a, $a + \mathrm{d}a$ and the distribution is truncated to zero outside the range $\bar{r} - 1.5\sigma < a < \bar{r} + 3.5\sigma$. The optical constants of graphite were taken from the data of Taft and Phillipp (1965) and the normalization is chosen such that

$$\Delta m = 0 \text{ at } \lambda^{-1} = 1.22\mu^{-1}$$
$$\Delta m = 1 \text{ at } \lambda^{-1} = 2.22\mu^{-1}. \tag{2}$$

Figure 1 shows normalized extinction curves for several combinations of $\bar{r}$, σ compared with the observational points of Johnson (1965) in the infrared for Cygnus, of Nandy (1964) in the optical for Cygnus, and of Stecher (1965) and Bless *et al.* (1968) in the rocket ultraviolet. All the curves in Figure 1 exhibit a pronounced hump at $\lambda \approx 2200$ Å. The best over-all agreement with the observations in the infrared, optical and near ultraviolet occurs for the case $\bar{r} = 0.045$–0.05μ, $\sigma \simeq 0.02\mu$, but it seems unavoidable that the theoretical curves for pure graphite flakes diverge significantly from the observations for $\lambda \lesssim 2000$ Å. Graphite core-ice mantle grains are capable of producing a general rise in the extinction curve further into the ultraviolet, but the graphite hump at ~ 2200 Å is then completely washed out (Wickramasinghe, 1967). While graphite core-solid H_2 grains could retain this hump, the formation and stability of such mantles may yet be open to question (Wickramasinghe and Nandy, 1968; Wickramasinghe and Krishna Swamy, 1969).

3. Graphite-Silicate Mixtures

We discuss here the possibility that graphite and silicate particles may exist as separate entities in the interstellar medium. Consider a size-distribution of graphite particles defined by (1) mixed with silicate particles of radii r_{sil}. The relative proportions of graphite to silicates is specified by a parameter x given by

$$x = \frac{\text{extinction by graphite component at } 4500\,\text{Å}}{\text{extinction by silicate component at } 4500\,\text{Å}} \,. \tag{3}$$

We carried out extinction calculations with various values of the size parameters $\bar{r}_{gr}$, σ, r_{sil} and with various values of x. The complex refractive index of the silicates was taken to be $m = 1.66$ over the entire wavelength range of interest. For r_{sil} in the range 0.07–0.09μ and $\bar{r}_{gr} \sim 0.05$–0.06μ the best fits emerged for $x \simeq 6$ – that is, for 6 times as much extinction at 4500 Å arising from the graphite component as from the silicate component (see also Hoyle and Wickramasinghe, 1969). The normalised extinction curves for

$$\bar{r}_{gr} = 0.05, 0.06\mu, \ \sigma = 0.02\mu; \quad r_{sil} = 0.07, 0.08\mu,$$

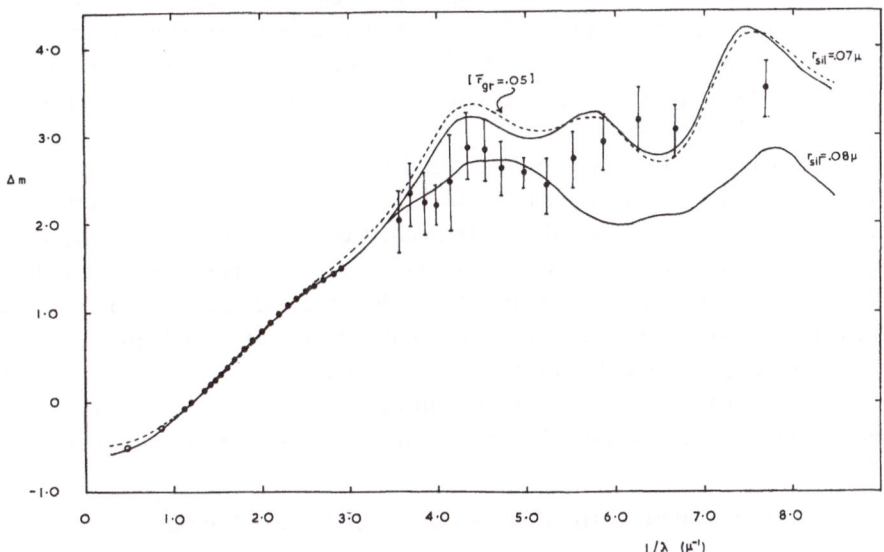

Fig. 2. Solid curves are the normalised extinction for size distributions of graphite particles defined by Equation (1) characterised by $\bar{r} = 0.06\mu$, $\sigma = 0.02\mu$ mixed together with silicate particles of radii $r_{sil} = 0.07\mu$ and 0.08μ. The particle types are mixed so that 6 times as much extinction at 4500 Å arises from the graphite as from the silicate. The dashed curve is for the case $\bar{r} = 0.05\mu$, $\sigma = 0.02\mu$ and $r_{sil} = 0.07\mu$ with $x = 6$. The observational points and the normalisation are the same as in Figure 1.

with $x=6$ are plotted in Figure 2. The agreement between the theoretical curves in Figure 2 and the observational points (the same as Figure 1) is seen to be quite satisfactory. The graphite hump at ~ 2200 Å is well reproduced; the other humps in the far ultraviolet may be expected to be smoothed out by a distribution of sizes for r_{sil}.

If n_{gr} and n_{sil} are the number densities of the graphite and silicate components of the mixture the above arguments of a fit to the extinction with $x=6$ would imply that

$$\frac{Q_{ext}(\text{graphite})\, \pi r_{gr}^2\, n_{gr}}{Q_{ext}(\text{silicate})\, \pi r_{sil}^2\, n_{sil}} \cong 6 \tag{4}$$

for extinction efficiencies evaluated at 4500 Å. With the adopted radii $\bar{r}_{gr} \simeq 0.06\mu$, $r_{sil} \simeq 0.07\mu$ and the computed values Q_{ext} (graphite) $= 2.7$, Q_{ext} (silicate) $= 0.37$ it follows from (4) that

$$n_{gr} \sim n_{sil}.$$

Since graphite and silicates have comparable specific gravities it follows that comparable mass densities of these materials are required to explain the extinction.

4. Accretion of Volatile Mantles on Silicates

Graphite particles with radii $\sim 0.06\mu$ absorb optical and ultraviolet radiation with efficiency factors close to unity. If the particles are considered to be relatively pure – without the presence of low frequency impurity oscillators – the temperature taken up by such grains in a typical interstellar radiation field may be estimated at ~ 40 K. Grains at these temperatures may not possess mantles of any type. Solid H_2 mantles will not be maintained, and ice mantles could have difficulty in forming at these high temperatures (Williams, 1968).

Silicate particles, on the other hand, would tend to take up lower temperatures due to their very low absorptivity ($Q < 10^{-3}$) at optical and near ultraviolet wavelengths. If their impurity induced low frequency conductivity exceeds the optical conductivity it is possible the grains do not take up a temperature significantly higher than 2.7 K (Hoyle and Wickramasinghe, 1969); the condensation of solid H_2 mantles may then be possible. In any case it seems highly likely that their temperatures will be less than ~ 10 K, so that the formation of ice mantles could occur. We consider next the extinction properties of graphite/silicate mixtures if (a) ice mantles form around the silicates, and (b) solid H_2 mantles form around the silicate cores.

5. Graphite Particle/Silicate Core-Ice Mantle Mixture

As before we consider two component grain mixtures with the relative proportions of graphite grains and silicate core-ice mantle grains defined by the parameter

$$x = \frac{\text{extinction by graphite component at 4500 Å}}{\text{extinction by silicate core-mantle grains at 4500 Å}}. \tag{5}$$

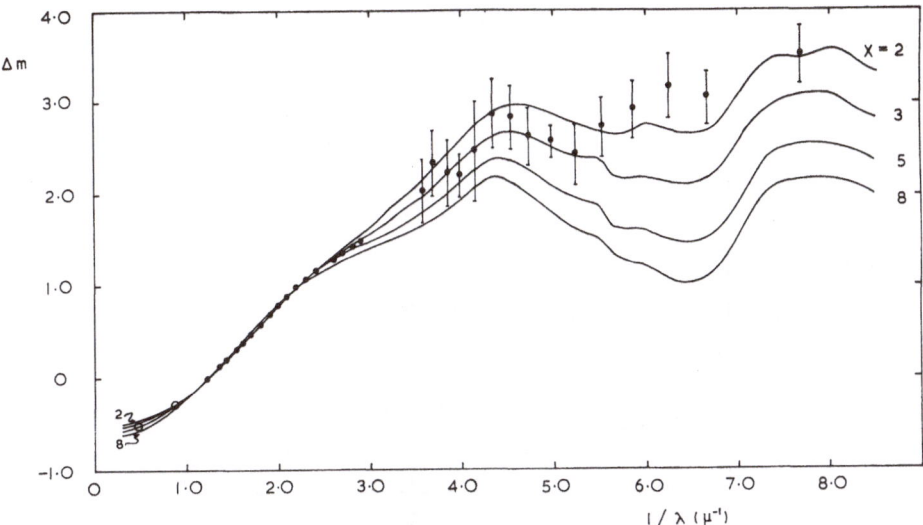

Fig. 3.   Normalised extinction curves for a graphite size distribution given by (1) with $\bar{r}_{gr} = 0.06\mu$, $\sigma = 0.02\mu$ mixed together with silicate core-ice mantle grains. The silicate core radius is 0.03μ, the ice mantle radius is 0.14μ and the several curves $x = 2 - 8$ represent mixtures in proportions defined by Equation (5). The observational points and the normalization are the same as in Figure 1.

We adopt $m = 1.33$ for the refractive index of ice at all wavelengths involved* and consider a distribution of graphite particles given by (1) with $\bar{r} = 0.06\mu$, $\sigma = 0.02\mu$. The normalised extinction for the case $r_{sil} = 0.03$, $r_{ice} = 0.14\mu$ with $x = 2, 3, 5$ and 8 are plotted in Figure 3, together with the observational points. The best agreement is seen to occur for the case $x \cong 2$. As before the agreement includes the graphite hump at 2200 Å.

If n_{gr}, n_{sil} represent the number densities of graphite particles and of silicate cores the fits obtained here imply

$$\frac{Q_{ext}(\text{graphite}) \, \pi r_{gr}^2 \, n_{gr}}{Q_{ext}(\text{sil. core/H}_2\text{O}) \, \pi r_{H_2O}^2 \, n_{sil}} \simeq 2 \tag{6}$$

with the Q_{ext} values computed at 4500 Å. For the appropriate values involved here it turns out that $n_{gr} \sim n_{sil}$; and since the assumed radii of the silicates is $\sim \frac{1}{2}$ that of the graphite it follows that the mass density of silicates is a factor ~ 10 lower than that of the graphite.

STRENGTH OF THE 3.1μ-ICE BAND

Attempts to detect the 3.1μ fundamental band of solid H_2O in the spectra of highly reddened stars have yielded negative results on two successive occasions (Danielson

---

* The replacement of $m = 1.33$ with the actual wavelength-dependent refractive index of H_2O-ice is likely to cause a slight lowering of the extinction curve for $\lambda < 1600$ Å in any given case. The general appearance of the curves including the 'hump' is likely to be unaltered, however.

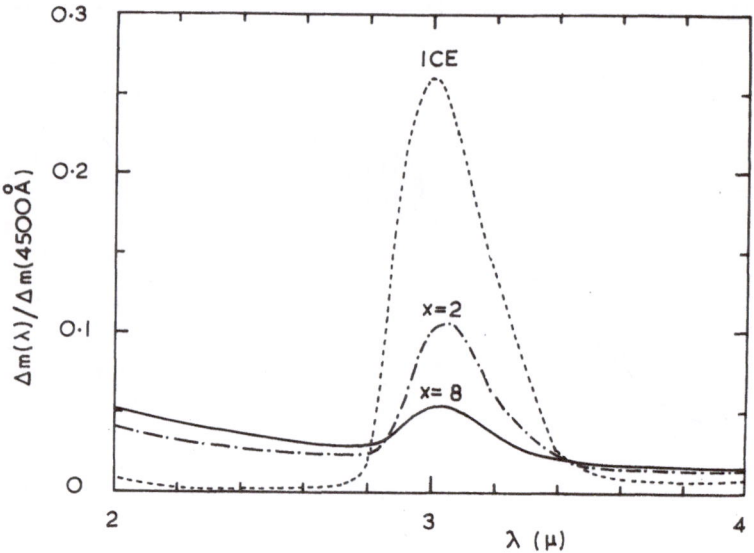

Fig. 4.   The extinction near the 3.1μ ice-band per unit extinction at 4500 Å for mixtures of silicate core-ice mantle grains and graphite particles with the same size parameters as in Figure 3. The uppermost dashed curve is appropriate to an ice sphere of radius 0.14μ, and the other curves $x = 2$, 8 correspond to 2 and 8 times as much extinction at 4500 Å arising from the graphite component of the mixture as from the non-graphite component.

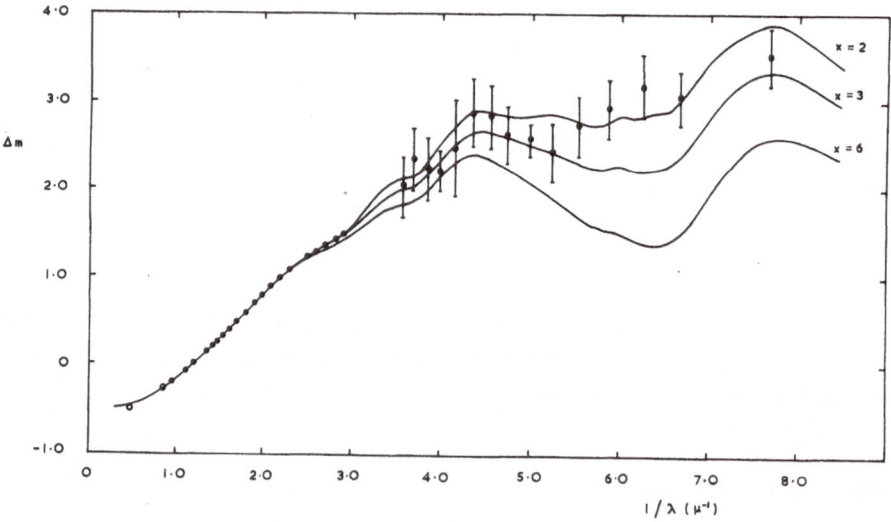

Fig. 5.   Normalised extinction curves for a graphite size-distribution given by (1) with $\bar{r}_{gr} = 0.05\mu$, $\sigma = 0.02\mu$ mixed with silicate core-solid H_2 mantle grains. The radius of the silicate cores is 0.05μ and the solid H_2 mantle radius is 0.35μ. The several curves marked $x = 2$, 3, 6 stand for mixtures with respectively 2, 3, 6 times as much extinction at 4500 Å arising from the graphite component as from the non-graphite component. The observational points and normalization are the same as in Figure 1.

et al., 1965; Knacke *et al.*, 1969). Although these results are yet to be regarded as somewhat preliminary, a quantity of obvious interest in any grain model involving H_2O-ice is the strength and shape of the 3.1μ ice band. For the silicate $- H_2O$/graphite mixtures discussed above we have computed the quantity

$$E(\lambda) = Q_{ext}(\lambda)/Q_{ext}(5000 \text{ Å}) \tag{7}$$

within and in the vicinity of this band. To calculate $Q_{ext}(\lambda)$ in the far infrared we replaced the composite silicate-ice grains by pure ice spheres of the same outer radius. In Figure 4 we have plotted the normalised extinction $E(\lambda)$ for mixtures of graphite particles characterised by $\bar{r}=0.06\mu$, $\sigma=0.02\mu$ and silicate$-H_2O$ grains of inner radius 0.03μ and outer radius 0.14μ. The relative proportions of the two separate components are as before defined by the parameter x (from (5)). The observational limits on $E(3.1\mu)$ set by the observations of Knacke *et al.* (1969) would seem to indicate that $x>2$ in the direction of HD 183143 and $x>8$ in the directions of Vl Cyg. No 12 and CIT 11. The possibility that much of the reddening in these objects occurs nearby to the stars may not be completely ruled out at the present moment, so that these values of x may well turn out to be untypical of the general interstellar medium.

6. Graphite Particle/Silicate Core-Solid H_2 Mantle Mixture

Finally we present extinction curves for the case when solid H_2 mantles may be present around the silicate particles in a mixture. Again we take a graphite particle size-distribution of the form (1) and prescribe the relative proportions of graphite and silicate core-mantle grains by the parameter x defined by (5). For $\bar{r}=0.05\mu$, $\sigma=0.02\mu$ as the graphite parameters and $r_{sil}=0.05\mu$, $r_{H_2}=0.35\mu$ the normalised extinction curves are given in Figure 5 for various values of x. The best agreement with the observational points is seen to occur for the case $x=2$. As in the previous cases it is seen that the graphite hump at 2200 Å is well fitted. The mass density of silicates in the interstellar medium is again at least an order of magnitude below that of graphite.

Acknowledgement

I thank Professor F. Hoyle for several discussions.

References

Bless, R. C., Code, A. D., and Houck, T. E.: 1968, *Astrophys. J.* **153**, 561.
Bless, R. C. and Savage, B. D.: 1969, This Symposium, p. 28.
Danielson, R. E., Woolf, N. J., and Gaustad, J. E.: 1965, *Astrophys. J.* **141**, 116.
Gilman, R. C.: 1969, *Astrophys. J.* **155**, L185.
Hoyle, F. and Wickramasinghe, N. C.: 1962, *Monthly Notices Roy. Astron. Soc.* **124**, 417.
Hoyle, F. and Wickramasinghe, N. C.: 1969, *Nature* **223**, 459.
Johnson, H. L.: 1965, *Astrophys. J.* **141**, 923.
Knacke, R. F., Gaustad, J. E., Gillett, F. C., and Stein, W. A.: 1969, *Astrophys. J.* **155**, L189.
Nandy, K.: 1964, *Publ. Roy. Observ. Edin.* **3**, No. 6.
Ney, E. P. and Allen, D. A.: 1969, *Astrophys. J.* **155**, L193.

Stecher, T. P.: 1965, *Astrophys. J.* **142**, 1683.
Stecher, T. P.: 1969, This Symposium, p. 24. *Astrophys. J.* **157**, L125.
Stecher, T. P. and Donn, B.: 1965, *Astrophys. J.* **142**, 1681.
Stein, W. A. and Gillett, F. C.: 1969, *Astrophys. J.* **155**, L197.
Taft, E. A. and Phillipp, H. R.: 1965, *Phys. Rev.* **138A**, 197.
Wickramasinghe, N. C.: 1967, *Interstellar Grains*, Chapman and Hall, London.
Wickramasinghe, N. C. and Guillaume, C.: 1965, *Nature* **207**, 366.
Wickramasinghe, N. C. and Krishna Swamy, K. S.: 1969, *Monthly Notices Roy. Astron. Soc.* **144**, 41.
Wickramasinghe, N. C. and Nandy, K.: 1968, *Nature* **219**, 1347.
Williams, D. A.: 1968, *Astrophys. J.* **151**, 935.
Woolf, N. J. and Ney, E. P.: 1969, *Astrophys. J.* **155**, L181.

MEASUREMENT OF INTERSTELLAR EXTINCTION
IN EMISSION LINE STARS

R. VIOTTI

Laboratorio di Astrofisica, Frascati, Italy

Usually selective interstellar extinction is derived by comparing the spectrum of a reddened star with an unreddened nearby star, preferably of the same spectral type. But in many cases this comparison is not possible, so that new methods have to be used.

A direct measure of this quantity is possible when a spectrum shows a number of emission lines as for the peculiar stars (novae, nova-like objects, symbiotic stars, etc.) and for nebulae. From a statistical point of view there are a small number of objects but they are of very high interest as far as stellar evolution is concerned.

The analysis of the different behaviour of the emission lines of the same ion in the different spectral regions may give the selective interstellar extinction provided that a suitable theory of the excitation of these lines is available. This may be the case for the ionized iron emission lines (permitted and forbidden) that have been observed in almost all the peculiar objects.

In fact, a detailed study of these lines in the optical spectrum of η Car and other astrophysical objects has shown that: (1) the Fe II levels are populated according to Boltzmann's law, giving a well determined excitation temperature, and that (2) the self-absorption in the strongest lines of Fe II is negligible (Viotti, 1969). Thus the intensity I must be proportional to the transition probability A and to the population of the upper level N_u; then the plot of I/AN_u against $1/\lambda$ gives immediately the colour excess for that spectrum.

From the [Fe II] emission lines Pagel (1969) obtained a very high $E(B-V)$ for η Car (about 1.2) and this value is confirmed by the writer's computations. This result was used to derive the absolute magnitude of η Car and its corrected continuum. The latter agrees fairly well with the continuum that has been directly computed by means of a similar analysis of the equivalent widths of the Fe II and [Fe II] emission lines. More recently, the method has been used at the Astrophysical Laboratory of Frascati to derive the true continuum of other peculiar stars, such as AG Peg and MHα 328–116 (Caputo *et al.*, 1970).

In conclusion, we stress the importance of obtaining the ultraviolet spectra of peculiar objects, at a sufficient dispersion to give a measure of the colour excess and of the true continuum. The former may give information on the absorption by the circumstellar matter, as in the case of η Car.

Both are very important for any theoretical study on the spectral formation in extended envelopes. Obviously, the method will be of the widest use as soon as more satisfactory theories become available on the formation of emission lines of hydrogen, helium and other elements in astrophysical objects.

Houziaux and Butler (eds.), Ultraviolet Stellar Spectra and Ground-Based Observations, 50–51.

References

Caputo, F., Panagia, N., and Gerola, H.: 1970, *Astrophys. J.* (in press).
Pagel, B. E. J.: 1969, *Nature* **221**, 325.
Viotti, R.: 1969, Thesis, Scuola di Perfezionamento in Fisica, Rome University, Italy.

THE INTERSTELLAR EXTINCTION CURVE
FROM 4000 Å TO 6500 Å

G. A. H. WALKER

Geophysics Dept., University of British Columbia, Vancouver, B.C., Canada

and

J. B. HUTCHINGS and P. F. YOUNGER

Dominion Astrophysical Observatory, Victoria, B.C., Canada

Abstract. Interstellar extinction curves (m_{ext} vs. $1/\lambda$) of 20 Å resolution have been obtained at the DAO from photoelectric scanner observations in the range 4000 Å to 5000 Å for five stars, and of 50 Å resolution for four stars in the range 4000 Å to 6500 Å from Willstrop's photoelectric data. There is a closely linear section between 4900 Å and 5800 Å for all of the curves. There are changes of gradient or discontinuities associated with the broadest diffuse interstellar bands at 6180 Å, 4882 Å, 4761 Å and 4430 Å. There is a marked discontinuity near 5800 Å and for some stars a broad absorption near 4200 Å. The 4430 Å band lies between two unequal wings of anomalously low extinction (one of which has been detected at Edinburgh). The irregularities vary from star to star, and those in the neighbourhood of the 4430 Å band seem to have the same form as those in the region of the absorption peak at 2200 Å

Interstellar extinction curves were derived for five early-type reddened stars in the range 4000 Å to 5000 Å from digitized scans made with a low resolution scanner on the D.A.O. 72-inch telescope in an attempt to define the form of the curve in the region of the discontinuities found by Nandy (1964, 1967). The resolution was 20 Å in the

TABLE I
Pairs of interstellar reddened and unreddened stars (reddened star first)

Star	Sp.	$\Delta E(B-V)$	Data	Curve
37022	O6	0.35	DAO	f
Model 98	O9		Underhill	
46711	B3 II	1.05	DAO	c
74280	B3 V			
46711	B3 II	1.05	DAO	d
Model 63	B2 V		Underhill	
154043	B1 I	0.68	Willstrop	h
165024	B0.5 II			
154368	O9.5 Iab	0.77	Willstrop	g
149438	B0 V			
160529	A2 Ia +	1.25	Willstrop	j
167356	A0 Ia			
167971	O8f	1.04	Willstrop	i
149438	B0 V			
183143	B7 Ia	1.23	DAO	b
197345	A2 Ia			
198478	B3 Ia	0.47	DAO	e
Model 63	B2 V		Underhill	
211971	A2 Ib	0.90	DAO	a
197345	A2 Ia			

Houziaux and Butler (eds.), Ultraviolet Stellar Spectra and Ground-Based Observations, 52–56.

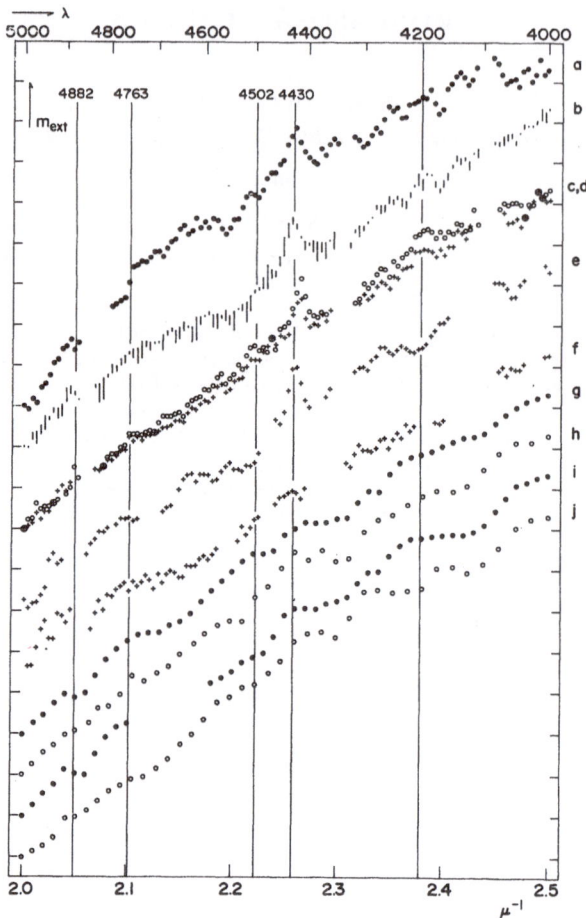

Fig. 1. Interstellar extinction curves (Δm_λ vs. $1/\lambda$) derived for the stars in Table I between 4000 Å and 5000 Å. Curves are normalized to $\Delta m_{5000} = 0$ and $\Delta m_{4000} = 1$. Curves (a) to (f) are based on D.A.O. scans of 20 Å resolution and (g) to (j) are based on Willstrop's (1965) scans of 50 Å resolution. Crosses indicate a comparison with appropriate model atmosphere fluxes. In (b) two curves based on different nights' observations have been superimposed and the points joined by bars. The positions of the diffuse absorption features at 4882 Å, 4763 Å, 4502 Å, and 4430 Å are shown. Further details are given in the text.

spectrum. Each scan was corrected for atmospheric extinction and sky background, and, in some cases calibrated in absolute flux units by reference to standards listed by Oke (1964). Two to four scans were taken and an average formed for each star. Wavelength calibration was applied using the stronger stellar absorption lines of H and He I. To derive an extinction curve spectral energy distributions of the reddened stars were ratioed with spectral energy distributions of unreddened stars or model atmosphere fluxes for stars of corresponding spectral type. The luminosities of the pairs of stars were never quite the same and consequently the members of the Balmer series which are strongly luminosity sensitive do not match and these regions have been

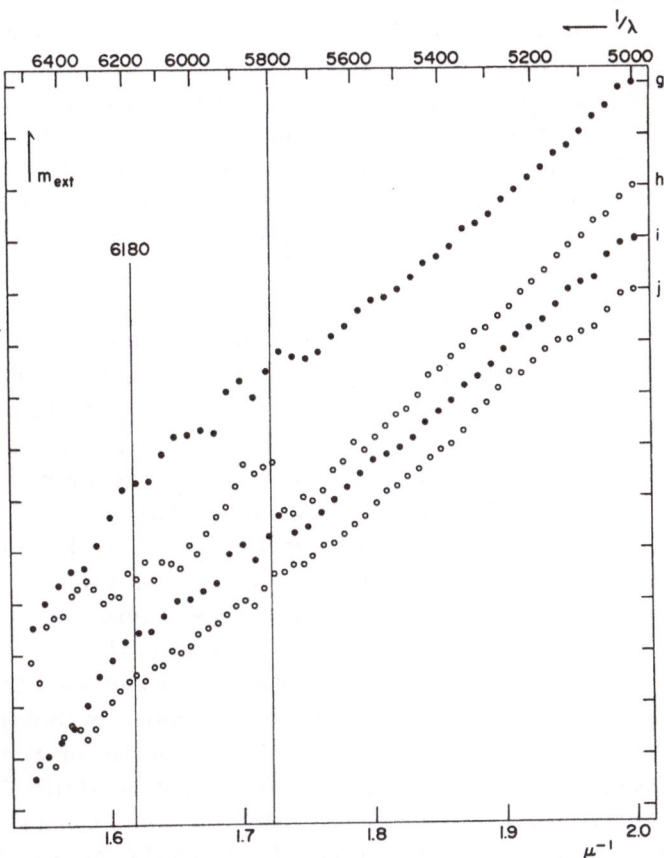

Fig. 2. Interstellar extinction curves between 5000 Å and 6500 Å at 50 Å resolution for stars observed by Willstrop (1965) in Table I. Curves normalized to $\Delta m_{5000} = 0$, $\Delta m_{4000} = 1$.

omitted from the curves. Only continuous opacity sources have been considered in calculating model atmosphere fluxes and consequently there is no matching in the region of either the Balmer series or the stronger He I lines which have been omitted from the curves.

The pairs of reddened stars and unreddened standards are listed in Table I. The extinction curves are shown in Figure 1. Each curve is normalised to give $m_{5000} = 0$, $m_{4000} = 1$. Two extinction curves derived from scans taken on different nights for HD183143 are shown superimposed with the pairs of points joined by bars. In the case of HD46711 curves have been derived from comparisons both with an unreddened standard star and with model atmosphere flux predictions. The close agreement of the two sets of results for HD183143 shows that none of the irregularities discussed below can be ascribed to observational error. The very close agreement of the two curves for HD46711 indicates that none of the larger observed irregularities in the extinction curves is caused by incomplete compensation for the stellar absorption lines between the standard and reddened star spectra since in general this should lead to oppositely directed errors in these two comparisons.

The data published by Willstrop (1965) for four pairs of reddened and unreddened early-type stars has been treated in the same manner as the D.A.O. observations to yield extinction curves at 50 Å resolution between 4000 Å and 6500 Å. The stars are listed in Table I and the extinction curves are shown in Figures 1 and 2. The region of the stronger 'f' emission has been omitted for HD167971. The positions of the broad diffuse interstellar absorption bands at 6180 Å, 4882 Å, 4763 Å, 4430 Å are shown (Wilson, 1958).

The most significant features of the curves in Figures 1 and 2 seem to be the following: the 4882 Å diffuse band appears as a discontinuity rather than an enhanced absorption; the 4763 Å diffuse band shows as a change of slope or discontinuity in all cases; the stars in Cygnus, HD183143, 198478, and 211971, show the 4430 Å band between unequal low absorption wings and the short wavelength wing can be seen for HD46711, and, weakly, both wings for several other stars. A broad absorption or change of slope can be seen near 4200 Å for HD46711 and to a lesser extent for the other stars. If it is real, however, the feature is not at the same wavelength for all of the stars. The star HD37022, the brightest member of the Trapezium, shows the most pronounced change of slope near 4763 Å and virtually no 4430 Å band absorption confirming Morgan's (1944) observation. The short wavelength low absorption wing associated with the 4430 Å band has already been reported by Brück and Nandy (1968). There is a change of gradient at 6180 Å for all of the curves and there is a marked discontinuity for all of the curves at 5800 Å except for HD160529. The difference between the curves for HD160529 and 154043 in Figure 2 is not as extreme as it appears. When the two curves between 5000 Å and 6500 Å are superimposed the curve for HD165024 has a smaller slope between 5000 Å and 5800 Å and the effect of the large discontinuity of slope at 5800 Å for HD154043 is to bring the curves into coincidence.

Although different in scale, the form of the curves in Figure 1 in the region of the 4430 Å interstellar band resembles the ultraviolet extinction curve in the region of the absorption peak near 2200 Å found by Stecher, and Bless and Savage, which they have presented at this symposium. Their results seem to show a change of slope of the extinction curve on passing through the region of the absorption peak and there appear to be wings of low absorption on either side of the peak.

References

Brück, M. T. and Nandy, K. N.: 1968, *Nature* **220**, 46.
Morgan, W. W.: 1944, *Astron. J.* **51**, 21.
Nandy, K. N.: 1964, *Publ. Roy. Obs. Edinburgh* **3**, 141.
Nandy, K. N.: 1967, *Publ. Roy. Obs. Edinburgh* **5**, 13.
Oke, J. B.: 1964, *Astrophys. J.* **140**, 689.
Underhill, A. B. and Walker, G. A. H.: 1966, *Monthly Notices Roy. Astron. Soc.* **131**, 475.
Willstrop, R. V.: 1965, *Mem. Roy. Astron. Soc.* **69**, 83.
Wilson, R.: 1958, *Astrophys. J.* **128**, 57.

Discussion

Underhill: You mentioned a hump (weaker extinction) just longward of 4430 Å. Do you think this is related to the emission wings of 4430 Å which Wampler thought he had detected in a few stars?

Walker: I don't think so. Wampler considered a smaller baseline in wavelength. I think he did find an asymmetry of profile.

Greenberg: How deep are the emission wings? Are they comparable to the height of the absorption peaks relative to the continuum?

Walker: I would prefer to consider these wings just as regions of lower extinction rather than 'emission' and their depth of course depends on the 'normal' level assumed for the extinction. However, one can extrapolate the linear portions of the extinction curve from 4882 Å and 4200 Å and they tend to intersect close to the bottom of the 4430 Å band. So, to this extent, the wings have the same depth as the absorption band.

Morton: To what extent are these effects general properties of the interstellar medium and how much do they depend on the particular stars used for the comparison? In other words are any of these effects proportional to $E(B-V)$?

Walker: The curves do vary from star to star although they are all normalized to the same color excess between 4000 Å and 5000 Å. There are not enough to separate possible regional differences from those for individual stars.

Wickramasinghe: Do you find that the discontinuity in the slope of the extinction curve is in any way dependent on the strength of the 4430 Å band? For example is the sharpness, or the ratio of the slopes any different in the case of stars with very weak 4430 Å?

Walker: I don't think so. HD37022 has virtually no 4430 Å absorption but it does show a marked change of slope near 4763 Å.

Borgman: Are the features only found when comparing observations with model atmospheres?

Walker: No, some of the presented graphs are based on comparison of reddened and unreddened stars.

Greenberg: I do not want to overemphasize the following point. I should merely like to point out that an 'apparent' discontinuity in the slope of the extinction curve at about $\lambda^{-1} = 2.2\,\mu^{-1}$ occurs in my theoretical calculations for cylinders which give a good match to extinction. I believe that this manifestation is due essentially to the fact that we are reproducing a curve which has a change from one sign of curvature to another occurring just about at this point. On my theoretical calculations there are no intrinsic discontinuities in the optical properties of the grains.

(Note added in subsequent discussion.) I should like to suggest that a possible way of establishing the emission and absorption features around 4430 Å would be to draw a 'theoretically' derived continuum rather than try to guess where the continuum runs from the observations. By a theoretically derived continuum I mean one which fits the extinction curve in the other spectral regions and is derived perhaps like the one which I presented.

C. THEORETICAL MODELS FOR STELLAR FLUXES

THE EFFECTIVE TEMPERATURES OF THE O STARS

DONALD C. MORTON

Princeton University Observatory, Princeton, N.J., U.S.A.

Abstract. Effective temperatures of O-type stars imbedded in diffuse nebulae are derived from measurements of Hα and radio fluxes from the nebulae and the apparent magnitudes of the stars. Accurate model atmospheres, with ultraviolet line blanketing where appropriate, are used for the theoretical relation between effective temperature and the ratio of Lyman continuum to visual stellar fluxes. Although there is considerable scatter in the results, an average temperature of 48000 K is found for spectral type O5, 40000 K for O6, and 35000 K for O7.

Two years ago Morton and Adams (1968) proposed a scale of effective temperatures for O and B stars by comparing the observations of the Balmer jump by Chalonge and Divan (1952) with theoretical model atmospheres which included line blanketing in the ultraviolet. For the O stars Morton and Adams adopted the scale in the third column of Table I, but the temperatures were very uncertain for the hottest ones because their Balmer jumps are very small. Also the temperature at O5 was an extrapolation because Chalonge and Divan had no observations earlier than O6 and no weight was given to the ultraviolet fluxes which dominate the emission of these stars.

A better procedure uses the nebula surrounding an O star to count the number of photons emitted in the stellar Lyman continuum, as described by Pottasch (1965) and Hjellming (1968). Each Lyman-continuum photon produces a Balmer photon so that it is possible to obtain for the exciting star the ratio

$$\frac{\text{number of Lyman continuum photons cm}^{-2} \text{ sec}^{-1}}{\text{energy in the } V \text{ magnitude band cm}^{-2} \text{ sec}^{-1}}$$

from the nebular Hα flux and the apparent visual magnitude of the star. This ratio also can be derived from the radio flux by equating the rates of ionizations and recombinations to give an expression for $\int n_e^2/T^{1/2} \, dV$ over the volume of the nebula and hence the rate of free-free emission. Therefore from observations of either the nebular Hα flux or radio emission it is possible to obtain the above ratio as a function of spectral type of the central star. This ratio then can be compared with theoretical models which include ultraviolet line blanketing, when important, to give a relation between effective temperature T_e and spectral type. A surface gravity of $g = 10^4$ cm sec^{-2} has been adopted in all cases.

This approach has the advantage of comparing fluxes over a very wide spectral range, but there are some difficulties. A few nebulae may be limited by density rather than by ionization so that all the Lyman-continuum photons are not converted to Balmer and radio emission, with the result that the derived temperatures are too low. In other nebulae with considerable obscuration all the exciting stars may not have been found so that the temperatures derived for the identified stars may be too high. In this investigation it has been assumed that these effects cancel on the average.

Houziaux and Butler (eds.), Ultraviolet Stellar Spectra and Ground-Based Observations, 59–63.

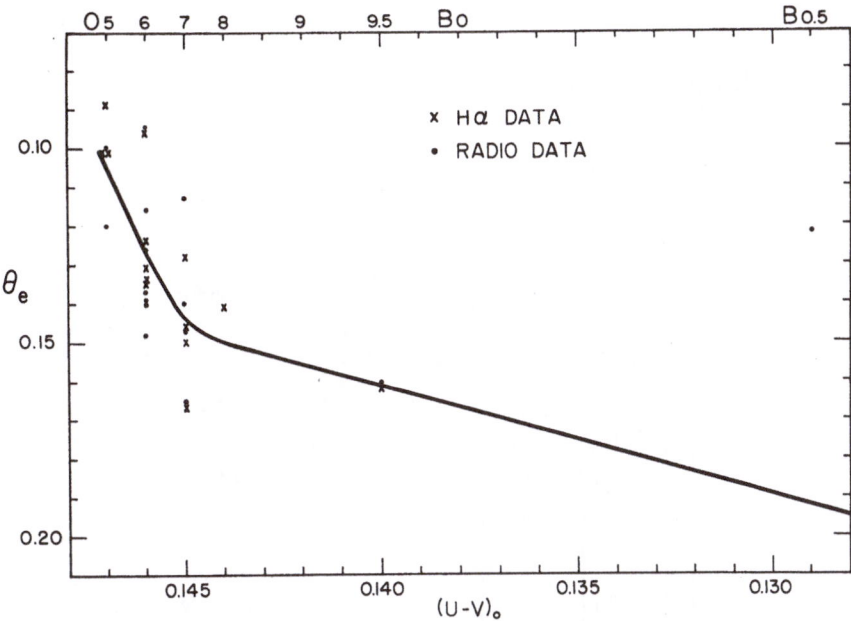

Fig. 1.   Reciprocal effective temperatures ($\theta_e = 5040.2/T_e$) of O stars in nebulae.

Figure 1 is a plot of the effective temperatures expressed as $\theta_e = 5040.2/T_e$ for nebulae where only one exciting star has been found. The line has been drawn through the average for each spectral type and fitted to the Morton-Adams scale for the B stars, since this agrees reasonably well with that proposed by Brown *et al.* (1967) from measurements with the intensity interferometer. The fourth column in Table I lists the adopted mean temperatures. The curve shows a sharp rise beginning at O7, and consequently the new temperatures for O5 and O6 are much hotter than the old scale. Full details of this research have been described elsewhere by Morton (1969).

The data on two nebulae with Wolf-Rayet stars suggest that HD 219460 (WN4.5 + BO) has a temperature around 50000 °K, while HD 168206 (WC8 + BO:) cannot be much hotter than 35000 °K.

TABLE I

Effective temperatures and bolometric corrections for O stars

Spectrum	$(U-V)_0$	T_e(K) Morton-Adams	T_e(K) from nebulae (adopted here)	B.C.
O5	−1.47	37500	48000	−4.32
O6	−1.46	36500	40000	−3.70
O7	−1.45	35700	35000	−3.27
O8	−1.44	35000	33500	−3.15
O9	−1.42	34300	32000	−3.06
O9.5	−1.40	32100	31000	−2.98

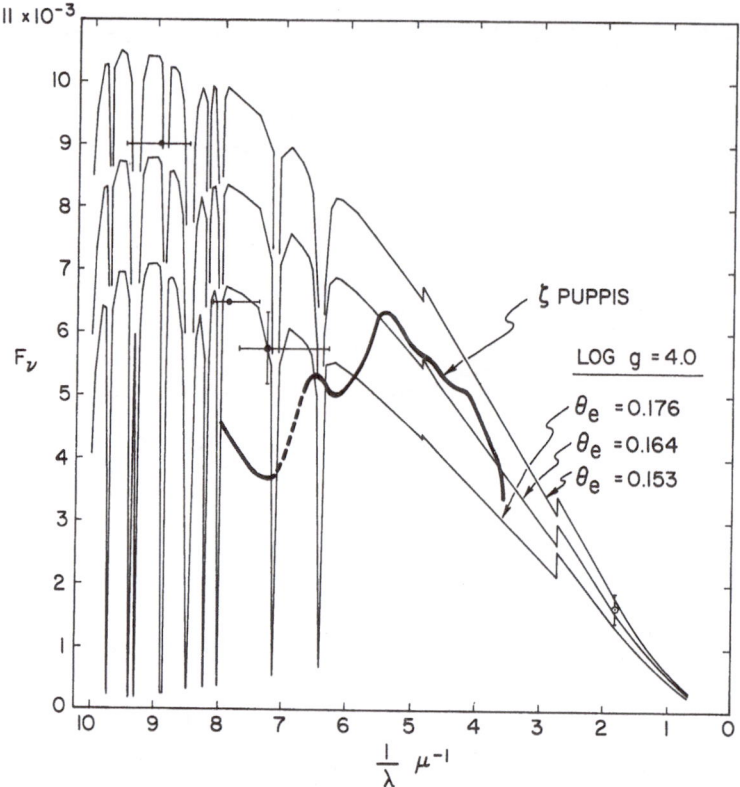

Fig. 2. Absolute surface fluxes of ζ Puppis in units of erg cm^{-2} sec^{-1} Hz^{-1} vs. reciprocal wavelengths in μ^{-1}. Here F_ν must be multiplied by π to obtain the physical flux. The heavy line is from Stecher's ultraviolet scan while the light lines represent three blanketed models. Broad-band measures by Carruthers and Smith are indicated by dots with horizontal lines to show the band widths. The point marked $\bigcirc$ was derived from the visual flux.

In the table, bolometric corrections calculated from the theoretical models are given for the new temperatures derived from the nebulae. These bolometric corrections have a zero point tied directly to the Sun as described by Bradley and Morton (1969). This procedure is to be preferred over the indirect method adopted by Morton and Adams who used the F-type models of Mihalas (1966) which may have serious errors in their ultraviolet fluxes, as will be described later in this symposium by Davis and Webb.

It is of interest to compare the new temperature scale with what we know about one hot O star, ζ Puppis, which is classed O5f. This star probably contributes to the ionization of the great Gum nebula, but unfortunately no Hα or radio measures are readily available for the integrated emission. However, the existing ultraviolet and visual data for the star seem to suggest that it is an exceptional case.

First Smith (1967) measured fluxes in a broad band around 1376 Å and found that the ratio $f_\lambda(1376)/f_\lambda(5475)$, corrected for interstellar extinction, was smaller for ζ Pup than for the O7 stars 15 Mon and ξ Per and the O9.5V star ζ Oph. Secondly, the angular diameter of ζ Pup measured with the intensity interferometer by Davis *et al.* (1969) gives an effective temperature of only 31 000 K for the sequence of model atmospheres used for the nebular temperature scale. Finally, Stecher's (1968) observed ultraviolet flux distribution from 1800 to 2200 Å, when placed on an absolute scale at the stellar surface by the angular diameter, corresponds to a model around 31 000 °K.

These absolute fluxes are shown in Figure 2, where F_ν must be multiplied by π to obtain the physical flux. The visual point derived from the apparent magnitude shows the error bars for the angular diameter, but the uncertainties from the absolute calibration of the V magnitude scale are of the same order. The ultraviolet part of the curve is Stecher's spectral scan with the strong emission and absorption lines smoothed from 1350 to 1750 Å. Also plotted in Figure 2 are blanketed models from Hickok and Morton (1968), and Bradley and Morton (1969) with $\log g = 4.0$ and $\theta_e = 0.153, 0.164$, 0.176, corresponding respectively to $T_e = 32940$, 30730 and 28640 K. (The model for $\theta_e = 0.153$, $\log g = 3.5$ lies between the curves for $\theta_e = 0.164$, $\log g = 4.0$ and $\theta_e = 0.153$ $\log g = 4.0$, so that a slightly hotter temperature would be needed to fit the data to a model with lower surface gravity.) The visual point and the UV flux longward of 1800 Å are reasonably consistent, but at shorter wavelengths there is a major decrease in flux. A check on Stecher's measures is provided by the broad-band observations of Smith (1967) at $1/\lambda = 7.3\mu^{-1}$ and Carruthers (1968, 1969) at 7.9 and $9.0\mu^{-1}$ which give $F_\nu = 5.7 \pm 0.6 \times 10^{-3}$, 6.5×10^{-3} and 9.0×10^{-3} erg cm^{-2} sec^{-1} Hz^{-1} respectively. The values at 7.3 and $7.9\mu^{-1}$ appear to confirm that there is a flux deficiency in ζ Pup below the blanketed model with $\theta_e = 0.164$, but the third point suggests the effect does not extend to $1/\lambda = 9.0\mu^{-1}$. However, the more important result is that for ζ Pup, $T_e \approx 31000$ K, which is much cooler than the temperatures derived from the Hα and radio observations of O5 and even O6 stars imbedded in nebulae.

Acknowledgement

I wish to thank John Davis of the University of Sydney for the opportunity to visit Australia to discuss the problems of stellar temperatures and to cooperate in the measurement of ζ Pup with the Narrabri interferometer.

References

Bradley, P. T. and Morton, D. C.: 1969, *Astrophys. J.* **156**, 687.
Brown, R. Hanbury, Davis, J., Allen, L. R., and Rome, J. M.: 1967, *Monthly Notices Roy. Astron. Soc.* **137**, 393.
Carruthers, G. R.: 1968, *Astrophys. J.* **151**, 269.
Carruthers, G. R.: 1969, *Astrophys. Space Sci.* **5**, 387.
Chalonge, D. and Divan, L.: 1952, *Ann. Astrophys.* **15**, 201.
Davis, J., Morton, D. C., Allen, L.R., and Brown, R. Hanbury: 1969, to be submitted to *Monthly Notices Roy. Astron. Soc.*
Hickok, F. R. and Morton, D. C.: 1968, *Astrophys. J.* **152**, 203.
Hjellming, R. M.: 1968, *Astrophys. J.* **154**, 533.
Mihalas, D.: 1966, *Astrophys. J. Suppl. Ser.* **13**, 1.
Morton, D. C. and Adams, T. F.: 1968, *Astrophys. J.* **151**, 611.

Morton, D. C.: 1969, *Astrophys. J.* **158**, 629.
Pottasch, S. R.: 1965, *Vistas in Astronomy*, Vol. 6 (ed. by A. Beer), Pergamon Press, Oxford, p. 149.
Smith, A. M.: 1967, *Astrophys. J.* **147**, 158, and private communication.
Stecher, T. P.: 1968, *Wolf-Rayet Stars* (ed. by K. B. Gebbie and R. N. Thomas), Nat. Bureau of Standards Special Publ. 307.

Discussion

Stecher: You have left out the emission lines in this consideration. When you add them in, the situation is somewhat improved.

Morton: For simplicity I have smoothed over both emission and absorption lines and they cancel each other to some extent so that the basic discrepancy remains.

Underhill: The O spectral types are assigned according to line ratios, chiefly HeI/HeII. The apparent strengths of these lines are not sensitively affected by T_{eff} and the continuous spectrum of the model. Thus any correlation between models and stars (types) based on identifying stars with models *via* the continuous spectrum will be very uncertain at type O.

Martynov: The temperature for W-R stars given in the paper is much too high: in the eclipsing binary V444 Cygni the surface brightness (in the optical part of spectrum) of the W-R component is much smaller than that of the O-component.

Morton: The Wolf-Rayet temperatures are not very accurate because they depend on only one star each, but a large range of temperatures is indicated. In a more recent investigation of these stars in nebulae I have found one at 31 000 K and another possibly as cool as 25 000 K.

Sunayev: I think the main contribution to the heating of gas in nebulae near hot stars may come from the stellar wind connected with mass loss from the stars. Dr. Tscheglov (Sternberg Astronomical Institute) has found from Hα observations regions in nebulae, with very high velocities (~ 100 km sec^{-1}). Prof. Pikelner has calculated the interaction of a stellar wind with the interstellar medium and showed that in some nebulae there must exist shock waves, which obviously must heat the interstellar gas. The energy in the stellar wind is very high in the case of O and B stars and it may be that its interaction with gas gives the main contribution to heating, and the photon flux in the Lyman continuum region may be not so high.

Morton: Certainly the wind from some O stars will have an important effect on the surrounding interstellar gas, but it is not certain that all stars in my list have significant mass loss.

THE EFFECT OF SILICON AND CARBON OPACITY ON
ULTRAVIOLET STELLAR SPECTRA

OWEN GINGERICH and DAVID LATHAM

Smithsonian Astrophysical Observatory and Harvard College Observatory,
Cambridge, Mass., U.S.A.

Abstract. Silicon and especially carbon bound-free absorptions considerably reduce the emergent flux in the ultraviolet for stars near spectral type A0 ($T_{eff} = 10000 \, \text{K}$). An overabundance of silicon and/or an underabundance of carbon can affect the Balmer discontinuity and the Paschen continuum by a few per cent at most. However, the abundances of these ultraviolet absorbers will have little effect on the temperature distribution calculated for a star if the model is chosen to match the visual spectrum. An examination of the ultraviolet spectrum of Sirius shows that still more opacity is needed; part of this absorption can be supplied by line blanketing.

The importance of bound-free silicon opacity in the ultraviolet spectra of solar-type stars has been appreciated for at least half a dozen years. After the identification in the solar spectrum of the ground-state emission edge by Tousey (1963) and the absorption edge from the first excited level by Gingerich and Rich (1966), computations made at the Smithsonian Astrophysical Observatory predicted that these edges should be visible in hotter stars in spite of the fact that over 99% of the silicon was ionized. Because a negligible amount of the solar flux emerges in the ultraviolet, the silicon absorption has comparatively little effect on the radiation balance. For hotter stars, however, the silicon can play an important role, especially when its abundance is enhanced.

The upper portion of Figure 1 shows the part of the continuous spectrum that can be observed from the ground for a main-sequence A0 star. Two observational parameters are shown: the Balmer discontinuity D_B and the slope of the Paschen continuum S_P. In spite of the different names attached to these parameters, they are both essentially color indices. In effect, they give us a means of comparing temperatures at different depths in the stellar atmosphere. This can be seen more readily from the lower part of the graph, where we have plotted nonobservable information available from the model-atmosphere calculation. On this graph we have plotted, as a function of wavelength, the relative location of optical depth unity. Where the atmosphere is comparatively transparent (for example, just redward of the Balmer discontinuity), we can see to deep, hot layers. In the more opaque regions (such as just to the violet of the Balmer discontinuity), the radiation arises from the higher, cooler layers. Hence, the Balmer decrement gives information about the temperature difference between a deep layer and a shallow layer in the atmosphere, while the slope of the Paschen continuum measures the relative temperature of an intermediate layer.

In Figure 2, this same graph has been extended into the ultraviolet. When the silicon is increased 10 times over the solar abundance, the augmented ultraviolet opacity raises the depth of formation of the continuum below 1527 Å. Now, in this ultraviolet region the tail of the Planck curve is rising increasingly rapidly with increasing temper-

Houziaux and Butler (eds.), Ultraviolet Stellar Spectra and Ground-Based Observations, 64–70.

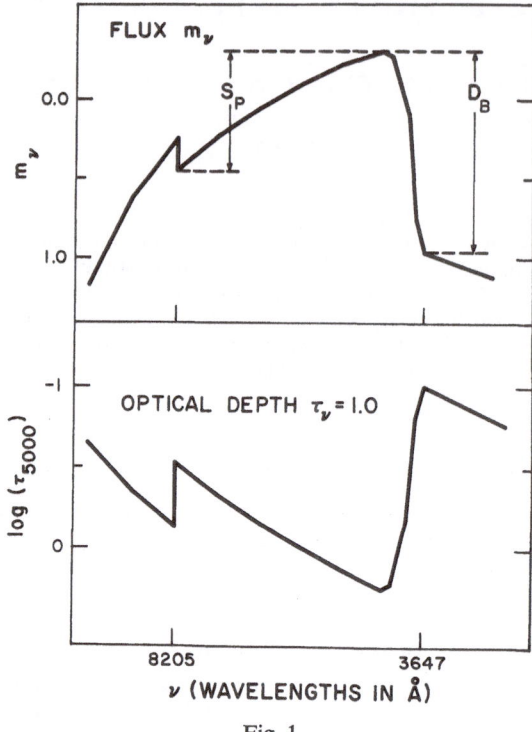

Fig. 1.

ature, and therefore with increasing depth, so that an appreciable fraction of the energy flowing into the outer layers of the atmosphere actually comes from much deeper layers than the optical-depth-unity curve suggests. Increasing the silicon opacity in this part of the ultraviolet effectively blocks this source of energy, and consequently the temperature of the overlying layers must increase in order to maintain the same total flux. With the outermost layers about 200° warmer, the temperature difference on the two sides of the Balmer discontinuity becomes less, and the discontinuity, as measured in magnitudes, becomes less negative. This effect from the silicon opacity alone has been discussed by Strom and Strom (1969).

However, when we include bound-free carbon opacity, which is far more important than silicon at these temperatures, the situation is no longer so straightforward. Figure 2 shows how a normal carbon abundance raises the depth of formation of the ultraviolet spectrum below 1239 Å. The opacity from a normal carbon abundance is so severe that in this wavelength region any given atmospheric layer is not much affected by radiation from adjacent layers. For the same reasons as before, the temperature in the outer layers must increase, but this time the effect reaches deeper and the Paschen continuum as well as the Balmer discontinuity will register a change. Even with the carbon abundance reduced to $\frac{1}{10}$ the solar value, it still dominates the ultraviolet opacity, but the effect on the Balmer discontinuity and Paschen continuum is comparable to enhancing the silicon opacity by a factor of 10.

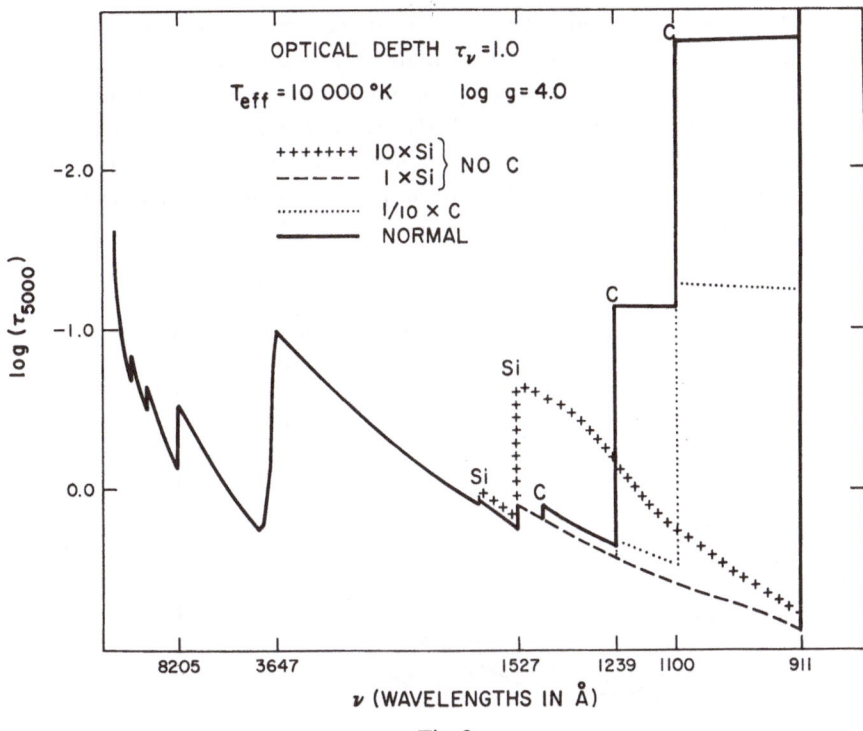

Fig. 2.

The observational effects of including carbon as well as silicon opacity are shown in Figure 3, where the Balmer discontinuity is plotted vs. the slope of the Paschen continuum. In the absence of the carbon opacity, we find the effect reported by the Stroms. When normal carbon is added, the effect of increasing the silicon abundance is halved, so that it becomes only marginally observable. On the basis of these results, we believe that it would be premature to attempt to deduce carbon or silicon abundances from ground-based observations of the continuous spectrum. An additional uncertainty exists on account of the line blocking due to the many strong ultraviolet resonance lines, not included in these models.

On the other hand, the ultraviolet continuum of late B- and A-type stars should be comparatively sensitive to enhanced abundances of silicon or reduced abundances of carbon. Predicted ultraviolet continuum fluxes for several cases are shown in Figure 4. The size of the discontinuity at 1527 Å should provide some indication of any overabundance of silicon, while those at 1100 and 1239 Å could indicate any underabundance of carbon. Notice the enormous change in the flux below 1200 Å in the various models.

We wish to make two separate but intimately related points: (1) If the opacity from a normal carbon abundance is omitted and the model is fitted to the visual continuum, then the total predicted flux would be about 10% too high, which is 250° in effective temperature. (2) Nevertheless, the temperature distribution established by fitting the visual continuum will have only a weak dependence on the detailed choice of the ultraviolet opacities, as we have just shown, and hence abundance analyses based on visual lines will not be critically affected by the amount of carbon or silicon absorption. In other words, *a correct fit of the models in the visual does not necessarily yield the true bolometric correction*; conversely, *a change in the ultraviolet opacities* (and hence in the bolometric correction) *will not vitiate the fit in the visual if the effective temperature is adjusted appropriately.*

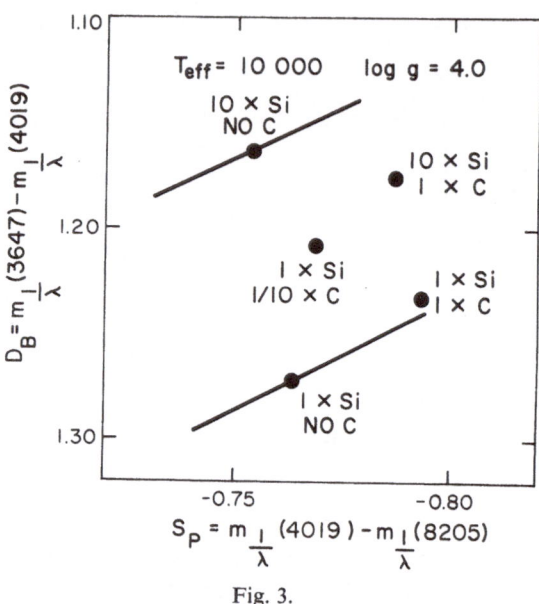

Fig. 3.

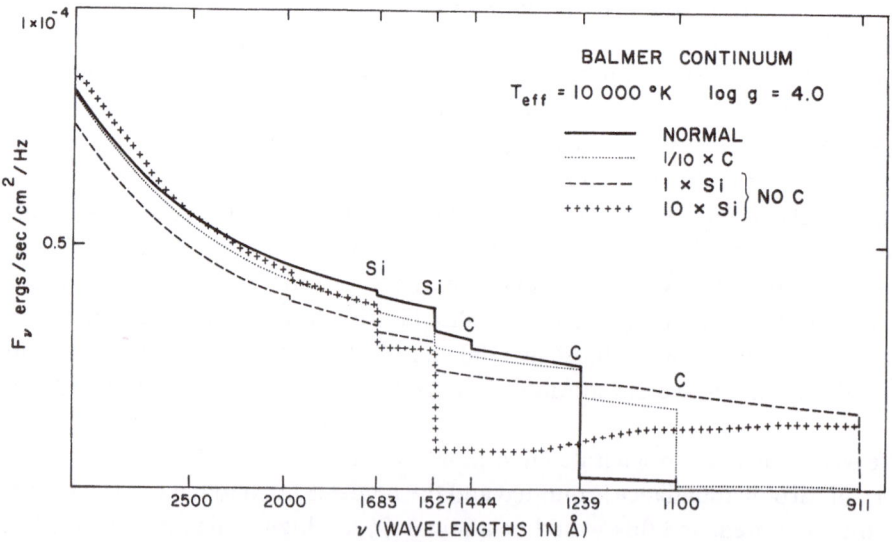

Fig. 4.

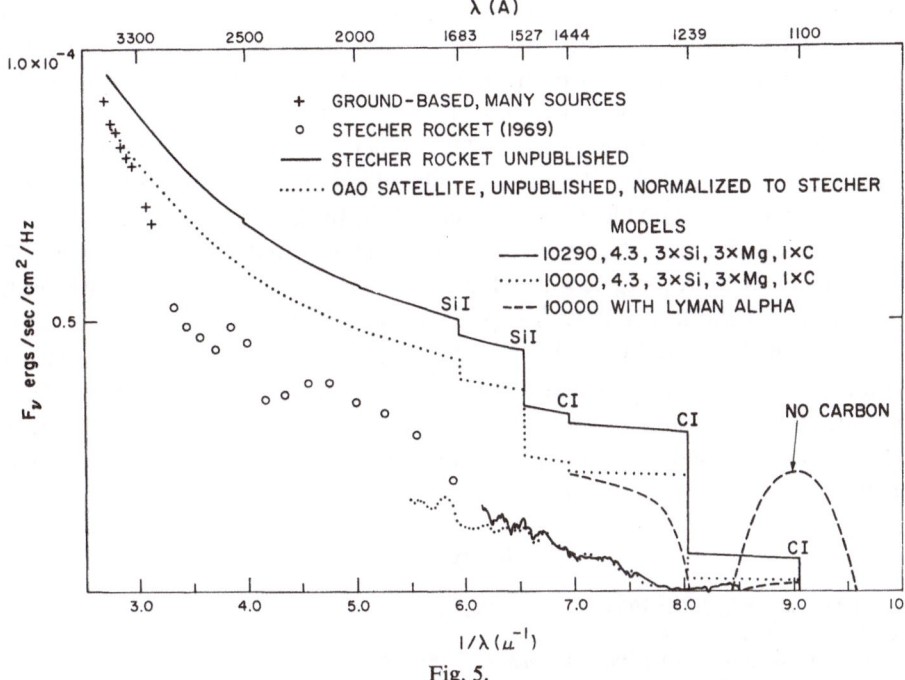

Fig. 5.

We must now inquire to what extent actual observations can be compared against these models. As yet, we have access to comparatively few data, either from direct ultraviolet spectrum scans or from the Celescope experiment. The Celescope filters overlap in just such a way as to make the color index U4-U3 a measure of the flux blocked by the carbon opacity, and U3-U2 a measure of the silicon. However, for the models shown here, the maximum effect on U4-U3 would be about 10%, and on U3-U2, about 30%. Whether the final reductions of the Celescope data will allow us to deduce silicon or carbon abundance differences remains to be seen.

In Figure 5 we compare two models with some recent ultraviolet observations of Sirius. We wish to thank Arthur Code and Theodore Stecher for permission to use some of their data in advance of publication. The slightly enhanced silicon and magnesium abundances of these models have been deduced in Latham's thesis research on the visible spectrum of Sirius. The effective temperatures of the two models differ by only 290°, but at the shortest wavelength this results in a rather dramatic difference in the predicted continuous spectrum. However, the 10000° model is at the cooler limit of matching the visual observations; that is, it just barely matches both the monochromatic surface flux in the visual and the shape of the spectral-energy distribution from 11000 to 3000 Å. A cooler model would require the error of both these observations to exceed 5%. The absolute calibrations of the ground-based observations and Stecher's rocket measurements are independent (Stecher, 1969); but if Stecher's calibration in the middle ultraviolet turns out to be 10 to 20% too low, then the rocket observations would match the ground-based data better at 3000 Å, where the two sets of measurements join. However, the observations would still lie well below the predicted curves throughout the ultraviolet regions. We have adjusted the relative photometry of the Wisconsin Orbiting Astronomical Observatory to match Stecher's curve, but have adjusted his wavelength scale to match the Wisconsin features.

Just longward of the carbon edge at 1239 Å, there are numerous strong resonance lines of CI, SiII, SI, and SII. We have calculated profiles for about 30 of them; we find that many of the equivalent widths exceed 1 Å. In the first 70 Å longward of the carbon edge, our incomplete set of calculations shows that about 50% of the flux would be blocked by resonance lines. Thus, we conclude that the strong bound-free absorption by carbon below 1239 Å is present, but that the sharp absorption edge is entirely smoothed out by lines. On the other hand, we have no explanation for the apparent absence of the silicon edge at 1527 Å. Until many more detailed calculations taking lines into account are available, it will probably be rather difficult to deduce carbon or silicon abundances solely from the approximate level of the far-ultraviolet spectrum.

References

Gingerich, O. and Rich, J. C.: 1966, *Astron. J.* **71**, 161.
Stecher, T. P.: 1969, *Astron. J.* **74**, 98.
Strom, S. E. and Strom, K. M.: 1969, *Astrophys. J.* **155**, 17.
Tousey, R.: 1963, *Space Sci. Rev.* **2**, 3.

Discussion

Stecher: I would believe that hydrogen Lyman-α and Lyman-β would completely obscure the effects of CI since the Lyman lines should be the strongest lines in the spectrum of an A star.

Gingerich: D. Peterson's calculation of the Lyman-α profile, based on the Griem Theory, indicated that carbon would be a much more important cause of opacity since it covers a wider wavelength region. Consequently we identified the Lyman-α core with a smaller feature on your spectrum, and Latham found that a good agreement with individual features of the Wisconsin OAO data could be obtained by sliding your spectrum by 30 Å.

Underhill: It is quite possible that if you included, in a schematic manner, the opacity due to the series of resonance lines approaching the CI and SiI limits you would get a model with the same structure (T, P_e) as for one without lines. This is because the gf value of the lines together is 3 or so times that of the continuum, at least. Your total opacity depends on Ngf and if you increase gf it is not necessary to increase N.

Gingerich: I agree that line blocking can be represented fairly well in a schematic way, as we are doing in the new Smithsonian grid of models. However, if I understand the reason for your remarks, you are suggesting that we use line blocking to replace the extra opacity from the enhanced silicon. This abundance is based on Latham's line analysis in the visual spectrum, not on an attempt to fit the UV. In any event, we need *more* UV opacity, which could be supplied by resonance line blocking.

Praderie: One must stress the urgent need for space ultraviolet observations for late A, F, and G type stars. A proposal for the observation of metallic discontinuities in such stars has been made by Dr. Bonnet and myself as guest observers on the OAO 2. The basis for such a proposal is twofold: firstly to study the variations with temperature and electron pressure of the 2076 Å discontinuity observed by Bonnet (1968, *Ann. Astrophys.* **31**) in the solar spectrum, which is sensitive to temperature as shown from the center to limb variation; secondly to study the variation of the silicon edges at 1520 Å and 1680 Å with spectral type. The magnitude of these predicted discontinuities varies strongly with T_{eff}: at 8000 K, $\log g = 3.9$ and with a normal silicon abundance, $\Delta \log F_\nu = 1.26$ at 1520 Å and $\Delta \log F_\nu = 0.94$ at 1680 Å. These values should be compared with the values computed by Gingerich for a 10000 K star. Moreover, the ratio of the predicted silicon steps depends on the silicon abundance (Praderie, 1968, Third Harvard Conference).

The aspect of the solar silicon edges cannot be interpreted using a classical radiative equilibrium solar model atmosphere; even with a realistic low chromosphere model, an extra absorber must be invoked to account for the 1680 Å discontinuity. Coming back to the observed ultraviolet spectrum of Sirius, one may wonder if such an absorber is still present, due to the absence of the 1520 Å discontinuity. Considering also that the theoretical flux computed by Gingerich is higher than the one observed, the chosen model for Sirius may be questioned as far as the outer layers are concerned; the log (τ_{5000}) vs. λ curve for $\tau_\lambda = 1$ seems to exclude this possibility for an A0 star. But for late A type stars, the UV spectrum emerges from higher layers than in A0 stars and chromospheres are not improbable. A recent and preliminary result on the K line of γ Boo (A7III), observed at 2.5 Å mm^{-1} by Le Contel and myself, favours a chromospheric temperature rise: the K line shows a small reversal.

Bonnet: I would like to stress that two features observed in the Sirius spectrum do appear also in the solar spectrum:

(1) For wavelength longer than 2000 Å a difference appears between computed and measured value of the intensities the latter lying below the former. This could be due either to a new source of opacity or to crowded absorption lines. This last possibility might be settled easily since many *gf* values are now available and the spectra can be computed taking this effect into account.

(2) The main discontinuities are almost completely absent in the spectra of the sun and Sirius. This might be due either to LTE departures or to the existence of a lower temperature gradient than of the models used in the computations.

Gingerich answers to both Praderie and Bonnet:

Mrs Praderie's calculations remind us of the fact that for F stars the ultraviolet opacity (and hence the depth of formation) varies over a much greater range than for A0, and therefore we have the possibility of constructing empirical temperature distributions for F stars from observed UV intensities, just as we can do for the Sun. On the other hand, at 10000° the 1520 Å Si discontinuity is formed at about the same depth as the Balmer discontinuity, so we cannot introduce a lower temperature gradient to explain the absence of the Si edge in Sirius. Non-LTE will tend to wash out the discontinuities, but I believe that numerous absorption lines may be even more effective.

D. OBSERVED STELLAR FLUXES

REVIEW OF ULTRAVIOLET AND VISUAL CONTINUUM OBSERVATIONS AND COMPARISONS WITH MODELS

R. C. BLESS

Space Astronomy Laboratory, Washburn Observatory,
University of Wisconsin, Madison, Wis., U.S.A.

Abstract. This paper first briefly describes model atmosphere grids now available for comparison with observations. The recent recalibration of the absolute energy distribution of α Lyr substantially improves the agreement of models and observations in the visual. Temperature scales determined by various methods agree reasonably well except for the hottest stars. Recent ultraviolet results suggest that earlier observations of O- and B-type stars indicating large flux deficiencies were probably in error. However, late B- and A-type stars may emit less energy in the UV than that predicted by models which do not include the opacities caused by silicon, magnesium, and carbon.

1. Model Stellar Atmospheres

Before comparing the observed continua of stars with model atmospheres, it is appropriate first of all to discuss briefly the present state of model atmosphere calculations.

A few years ago a plateau was reached in the computation of standard, LTE models with continuous opacity sources. Mihalas (1964), Strom and Avrett (1965), and Underhill (1963) computed large grids of models in radiative equilibrium, generally with opacities due to bound-free transitions in H, H^-, H_2^+, He I, He II, free-free transitions, and electron and Rayleigh scattering. Since that time the problem has been to ensure that all of the important sources of continuous opacity have been included, to take into account the effects of the absorption lines, of convection and of rotation and to evaluate the effects of departures from LTE.

The most extensive grids of models which explicitly include absorption lines in the opacities have been calculated by Mihalas (1966), by Mihalas and Morton (1965), by Morton and his students (see, e.g., Hickok and Morton, 1968), and by Klinglesmith and Fischel (1969). Mihalas included the Balmer lines of hydrogen through H 20 in a grid with effective temperatures from 7200 K to 12600 K and with $\log g = 2$, 3, and 4. In this temperature range the Balmer lines are very strong and are located where much of the flux emerges. They are therefore expected to be the main source of line opacity. At higher temperatures the lines most strongly affecting the structure of the atmosphere are those located farther in the ultraviolet, where most of the star's radiation is emitted. In their calculations of hotter models, the Princeton group included approximately one hundred lines that were expected to have equivalent widths of 2 Å or greater. These lines lie shortward of 1600 Å. An approximate expression by Griem (1960, 1962) was used for the broadening in the hydrogen Lyman lines while for the other lines a Doppler-damping profile was taken with the damping constant assumed to be 10 times the classical value. Models appropriate to main sequence and near-main sequence stars with effective temperatures from 14400 K to 37500 K have now

Houziaux and Butler (eds.), Ultraviolet Stellar Spectra and Ground-Based Observations, 73–82.
All Rights Reserved. Copyright © 1970 by the IAU.

been computed in this way. The effect of the line blanketing is to redistribute the flux from the lines to the regions between the lines and to longer wavelengths extending into the visual spectrum. Thus at B0 the visual spectrum looks like that from an unblanketed model with an effective temperature some 1500 K higher.

A few attempts have been made to estimate the blanketing effect in the 2000 to 3000 Å region. Here there are not many strong lines and the opacity will apparently be produced by a large number of relatively weak lines rather than by a few strong lines plus weak lines. Hence the calculations are somewhat more difficult and more statistical in nature. Calculations by Elst (1967) show that at type B1 maximum blanketing for this region is reached near 1900 to 2300 Å where it is slightly more than 0.1 magnitudes per 100 Å. At longer wavelengths blanketing effects are even smaller.

Another large model atmosphere program has been constructed by Klinglesmith and Fischel at Goddard Space Flight Center. In addition to the usual opacities their program includes the first 40 lines of the Balmer and Lyman series of hydrogen. They will soon publish a Goddard Space Flight Center Note consisting of data for 120 models, including many which are hydrogen deficient. At the Smithsonian Astrophysical Observatory, Strom, Gingerich, and their colleagues have developed model atmosphere programs capable of representing stars over a large portion of the HR diagram. This grid will be published in the *Proceedings of the Third Harvard-Smithsonian Atmospheres Conference*. Thus it is now possible to compute reasonably sophisticated model atmospheres for stars of spectral type F and earlier, on and near the main sequence, in which at least some of the effects of line opacities are included.

Recently Gingerich, Strom, and others at the Smithsonian Astrophysical Observatory have recognized that the photoionization continua of magnesium, carbon, and silicon can in certain situations be important sources of opacity, and that possibly the wing of Lyman α may affect the structure of the atmosphere throughout the ultraviolet. Our understanding of this last source of opacity seems to be somewhat confused at this point, however. It is apparent that carbon is important and that the ^3P ground state photoionization limit of silicon at 1520 Å and the corresponding limit from the ^1D first excited state at 1677 Å cause considerable amount of energy to be redistributed at longer wavelengths up to the Balmer limit. An overabundance of Si is apparently the cause of the abnormally small Balmer jump observed in Sirius. A few years ago Mihalas (1965) computed two models appropriate to late A- or early F-type stars in which he included convection by means of mixing-length theory. Compared to purely radiative models, the convective models showed a substantial flux deficit shortward of about 1800 Å. This deficit increased with greater convective efficiency. It was hoped that observations in the ultraviolet would provide a test of these convection models. However, the silicon opacity in these stars may be large enough even with normal abundance to modify the atmospheric structure in such a way as to vitiate these convective models. Thus the observations may yield information on abundance rather than convection.

Recent calculations by Auer and Mihalas (1969) in which the first three levels of the hydrogen atom are permitted to depart from LTE (detailed balancing is not assumed as has been the case in the past) indicate no appreciable change in either the Paschen or Balmer continua compared to the LTE models. This result agrees with observations by Hayes (1967) who found no departures from the LTE values for the ratio of the Balmer and Paschen discontinuities. However, these calculations do show that for the non-LTE model the Lyman continuum is depressed by approximately a factor of 4 at $T_e = 15000$ K and by a factor of 2 at $T_e = 25000$ K. This result could affect considerations of the excitation of gaseous nebulae and in particular the effective temperatures of the exciting stars derived from observations of the nebular flux.

2. Visual Observations

On the observational side, a new spectrophotometric calibration of early-type stars by Hayes (1967) has considerably improved the comparison of continuous energy distributions with predictions of model atmospheres. Figure 1 compares the absolute calibrations of Hayes (1967), Code (1960), and Oke (1964). Oke's was not an absolute calibration in the usual sense; rather he used an unblanketed model computed by Mihalas ($T_e = 9500°$, $\log g = 4.44$), which best represented the data on Vega known in 1964, as an interpolative device. Code's calibration was a composite one, including observations made by several investigators, principally Kienle and his co-workers, Williams, and Hall (see Code, 1960), as well as observations by Whitford and Code in the red. The calibration he adopted was a mean of these measurements and agrees well with the recent measurements of Hayes except in the region from about 5000 Å shortward to the Balmer jump. In this region Kienle's observations were considerably higher than those of Williams, and apparently better represented the continuum of

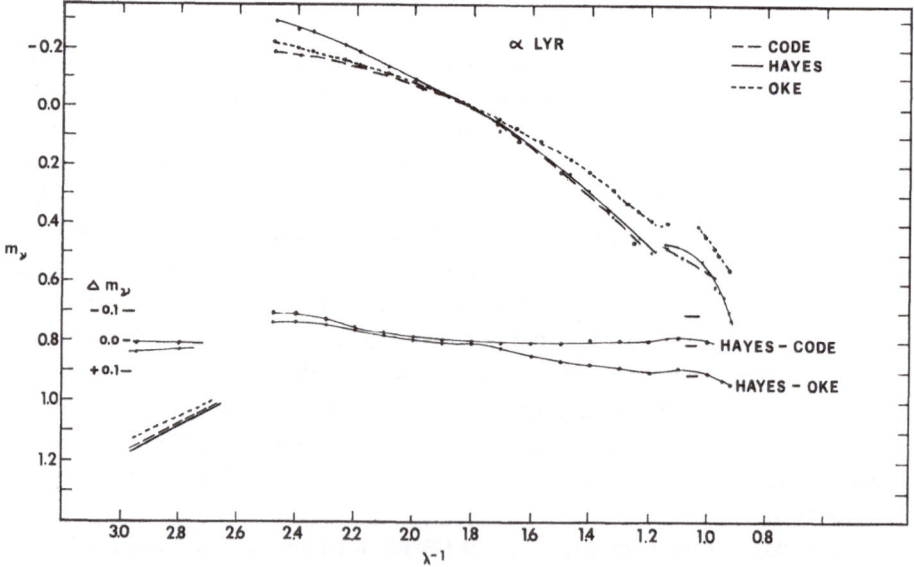

Fig. 1.   Absolute calibration of Vega.

Vega than did the mean Code drew between the two sets of observations. That the Balmer jump should be greater than that given by Code was also indicated by measurements of Bahner (1963), who found good agreement with Code's energy distribution, except for the Balmer jump, which Bahner increased by about $0^{m}.13$. With Code's calibration one cannot fit a model atmosphere to the Paschen continuum both shortward and longward of about 5500 Å. With Oke's energy curve one can fit the Paschen continuum, but the observed Balmer discontinuity is smaller than that predicted by the model. This led to effective temperatures, as determined from the Paschen continuum, sometimes as much as several thousand degrees higher than those found from the Balmer jump for late B- or early A-type stars. With Hayes' calibration the Paschen continuum is steeper, giving a Balmer discontinuity about $0^{m}.13$ larger than Code's. Also, the steeper Paschen continuum implies a higher effective temperature, which requires models with smaller Balmer jumps for stars earlier than A0 and helps to fit the discontinuity. Using this calibration, Wolff et al. (1968) have found that one can generally get good agreement between effective temperatures of A- and B-type stars found from the Paschen continuum and those found from the Balmer jump. Furthermore, the resulting temperature scale agrees well with that found by Hanbury Brown et al. (1967) from their interferometric observations. Earlier in this conference Oke and Schild (this volume, pp. 13–17) described their absolute calibration work at Palomar. Their preliminary results indicate substantial agreement with Hayes' calibration.

More work needs to be done on the absolute value of the flux from a star at a given wavelength. The value of 3.8×10^{-9} ergs cm^{-2} sec^{-1} Å^{-1} at an effective wavelength of 5560 Å for a star with $V = 0^{m}.0$ should be redetermined using a narrower bandpass than the approximately 100 Å width interference filter used by Code and Whitford. We hope to remeasure this quantity at Wisconsin this fall, using as absolute standards both tungsten ribbon filament lamps calibrated by the National Bureau of Standards and the synchrotron radiation from the storage ring near Madison, which is capable of giving absolute as well as relative flux energy distributions. The present quality of spectrophotometric observations and model atmosphere predictions is such that the uncertainty of about 5–10% in the absolute flux value and the possible uncertainty in the effective wavelength at which it is measured is now just beginning to be felt.

Recently there have been several rediscussions of the effective temperature scale, especially for the earlier type stars. Table I compares these temperature scales. The first was given by Hanbury Brown et al. (1967) from their interferometric measurements. The second temperature scale shown is that of Wolff et al. (1968), who compared the Balmer line blanketed models of Mihalas with photometric observations using Hayes' recalibration of Vega. The third scale is that of Morton (1969), and Morton and Adams (1968), who identified stars with blanketed models through the observed UBV colors or the Balmer jump. The last column gives Heintze's (1969) temperature scale. He used line profiles as well as continuous energy distributions for its determination. Going towards earlier type stars the agreement among the four scales is quite good up to $(B - V) \approx -0.15$, where fairly large differences begin to appear. The

TABLE I

Effective temperature scales

$(B-V)$	Hanbury Brown et al.	Wolff et al.	Morton and Adams	Heintze
−0.25				25 500
−0.23	26 600:		19 700	20 400
−0.22	22 900		19 100	19 000
−0.20	18 600	18 500	17 900	16 800
−0.15	14 250	14 800	15 100	13 800
−0.10	12 400	12 500	12 500	12 300
−0.05	11 150	11 000	10 400	10 700
0.00	10 250	10 000	9 600	9 700
+0.05	9 600	9 450	9 150	9 400
+0.10	9 100	8 970	8 780	
+0.15	8 650	8 600	8 430	
+0.20	8 250		8 120	
+0.25	7 850		7 850	
+0.30	7 500		7 600	
+0.40	6 850		7 000	
+0.50	6 325		6 420	
+0.60	5 875		5 920	

uncertainties in the temperature determinations are greater for the hotter stars because the visual energy distributions are insensitive to effective temperature; ultraviolet observations are crucial here. The consistency of the scales determined in the visual is reassuring, but does not guarantee that this is the correct scale since all the models used in the temperature determinations could be incorrect in the ultraviolet. Nearly all the temperature scales may be systematically too hot for stars of earliest type, since model atmospheres with far-UV line blanketing were not generally used, and the effect of this blanketing is to lower θ_e by a few hundredths, as noted earlier.

3. Ultraviolet Observations

Bless et al. (1968) reviewed the existing ultraviolet filter observations and compared them with model atmosphere calculations. We found that for some dozen stars observed by both Wisconsin and Boggess (1967) in the 2000 Å–3000 Å region, the agreement in the absolute fluxes was quite good, ~0^{m}25. This probably represents a reasonable estimate of the accuracy attainable by unstabilized rocket photometry. Shortward of 2000 Å, however, the accuracy is often much poorer than this.

Observations of 35 stars at several wavelengths shortward and longward of 2000 Å, taken from various authors, were compared with models chosen according to their (B−V) colors and the Morton-Adams (1968) temperature scale, using blanketed models wherever possible. Most of the observations fell within ±0^{m}5 of the model predictions with more observations falling below the predicted flux level than above. This result may now be weakened somewhat since OAO and rocket observations suggest that blanketing may not be as great as predicted by the blanketed models. However, these models do not include the effects of weak lines. Flux measurements at 1427 Å and 1115 Å fell far below the model predictions, even with line-blanketed models. Since observations at 1376 Å and 1314 Å were in reasonable agreement with the models, it was suspected that 1427 Å measurements and possibly the 1115 Å measurements as well were in error rather than the model predictions. Recent observations by Carruthers (1969) have shown that the earlier NRL measurements at 1115 Å should be increased by as much as an order of magnitude. I think a cautionary note is appropriate here. A few years ago there were thought to be very serious deficiencies in the observed UV fluxes below about 2500 Å, ranging up to factors of 10 or even 20. Similarly at very short wavelengths, around 1100 to 1400 Å, very large deficiencies were thought to exist. It now appears that the observations around 2500 Å were biased by calibration problems and that the large deficiencies are not real. The same is at least partially true for the shorter wavelength observations. Changes in the temperature scale and the inclusion of some line blanketing in the models also improved the agreement. One must be fully aware of the difficulties of making accurate measurements in this wavelength region. Scattered light often affects the accuracy of laboratory calibrations and may result in an underestimate of observed fluxes. The sensitivity of some detectors decreases with time so that calibrations must be made as near to flight time as possible. Also, in some cases residual atmospheric extinction may not have been fully taken into account, again resulting in an underestimate of the real flux. Ultraviolet deficiencies (with respect to presently published models) may in fact exist, especially at short wavelengths for late B- to F-type stars; however, one must interpret the observations with caution.

Since the review by Bless *et al.* (1968) was written, a few other observations have been published. Yamashita (1968) observed half a dozen early-type stars in three spectral regions from 1060 to 1480 Å. In general, these observations appear to show flux deficiencies up to 1 to 2 magnitudes. However, the uncertainties in the measurements are very large, the smallest being quoted as $0^{m}.30$, the largest as $1^{m}.3$. Thus it is not clear whether real deficiencies exist here. Stecher (1969a) has recently completed data reduction of observations obtained from several rocket flights. He flew an objective grating system which used the rocket roll to scan the spectrum with approximately 100 Å resolution. About 25 stars were observed in this manner, ranging from an O5f star, ζ Pup, to an F5 IV star, α CMi. Stecher feels that his observations indicate considerable line blanketing in the 1800 to 2100 Å range in hotter stars, and near 2300 Å in the cooler stars. Stecher (1969b) has also completed the reduction of data obtained with a 10 Å resolution photoelectric spectrum scanner which he has flown on stabilized Aerobee rockets. These observations cover the spectral range from 3000 Å to Lyman α, and contain a wealth of information on the stronger spectral lines, as well as on the continuous energy distribution. Opal *et al.* (1968) have observed η UMa with an objective prism telescope over the wavelength region 1175 to 1800 Å. The shape of the energy distribution agrees nicely with the model atmosphere predictions. However, the absolute flux is lower than predicted by slightly less than a factor of 2.

This is just within the quoted calibration error and it seems likely to me that the discrepancy is caused by calibration difficulties rather than by an unknown opacity source, especially since this unknown absorber would have to be grey over the spectral region 1175 to 1800 Å.

Additional ultraviolet observations have been made by Henize *et al.* (1968) in the wavelength region 2300 to 5000 Å, with instrumentation carried on the Gemini 10, 11, and 12 spacecrafts. They used a camera with a 30° field of view to which was attached either an objective prism or an objective grating. The former gives a dispersion of 1400 Å per mm at 2500 Å, while the latter gives a dispersion of 184 Å per mm. Several hundred stars have been observed in the Orion and Scorpius regions with the objective prism system. Henize is currently reducing observations of about 15 stars taken with the grating system in Scorpius and Orion. The group at Geneva, under the direction of Golay, is continuing its program of balloon observations with an objective prism camera. Recently they had a successful flight at an altitude of about 40 kilometers. They have recorded the spectra of many stars down to about 2500 Å with resolution as high as 50 Å and are now reducing this large amount of data. Filter photometry in the 2000 to 3000 Å region and 100 Å resolution spectrophotometry have also been carried out by Campbell, Sudbury, and their colleagues at Edinburgh, who flew small telescopes on sounding rockets. These observations are also in the reduction stage.

Later we will hear about additional ultraviolet photometry being done by Soviet, U.S., and French experimenters.

Finally, I would like to say a little about OAO observations. In the paper on interstellar extinction we mentioned one result of wider interest, namely, that unreddened early-type stars of the same visual spectral type have the same continuous spectrum in

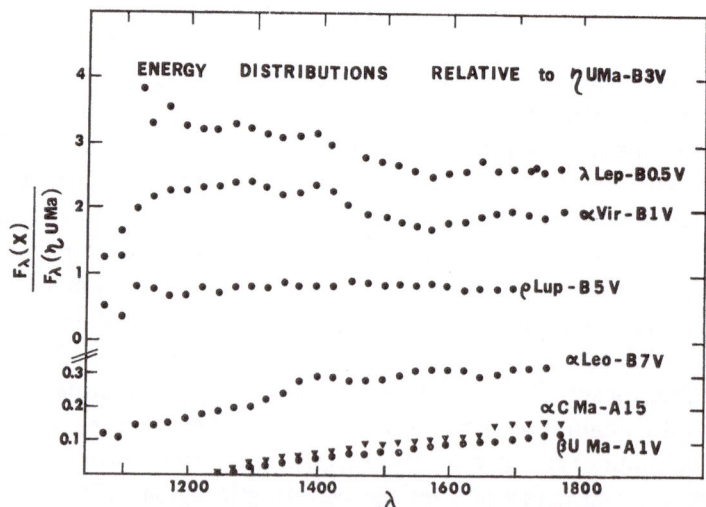

Fig. 2.   Preliminary relative energy distributions in the far ultraviolet derived from Orbiting Astronomical Observatory scans.

the ultraviolet. If this is found to be generally true it will greatly simplify the interpretation and analysis of stellar spectra.

Figure 2 displays preliminary scanner data of seven stars from B0.5 V to A1V, normalized to η UMa, a B3 V star, which was assumed to have $T_e = 17800$ K. The observations were compared with blanketed models with $\log g = 4.0$. Effective temperatures so determined for the B5 and earlier stars are a little hotter by 1000–2000 K than those given by Morton and Adams (1968). Also, the observed fluxes from the cooler stars are less than those predicted by the models by $0^m.5$ to $1^m.0$. Possibly the inclusion of carbon opacity in the models would decrease this apparent discrepancy. Notice particularly the scan of Sirius in Figure 2. A noteworthy feature of this scan is the distinct discontinuity at about 1680 Å although none appears at 1520 Å. The former wavelength corresponds to photoionization from the ^1D level of silicon, about one volt above the ^3P ground state, whose absorption edge lies at 1520 Å. If the jump at 1680 Å is real it is difficult to understand the absence of a discontinuity corresponding to ionization from the ground state.

In addition to the two spectrum scanners mentioned in the extinction paper, the Wisconsin OAO package also includes four telescopes of 20 cm aperture for filter photometry in nine different wavelength bands from Lyman α to the blue. The filter half-widths are about 300 Å. Typically our limit for early-type stars is about 7^m visual, although we have gone considerably fainter. This limit is set primarily by the confusion of other stars in the field for the 10′ field stop we use.

With these photometers we have so far observed about 450 stars having a wide range of spectral type, luminosity class, reddening, and peculiarities. As with the spectral scans, the photometric quality is quite good. Figure 3 shows photometry at

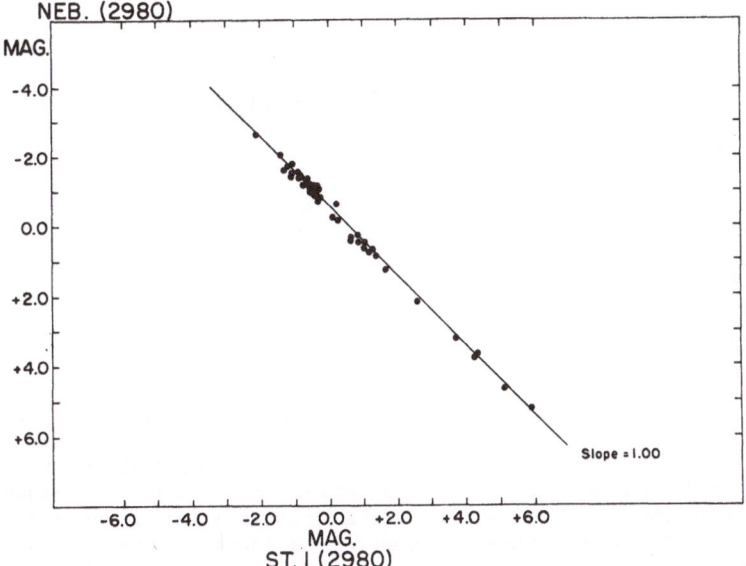

Fig. 3.   2980 Å filter photometry of the same stars with two different OAO instruments.

2980 Å done simultaneously with two different instruments over a period of several weeks. These instruments have nearly identical bandpasses. Over the 7^m range of the observations, the average deviation of an observation is 0^m.02–0^m.03.

Soon, hopefully, already existing OAO data will be put in a form suitable for machine reduction so that the various small corrections can be made where necessary. This large amount of homogeneous data for many stars should enable us to begin answering some of the questions raised here.

References

Auer, L. A. and Mihalas, D.: 1969, *Astrophys. J.* **156**, 681.
Bahner, K.: 1963, *Astrophys. J.* **138**, 1314.
Bless, R. C., Code, A. D., and Houck, T. E.: 1968, *Astrophys. J.* **153**, 561.
Boggess, A.: 1967, private communication.
Brown, R. Hanbury, Davis, J., Allen, L. R., and Rome, J. M.: 1967, *Monthly Notices Roy. Astron. Soc.* **137**, 393.
Carruthers, G. R.: 1969, *Astrophys. Space Sci.* **5**, 387.
Code, A. D.: 1960, in *Stars and Stellar Systems*, Vol. VI, *Stellar Atmospheres* (ed. by J. L. Greenstein), University of Chicago Press, Chicago, Chapter 2, p. 50.
Elst, E. W.: 1967, *Bull. Astron. Inst. Netherl.* **19**, 90.
Griem, H. R.: 1960, *Astrophys. J.* **132**, 883.
Griem, H. R.: 1962, *Astrophys. J.* **136**, 422.
Hayes, D.: 1967, unpublished dissertation, University of Calif., Los Angeles.
Heintze, J. R. W.: 1969, *Bull. Astron. Inst. Netherl.* **20**, 154.
Henize, K. G., Wray, J. D., and Wackerling, L. R.: 1968, *Bull. Astron. Inst. Csl.* **19**, 279.
Hickok, F. R. and Morton, D. C.: 1968, *Astrophys. J.* **152**, 203.
Klinglesmith, D. A. and Fischel, D.: 1969, private communication.
Mihalas, D.: 1964, *Astrophys. J. Suppl. Ser.* **9**, 321.
Mihalas, D.: 1965, *Astrophys. J.* **141**, 564.
Mihalas, D.: 1966, *Astrophys. J. Suppl. Ser.* **13**, 1.
Mihalas, D. and Morton, D. C.: 1965, *Astrophys. J.* **142**, 253.
Morton, D. C.: 1969, *Astrophys. J.*, **158**, 629.
Morton, D. C. and Adams, T. F.: 1968, *Astrophys. J.* **151**, 611.
Oke, J. B.: 1964, *Astrophys. J.* **140**, 689.
Opal, C. B., Moos, H. W., Fastie, W. G., Bottema, M., and Henry, R. C.: 1968, *Astrophys. J.* **153**, L179.
Stecher, T. P.: 1969a, *Astrophys. J.* **74**, 98.
Stecher, T. P.: 1969b *Astrophys. J.*, in press.
Strom, S. E. and Avrett, E. H.: 1965, *Astrophys. J. Suppl. Ser.* **12**, 1.
Underhill, A. B.: 1963, *Publ. Dom. Astrophys. Obs. Victoria* **11**, 467.
Wolff, S. C., Kuhi, L. V., and Hayes, D.: 1968, *Astrophys. J.* **152**, 871.
Yamashita, K.: 1968, *Astrophys. Space Sci.* **2**, 4.

Discussion

Morton: The non-LTE effects which may reduce the Lyman continuum appear to decrease with increasing effective temperature so that we have ignored this effect for the hottest models.

Bless: Two models have so far been published, the hotter one just getting into the T_e range you are interested in. It would be nice to have an even hotter model in order to check this point.

Underhill: Feautrier has constructed a few model atmospheres taking fully into account the effects of non-LTE in the hydrogen spectrum. The result is a rise in the temperature in the outermost part of the atmosphere. This hot plateau is more significant for type A0 than for B0, as I recall, and it may influence the formation of strong lines.

Bless: I think Auer and Mihalas found the same effect.

Morton: I suspect that non-LTE effects will be important in the cores of the lines, but not in the damping wings which are formed deep in the atmosphere and are the main contribution to the blanketing.

Underhill: The rather large differences in fluxes near 2000 Å predicted from A type models when different sources of opacity are used arise chiefly because at this wavelength one is far in the shortward wing of the black body curve for the relevant temperatures. Consequently a change in temperature of a few hundred degrees causes a large percentage change in the small flux emitted. To interpret the absolute fluxes in the region accurately one would need to be very sure of the temperature of the atmosphere.

Gingerich: In the process of computing an extensive grid of model stellar atmospheres (to be published in the *Proceedings of the Third Harvard Smithsonian Stellar Atmosphere Conference*), D. Carbon has investigated the observational effects of a mixing-length theory of convection. These models include opacity from silicon, carbon, and schematic line-blocking. The convective effects are greatest for a main-sequence star with $T_{eff} = 7500°$, and appear in the Balmer discontinuity as well as at 1683^+ Å. The effects are small in the UV, and occur in a region of uncertain opacity. They are essentially independent of the mixing-length chosen. We conclude that UV observations will not be able to distinguish between various amounts of convection.

Henize: Have you observed the predicted flux differences between pole-on and equator-on rapidly rotating stars?

Bless: We have filter photometry observations of rapidly rotating stars, but have not yet reduced them.

Stecher: The star β Lup B5 V looked the same as η UMa B3 V. Is this a property of rotation or is it some other effect such as a classification error?

Bless: I don't think it has to do with rotation, nor do I know how certain the classification is.

THE STELLAR TEMPERATURE SCALE FROM O5 TO A0

D. S. HAYES

Dept. of Physics and Astronomy, Rensselaer Polytechnic Institute, Troy, N.Y., U.S.A.

Abstract. An absolute spectrophotometric calibration of Vega and eleven other standard stars was completed by the author in 1967. This calibration extends from 3200 to 10870 Å, and therefore gives a useful calibration of the Paschen discontinuity as well as the Balmer discontinuity, for which calibrations have been available for some time.

The measured values of the ratio of the size of the Paschen discontinuity to the size of the Balmer discontinuity have been compared with corresponding values predicted by model atmospheres. The comparison shows that LTE models satisfactorily predict the continuum in the wavelength range given above.

A stellar temperature scale has been determined by fitting the measured sizes of the Balmer discontinuity of 43 stars to blanketed model atmospheres. This temperature scale agrees very well with Morton's modified form of the Morton-Adams temperature scale for all except the latest B-stars, for which higher temperatures are predicted.

1. Introduction

In 1967 the author completed a recalibration of the energy distribution of Vega (Hayes, 1967, 1969). It has been shown (Wolff *et al.*, 1968, hereafter referred to as WKH) that this new calibration allows the determination of the temperatures of early-type stars through measurement of the Balmer discontinuity. That paper included stars extending only as early as B3V, and used unblanketed LTE model atmospheres. This paper reports a rediscussion of the WKH approach with the addition of a number of earlier stars, extending the scale to O5, and with the use of the blanketed model atmospheres calculated by the Princeton group (Mihalas and Morton, 1965; Adams and Morton, 1968; Hickok and Morton, 1968; Bradley and Morton, 1969).

In view of the recent discussion about the possible importance of departures from LTE on the Lyman continuum of model atmospheres, it should be made clear that a basic assumption of this paper is that the LTE blanketed models are correct. This means that 'T_e' (or $\theta_e = 5040/T_e$) will be here taken to be a parameter defined by the LTE models. The physical significance of this parameter may still be open to some re-interpretation once accurate non-LTE models become available.

2. Observations

The 43 stars used in this temperature scale have been selected from scanner observations by the author (the data from 3200 to 7350 Å previously published in WKH), from the 13-color photometry by Mitchell and Johnson (1969), and from the list of scanner observations by Whiteoak (1966). In the latter two cases, the observations have been reduced to the calibration of Vega by the author (Hayes, 1967, 1970). Further, they have been selected to represent single, normal near main-sequence stars,

Houziaux and Butler (eds.), Ultraviolet Stellar Spectra and Ground-Based Observations, 83–89.

as much as possible. Exceptions are found in the O-stars, some of which are multiple. Of stars have been excluded. That the stars are single means that double-lined spectroscopic binaries and visual doubles with $\Delta m < 5^{\text{m}}0$ and a separation less than $1'$ have been avoided. Among the stars later than B0, only little-reddened stars have been used.

3. Interpretation of the Observations

The quantity B is defined as being $-2.5\log[F(1/\lambda=2.74^+)/F(1/\lambda=2.74^-)]$, a measure of the Balmer discontinuity which is directly comparable with the models. It is to be distinguished from D, defined by Barbier and Chalonge (Barbier, 1955), which is $\log[F(1/\lambda=2.7^-)/F(1/\lambda=2.7^+)]$. It is also distinguished from D by the method of

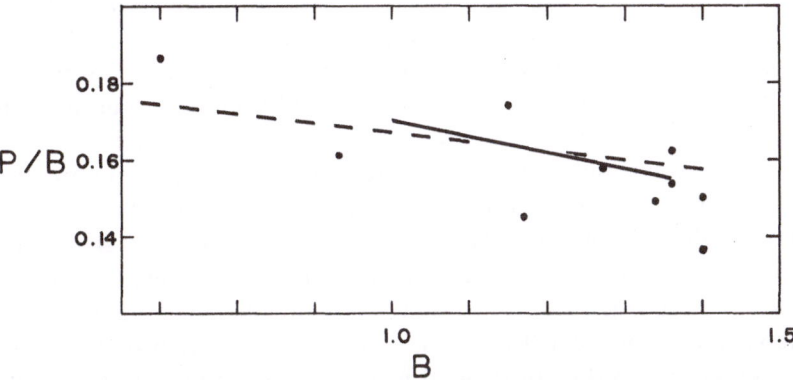

Fig. 1. The ratio of the logarithmic sizes of the Paschen and Balmer discontinuities plotted with respect to the Balmer discontinuity. The dots represent the observations. The dashed line is the locus of LTE models by Mihalas, and the solid line the locus of non-LTE models by Mihalas.

extrapolating to the wavelength of reference from longer wavelengths. A model atmosphere is selected which best fits the Paschen continuum from 4000 to 8000 Å. The model then allows a reproduceable extrapolation to $1/\lambda=2.74^-$. Recall that D, however, uses a linear extrapolation from the blue region, with a much greater chance of errors of judgment. The lower point (2.74^+) is defined by the observations.

The quantity B_c is simply the Balmer discontinuity of the model above, best fitting the Paschen continuum. Finally, $P= -2.5 \log[F(1/\lambda=1.22^+)/F(1/\lambda=1.22^-)]$ is measured differentially with respect to the model.

Because of the linearity and range of photoelectric observations, we should not expect to find any errors in the measured values of B which are a function of spectral type; the only systematic error present should be constant with spectral type and be due only to the calibration of Vega.

The calibration of the Paschen discontinuity allows an evaluation of the success of the LTE models in predicting the visual spectrum. Figure 1 shows the ratio P/B plotted against B. Also shown are the corresponding results for LTE models by Mihalas (1964) and non-LTE models by Mihalas (1968). The non-LTE results of Mihalas indicate that the departures from LTE must be small in the second, third, and fourth levels of hydrogen. The observations confirm this. Thus, theory and observations agree that LTE models satisfactorily predict the stellar continuum in the region from 3200 to 11 000 Å. If the new results reported by Oke and Schild at this symposium (this volume p. 13–17) are taken to indicate a larger value of P, the points in the diagram will move up, but by no more than 0.02 or at most 0.03 in P/B, which will not vitiate our conclusion.

The value of B for Vega, that is the primary calibration, has been changed for this investigation by 0.04, in the sense of making B smaller. This has been done for several reasons, most arising because, in essence, we are performing an interpolation among other, more fundamental, temperature determinations. In particular, the size of the change, 0.04, has been obtained by comparison with the results from O-stars reported by Morton at this symposium (this volume, p. 59–63).

The reasons for the change are as follows. One is that the calibrations by Bahner (1963), Divan (1966) and the new results of Oke and Schild indicate that B for Vega should be smaller by about this amount in the former two cases, and more in the latter case. Another reason is that only if B is decreased by about 0.03 do we obtain agreement between B and B_c; that is, the model which best fits the Paschen continuum should have a value of B_c which is the same as the measured value of B for the star.

A third reason for reducing B for Vega arises because the temperatures are only defined with respect to the LTE models. The largest value of B predicted by the models is about 1.40 at $\theta_e = 0.530$ for $\log g = 4.0$. The calibration of Vega itself gives $B = 1.44$, and the same value for θ Vir. It is not possible to fit these values of B using lower gravities and still have the Paschen continuum fit the lower gravity model. Thus, the calibration and the models are fundamentally incompatible. By reducing B of Vega of 0.04, we obtain agreement. Otherwise, one should not even use these models for determining temperatures, because they would be undefined near A0.

As the classification parameter we have used the color $(U-V)$, instead of the MK type because rotation changes the apparent T_e, without changing the MK type seriously (Collins, 1968), thus giving a spread in T_e at a given MK type. Increasing rotation causes a hypothetical star to move along the mean T_e vs. color line. The color $(U-V)$ is probably the best to use for early type stars because of its long baseline.

4. Auxiliary Color Systems

Since only MK types are available for the O-stars from Whiteoak's list, and since they are all reddened to some degree, it was necessary to find some means to obtain their unreddened colors. It was hoped that an unreddened color for each star could be obtained, rather than depending upon the mean intrinsic colors published by Johnson (1966). In order to obtain the maximum sensitivity to reddening, an auxiliary monochromatic color system was defined. The magnitudes UV, G, and IR represent the scanner points at $1/\lambda = 2.80$, 2.02, and 1.136, respectively. For stars taken from Mitchell and Johnson's 13-color photometry, the following transformations were used:

$$(UV - G) = 1.03 \, (35) - 0.03 \, (33) - 0.04 \, (45)$$
$$(G - IR) = -0.65 + 0.935 \, (52 - 86)$$

where **33, 35, 45, 52**, and **86** denote 5 of the magnitudes in the 13-color system.

A color-color diagram was plotted for the stars, with the reddening line defined by five O8 stars with nearly the same value of B. The unreddened main sequence was defined by the models by Mihalas (1964). The unreddened value $(UV-G)_0$ was obtained for each star. A relation between $(UV-G)$ and $(U-V)$ was established by determining $(UV-G)$ for a large number of early-type stars for which thirteen color photometry was available. All of these stars also had been observed in the UBV system (Johnson *et al.*, 1966). $(UV-G)$ was plotted vs. $(U-V)$ for 78 stars and the mean transformation line determined. With this transformation line the values of $(U-V)_0$ for all of the program stars was obtained. It should be noted that, based upon only a small sample, mean intrinsic colors for the O-stars could be determined, and they differ from those published by Johnson (1966).

5. The Temperature Scale

The values of B for each star can be associated with values of the parameter T_e through the blanketed LTE models of the Princeton group. (In practice, the unblanketed LTE models by Mihalas (1964) were used, and the resulting values of T_e corrected to agree

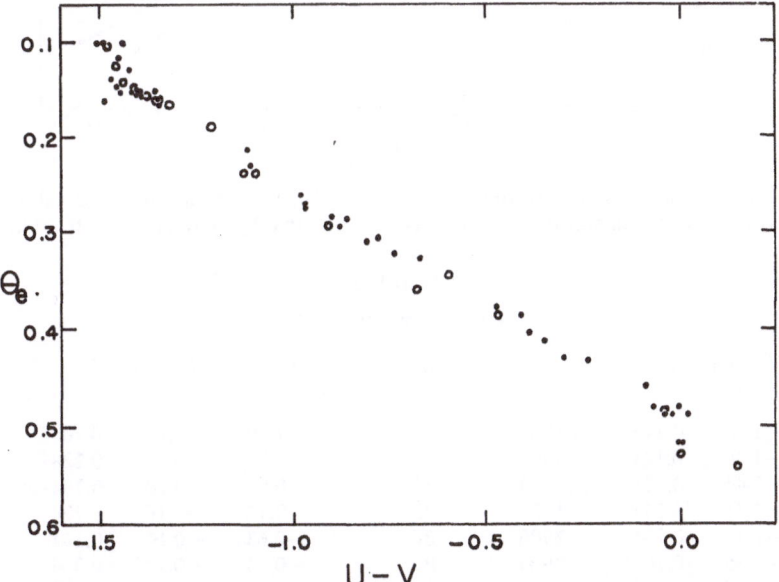

Fig. 2.   The value of $\theta_e = 5040/T_e$ plotted with respect to $(U-V)$. The dots represent the observations from this paper. The circles for $(U-V) < -1.25$ represent the results by Morton, and those for $(U-V) > -1.25$ the Sidney interferometer results.

with the blanketed models.) In Figure 2 are shown the values of θ_e plotted vs. $(U-V)_0$.
Also shown are the Sidney interferometer results (Brown *et al.*, 1967). The low point at
$(U-V) = -0.82$ is α Eri. It is discrepant probably because of an erroneous measurement of $(U-B)$. Morton's results are shown plotted with respect to the mean intrinsic
colors for the early spectral types derived above from the auxiliary color system. It is
quite clear that the present method has given temperatures in excellent agreement with
those obtained by the other two techniques.

A mean line has been drawn through the points shown in Figure 2; it is shown in
Figure 3. The corresponding values are listed in Table I. Although spectral types are
given in this table for convenience, it is to be remembered that the identification of a
color with a spectral type is to be regarded with great caution. Also shown in Figure 3

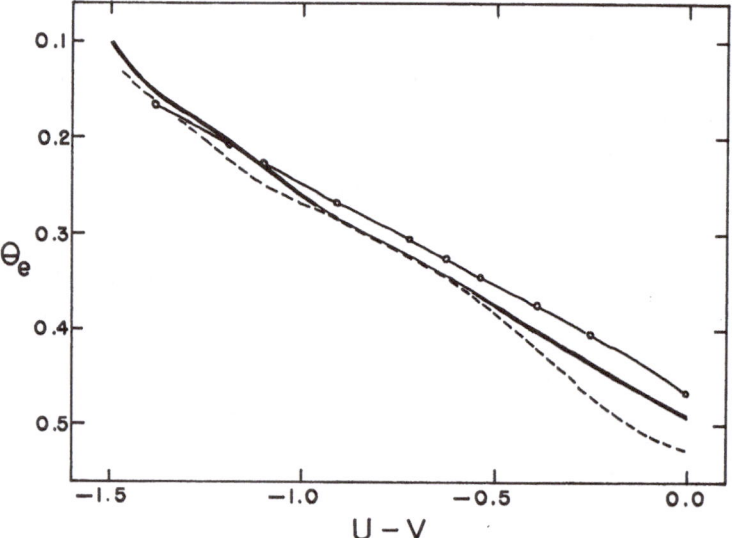

Fig. 3.   The mean line derived from the data shown in Figure 2 (solid line). Also shown are the
Morton-Adams temperature scale (dashed line) and that by Harris (line with circles).

TABLE I

The temperature scale

MK Type	$(U-V)$	θ_e	T_e	MK Type	$(U-V)$	$(B-V)$	θ_e	T_e
O5	−1.48	0.112	45000	B1	−1.19	−0.26	0.204	24700
O6	−1.46	0.123	40900	B2	−1.10	−0.24	0.228	22100
O7	−1.44	0.133	37900	B3	−0.91	−0.20	0.285	17700
O8	−1.41	0.145	34700	B5	−0.72	−0.16	0.324	15600
O9	−1.38	0.154	32700	B6	−0.63	−0.14	0.344	14700
O9.5	−1.35	0.163	30900	B7	−0.54	−0.12	0.364	13800
B0	−1.32	0.170	29600	B8	−0.39	−0.09	0.401	12600
B0.5	−1.28	0.180	28000	B9	−0.25	−0.06	0.435	11600
				A0	0.00	0.00	0.492	10200

are the temperature scales by Harris (1963) and Morton and Adams (1968). It is well-known that the temperatures given by Harris are too high in the middle and later B stars and this is clearly shown here. The agreement with the Morton-Adams temperatures is good in the earliest stars, excellent in the middle B-stars, but poor for the late B stars. The present temperatures should be superior to the Morton-Adams temperatures because the present ones rely upon photoelectric measurements of B, whereas the latter depend upon the older photographic spectrophotometry by Chalonge and Divan (1952).

A comparison of the observed ultraviolet fluxes with those predicted by models has already been given by Bless in his review paper at this symposium (this volume, p. 73) and also by Bless *et al.* (1968). Both of these papers refer to the Morton-Adams temperature scale. If we refer to the differences between the temperature scales cited above, we can conclude that for the early B stars the UV fluxes agree very well with model predictions when the present temperature scale is used. But for the late B stars, the observed fluxes fall significantly below the models.

Acknowledgement

I would like to thank D. C. Morton for allowing me to quote his results in advance of publication.

References

Adams, T. F. and Morton, D. C.: 1968, *Astrophys. J.* **152**, 195.
Bahner, K.: 1963, *Astrophys. J.* **138**, 1314.
Barbier, D.: 1955, 'Principes fondamentaux de classification stellaire', *Colloques Int. Centre Nat. Rech. Sc.*, Paris, **55**, 47.
Bless, R. C., Code, A. D., and Houck, T. E.: 1968, *Astrophys. J.* **153**, 561.
Bradley, P. T. and Morton, D. C.: 1969, *Astrophys. J.* **156**, 687.
Brown, R. Hanbury, Davis, J., Allen, L. R., and Rome, J. M.: 1967, *Monthly Notices Roy. Astron. Soc.* **137**, 393.
Chalonge, D. and Divan, L.: 1952, *Ann. Astrophys.* **15**, 201.
Collins II, G. W.: 1968, *Astrophys. J.* **152**, 847.
Divan, L.: 1966, *IAU Symposium* **24**: *Spectral Classification and Multicolour Photometry* (ed. by K. Loden, L. O. Loden, and U. Sinnerstad), Academic Press, London and New York, p. 311.
Harris III, D. L.: 1963, 'The Stellar Temperature Scale and Bolometric Corrections' in *Stars and Stellar Systems*, Vol. III, *Basic Astronomical Data* (ed. by K. Aa. Strand), University of Chicago Press, Chicago, p. 263.
Hayes, D. S.: 1967, Unpublished Ph.D. dissertation, University of California at Los Angeles.
Hayes, D. S.: 1970, *Astrophys. J.* **159**, 165.
Hickok, F. R. and Morton, D. C.: 1968, *Astrophys. J.* **152**, 203.
Johnson, H. L.: 1966, *Ann. Rev. Astron. Astrophys.* **4**, 193.
Johnson, H. L., Mitchell, R. I., Iriarte, B., and Wisniewski, W. Z.: 1966, *Comm. Lun. Planet. Lab.* **4**, 99.
Mihalas, D.: 1964, *Astrophys. J. Suppl. Ser.* **9**, 321.
Mihalas, D.: 1968, *Astrophys. J.* **153**, 317.
Mihalas, D. M. and Morton, D. C.: 1965, *Astrophys. J.* **142**, 253.
Mitchell, R. I. and Johnson, H. L.: 1969, *Comm. Lun. Planet. Lab.* **8**, 1.
Morton, D. C. and Adams, T. F.: 1968, *Astrophys. J.* **151**, 611.
Whiteoak, J. B.: 1966, *Astrophys. J.* **144**, 305.
Wolff, S. C., Kuhi, L. V., and Hayes, D.: 1968, *Astrophys. J.* **152**, 871.

Discussion

Houziaux: What is the reason for defining quantities such as B, B_c, P etc., which in fact refer to quantities that do not exist? There are no such things as 'discontinuities'. What you have is a lot of close lines which overlap and gradually merge to reach the continuous spectrum on the short wavelength side of the 'discontinuity'. In certain cases, especially when the 'discontinuities' are small, such an extrapolation may be an additional source of error. Is it not just as good to take the flux ratio at two points where we believe we see the true continuum?

Hayes: Such a flux ratio would be sensitive not only to the size of the Balmer discontinuity, for example, but also to contributions to the slope of the continuum, such as errors in the calibration or interstellar reddening. I wish to isolate the effects of the Balmer discontinuity from those of the slope.

Kuhi: Does this mean that you are changing the absolute calibration again? And if so, what went wrong with the original experiment?

Hayes: Not really. Remember that I am using this technique as a form of interpolating device between the temperatures given by Morton at the high end and those from the Sidney interferometer results at the other end. I also want to emphasize that the temperature scale is constructed with the LTE models, and that one reason for changing the value of B consistently by $0.^m04$ is to make the observations and the models compatible at the low end. I am not recommending this as a general change in my calibration, although it is within the range of the normal errors of measurement of that calibration.

Martynov: The use of symbol T_e for the temperature of the star determined from the measures of outgoing flux only is misleading because T_e usually denotes the effective temperature which can be determined only for stars with known dimensions and distance (Sun and the components of some eclipsing binaries), or stars with known angular diameters. The temperature discussed in the paper by Hayes (as well as by Bless) is 'colour temperature', or 'distribution temperature'.

Hayes: This use of the symbol 'T_e' is an interpretation with respect to the models. Its relation to a physical effective temperature depends upon the accuracy of the models. It is an effective temperature, however, because it depends upon the total radiation predicted by the model.

ON ULTRAVIOLET FLUXES, BOLOMETRIC
CORRECTIONS AND EFFECTIVE TEMPERATURES
OF LATE B TO F STARS

JOHN DAVIS and ROBERT J. WEBB

Cornell-Sydney University Astronomy Centre, University of Sydney, Sydney, Australia

Abstract. Observed ultraviolet fluxes have been compared with the predictions of line blanketed model atmospheres for stars of spectral type later than B5. The comparison reveals significant differences which increase towards later spectral type and which lead to differences between empirical and theoretical bolometric corrections. Empirical bolometric corrections for the spectral range B8–F5 have been computed from a combination of ultraviolet, visual and infrared observational data. These have been used to derive empirically based effective temperatures for five A and F type stars for which angular diameter measurements are available.

1. Introduction

The bolometric correction and effective temperature scales for early type stars have been based on the predictions of theoretical model stellar atmospheres in the past. This is because a large fraction of the emergent flux for hot stars is in the ultraviolet and, without measurements in this region of the spectrum, recourse to theoretical estimates was necessary. Since bolometric corrections and effective temperatures play an essential role in the comparison of observations with the results of stellar-interior calculations it would be desirable to derive them directly from observable quantities wherever possible. The empirical data now available in the form of visual spectral energy distributions, infrared photometry, and ultraviolet flux and angular diameter measurements are adequate for this purpose for main sequence stars in the spectral range B8–F5.

The bolometric corrections and effective temperatures derived from observational data for B8–F5 type stars differ significantly from those based on theoretical models and a comparison of the observed ultraviolet fluxes with the model predictions has revealed the reason for the differences.

2. Ultraviolet Fluxes

A. OBSERVED AND THEORETICAL ULTRAVIOLET FLUXES

Figure 1 shows a comparison of observed ultraviolet fluxes with the predictions of model atmospheres for stars of luminosity classes IV and V and spectral type B5 and later at 2800 Å, 2100 Å and 1376 Å.

In selecting observational data for the comparison, peculiar stars and double and variable stars for which the observed fluxes might be significantly affected by their nature, have been excluded. All the observational data have been converted to fluxes

Houziaux and Butler (eds.), Ultraviolet Stellar Spectra and Ground-Based Observations, 90–99.
All Rights Reserved. Copyright © 1970 by the IAU.

per unit frequency interval, reduced to $V=0.00$, and corrected for reddening where necessary.

The theoretical models chosen for comparison with the observations are the Balmer line blanketed models of Mihalas (1966) with $\log g = 4$. Before the theoretical fluxes could be plotted in Figure 1 it was necessary to identify the models with real stars and to reduce the predicted emergent fluxes to $V=0.00$. The effective temperature scales of Wolff *et al.* (1968) and Hanbury Brown *et al.* (1967) which were based on the Mihalas Balmer line blanketed models may be regarded as an identification of the models

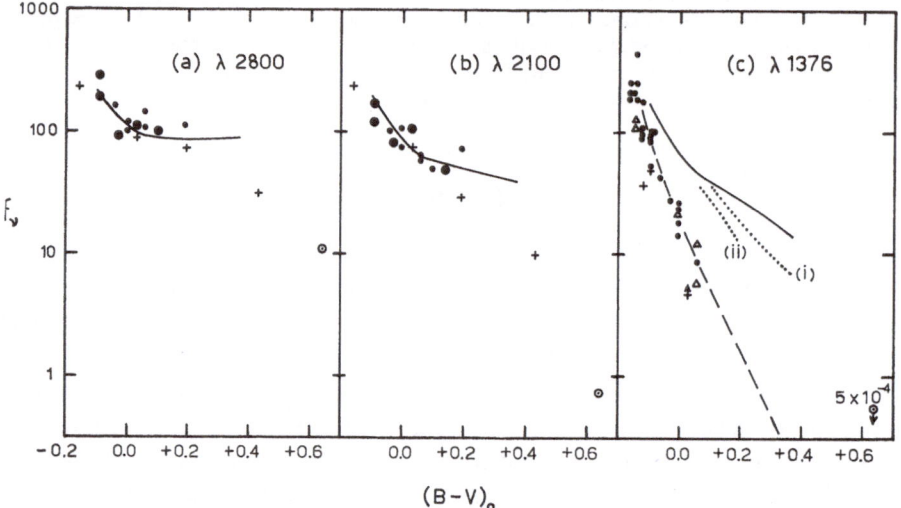

Fig. 1.   Ultraviolet fluxes for main sequence stars and model atmospheres for (a) 2800 Å, (b) 2100 Å and (c) 1376 Å. The full curves represent the predicted fluxes for the Balmer line blanketed models of Mihalas. All data have been reduced to $V=0.00$ (3.78×10^{-20} erg cm^{-2} sec^{-1} Hz^{-1} at $1.83\mu^{-1}$). The ordinate is in units of 10^{-22} erg cm^{-2} sec^{-1} Hz^{-1}. – (a) and (b). Large dots: observations by Bless *et al.*; small dots: observations by Bless *et al.* which were followed by a colon; crosses: observations by Stecher. The solar fluxes are those given by Labs and Neckel. – (c) Dots: observations by Smith at 1376 Å; triangles: observations by Chubb and Byram (1963) at 1427 Å; crosses: observations by Chubb and Byram at 1314 Å. The solar flux is by Detwiler *et al.* (1961). The dotted curves represent the predicted fluxes for the convective models of Mihalas. The curve marked (i) is for $1/H = 1$ and the curve (ii) is for $1/H = 2$. The dashed curve passes through the mean fluxes of Smith at $B-V = -0.09$ and $B-V = 0.00$ and is extrapolated to the solar point.

with real stars and these have been used in conjunction with Mihalas' own identification of his models with stars. To normalise the fluxes to $V=0.00$ the flux at the constant energy reciprocal wavelength (Code, 1960) of the V magnitude system ($1.83\mu^{-1}$) was taken to be 3.78×10^{-20} erg cm^{-2} sec^{-1} Hz^{-1}, a value based on the work of Code (1960), Willstrop (1965), and Labs and Neckel (1968) (see Davis and Webb, 1970). In addition to the fluxes for the Balmer line blanketed models, which are represented by the full curves in Figure 1, fluxes for the convective models of Mihalas (1965) with $1/H=1$ and $1/H=2$ are also plotted in Figure 1c for 1376 Å. The fluxes for the convective models do not differ significantly from those for the Balmer line blanketed radiative models at 2800 Å and 2100 Å.

Figures 1a and 1b show that for 2800 Å and 2100 Å the observations and theoretical predictions are in good agreement for the earlier spectral types considered, but that in passing to later spectral types the observations fall significantly and increasingly below the theoretical curves, the breakaway occurring earlier for the shorter wavelength. The data by Bless *et al.* (1968) and Stecher (1969) at 2500 Å show effects intermediate to those at 2800 Å and 2100 Å. The fact that the observations can be extrapolated smoothly to the solar fluxes confirms that the fall away from the theoretical curves is a real effect.

At 1376 Å the observed fluxes all lie below the theoretical curves and it is noted that even if Smith's absolute calibration of his observational data is wrong, and it appears unlikely that it can be seriously in error, there is a difference in slope between the observations and the theoretical curves. The solar flux falls well below Figure 1c but the dashed curve shows that the observations can be extrapolated smoothly to it. It is concluded that for 1376 Å the fluxes are less than predicted by the Mihalas models and that the discrepancy increases towards later spectral types just as it does at longer ultraviolet wavelengths. The difference between theory and observation is ~ 1.2 mag. at $(B-V)_0 = 0.00$ and the extrapolated curve suggests that it is of the order of 5 mag. at $(B-V)_0 = +0.40$.

The discrepancies between the Mihalas model predictions and the observed ultraviolet fluxes would be greatly reduced by the inclusion in the models of additional sources of ultraviolet opacity such as bound-free absorption by Mg I and Si II (Strom and Strom, 1969) and by C I (Gingerich, 1969), the more important bound-bound metallic and hydrogen absorptions (see for example Mihalas and Morton (1965), Van Citters and Morton (1969)) and, to a lesser extent, by the effects of convection, Mihalas (1965) as shown in Figure 1c. The inclusion of any of these features will affect the relative magnitude of the visual and ultraviolet fluxes. For example, for a given effective temperature the visual fluxes would be expected to increase relative to those predicted by the Mihalas models if the ultraviolet spectrum is depressed. Alternatively, for a given visual flux, the inclusion of additional ultraviolet opacities will lead to a model of lower effective temperature. It is important to take this into account in considering bolometric correction and effective temperature scales based on the identification of observable parameters with model predictions.

3. Bolometric Corrections

The bolometric correction (BC) is, conventionally, the correction required to reduce visual magnitudes to bolometric magnitudes and is defined by

$$BC = m_{bol} - V \tag{1}$$

which is equivalent to writing

$$BC = 2.5 \log \frac{\int_0^\infty f_\nu S_V \, d\nu}{\int_0^\infty f_\nu \, d\nu} + \text{constant} \qquad (2)$$

where f_ν is the flux per unit bandwidth received outside the atmosphere from a star at frequency ν, and S_V is the sensitivity function of the V magnitude system.

The constant in Equation (2) is uniquely defined if we adopt a value for the bolometric correction for a specified spectral distribution of f_ν. In practice the Sun is the only star for which f_ν is sufficiently well known to define the zero point of the bolometric correction scale. For a star of spectral type G2 V a value of $BC = -0.07$ has been adopted by Popper (1959), Harris (1963) and many others. We therefore adopt -0.07 for the bolometric correction for the Sun together with the solar spectral energy distribution tabulated by Labs and Neckel (1968) and the sensitivity function S_V, outside the atmosphere, tabulated by Matthews and Sandage (1963) (their ν_0).

Equation (2) can now be written

$$BC = 2.5 \log \frac{\int_0^\infty f_\nu S_V \, d\nu}{\int_0^\infty f_\nu \, d\nu} + \left\{ -0.07 - 2.5 \log \frac{\int_0^\infty f_{\nu\odot} S_V \, d\nu}{\int_0^\infty f_{\nu\odot} \, d\nu} \right\}, \qquad (3)$$

where $f_{\nu\odot}$ is the solar flux per unit bandwidth from Labs and Neckel (1968) (from their $H(\lambda)$).

In the case of a model stellar atmosphere the bolometric correction is given by Equation (3) if f_ν is replaced by πF_ν.

B. BOLOMETRIC CORRECTIONS FROM OBSERVATIONAL DATA

Bolometric corrections for stars of spectral type later than about F5 can be derived from ground-based observations alone as has been done by Kuiper (1938) and Popper (1959) using radiometric magnitudes, and by Johnson (1966) using multicolour photometry. For earlier type stars a large fraction of the flux is in the ultraviolet region of the spectrum and a reliable bolometric correction can only be derived if the absolute flux in the ultraviolet has been measured with reasonable precision. Because of the present uncertainty in the absolute calibration of observations for $1/\lambda \gtrsim 7.6\mu^{-1}$ it is not possible to derive accurate bolometric corrections for stars of spectral type earlier than about B8.

Empirical bolometric corrections have been computed for the four cases listed in Table I using Equation (3). Ultraviolet fluxes (Smith, 1967, 1969; Bless et al., 1968;

and Stecher, 1969), visual energy distributions (Bahner, 1963; Hayes, 1967; and Wolff *et al.*, 1968) and absolute infrared photometry (Johnson *et al.*, 1966, and Johnson, 1966) were combined to obtain the absolute energy distributions (i.e. the energy distributions in flux units) for the four cases by reducing all absolute fluxes to $V=0.00$ and normalising the visual energy distributions to $V=0.00$ by making $f(1/\lambda=1.83)=3.78 \times 10^{-20}$ erg cm^{-2} sec^{-1} Hz^{-1}.

As an example the normalised empirical fluxes for α CMi are plotted in Figure 2. The continuum flux distribution for a Balmer line blanketed model with $\theta_e=0.75$ and $\log g=4$, obtained by extrapolating the data for the Mihalas (1966) grid of models, has been included in the diagram for comparison with the empirical data. When corrections for line absorption effects between $2.75\mu^{-1}$ and $1.57\mu^{-1}$ (Oke and Conti, 1966) are made to the theoretical continuum fluxes, as shown by the dotted curve, the

TABLE I

Computed empirical bolometric corrections

Spectral type	$B-V$	Empirical BC
B8 V[a]	-0.09	-0.55 ± 0.10
A0 V[b]	0.00	-0.21 ± 0.08
A7 IV, V[c]	$+0.22$	$+0.01 \pm 0.06$
F5 IV–V[d]	$+0.42$	-0.04 ± 0.04

[a] Mean flux data for B8 V used for BC
[b] Mean flux data for A0 V used for BC
[c] Flux data for α Aql used for BC
[d] Flux data for α CMi used for BC

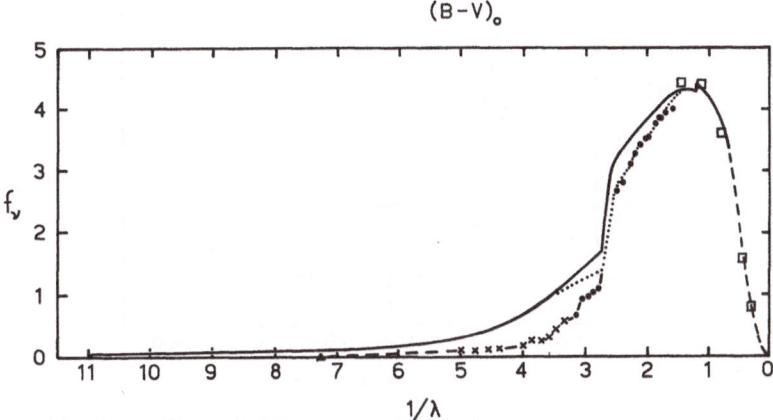

Fig. 2.   Empirical absolute flux distribution for α CMi (F5 IV–V) normalised to $V=0.00$. The data are represented by squares: Johnson (1966); dots: Bahner (1963); crosses: Stecher (1969); triangle: from Figure 1c. Full curve: continuum flux distribution for model with $\theta_e=0.75$ and $\log g=4$ (extrapolated from Mihalas, 1966) normalised to $V=0.00$; dotted curve: model corrected for line absorption; dashed curve: drawn through empirical points. Ordinate: in units of 10^{-20} erg cm^{-2} sec^{-1} Hz^{-1}; abscissa: in μ^{-1}.

agreement with the observational points is reasonably good on the long wavelength side of the Balmer jump. The differences between theory and observation in the ultra-violet are obvious. The empirical absolute energy curve has been completed by taking a linear interpolation between Stecher's point at $5.88\mu^{-1}$ and the point derived from Figure 1 at $7.27\mu^{-1}$.

Since the bandpasses for the observational data avoid the inclusion of Balmer lines, small corrections ($\leqslant 0.04$) have been made to the bolometric corrections derived from the empirical absolute energy curves. The uncertainties in the final values for the empirical bolometric corrections are set primarily by the uncertainties in the absolute calibration of the ultraviolet fluxes and in the far ultraviolet extrapolation, the latter dominating for the B8 V and A0 V cases.

C. DISCUSSION

The empirical bolometric corrections have been plotted in Figure 3 and a curve drawn through them to join smoothly to the point representing the Sun. For comparison, bolometric corrections for the Mihalas Balmer line blanketed models computed from the equivalent of Equation (3) (Davis and Webb, 1970*) are also shown. The empirical values fall below the curve through the theoretical points as would be expected from the differences in the ultraviolet fluxes.

The empirical bolometric corrections tabulated by Popper (1959), which give BC = -0.07 for a star of the same colour as the Sun, are also plotted in Figure 3 and the

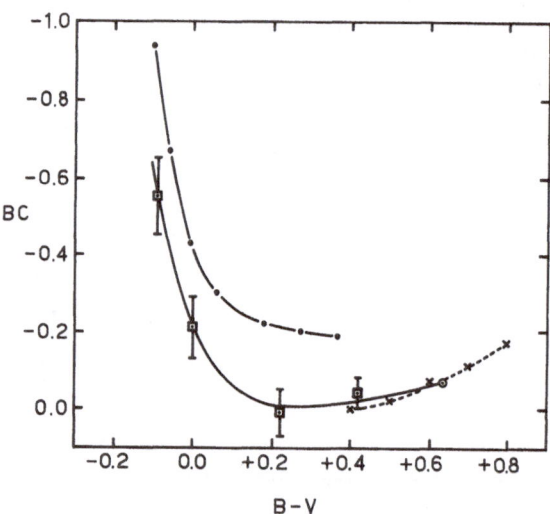

Fig. 3. Bolometric corrections for main sequence stars. Dots: Mihalas Balmer line blanketed model values from Davis and Webb (1970); crosses: mean empirical values (Popper, 1959, his Table II); squares: empirical values from Table I. The curves have been fitted to the points by eye in each case.

---

* This reference contains a detailed discussion of bolometric corrections derived for model atmospheres.

new results show good agreement with them. The small difference in slope in the overlapping region which is suggested by the curves drawn through the data can be explained by the fact that Popper's bolometric corrections do not include corrections for the ultraviolet flux for $1/\lambda > 3.3\mu^{-1}$.

Johnson (1966) has given empirical bolometric corrections which are appreciably greater than the values presented in Table I. This is expected since Johnson based his calculations for stars hotter than the Sun on estimates of the ultraviolet flux made from the data available at the end of 1965. These data were generally lower than the more recent data, on which the present values are based, and therefore led to less negative values for the bolometric corrections.

4. Effective Temperatures of A and F Type Stars

The new empirical bolometric corrections can be combined with measurements of stellar angular diameters to give empirical effective temperatures. Table II lists five stars in the spectral range A–F whose angular diameters have been measured by Hanbury Brown et al. (1967). The bolometric corrections given in Table II are, for α Aql and α CMi, the values derived for these stars; for α Lyr and α CMa, the value derived for $B-V=0.00$; and for α PsA, the interpolated value given by the empirical curve in Figure 3 for $B-V=+0.09$. In addition to the photometric data for these stars (Johnson et al., 1966) and their angular diameters (Hanbury Brown et al., 1967), the angular diameter (1919″), the V magnitude (-26.74 from Johnson (1965)), the effective temperature (5780 K from Labs and Neckel (1968)), and the bolometric correction (-0.07 by definition, Section 3a) of the Sun have been used to obtain the effective temperatures in column 5 of Table II. These effective temperatures are based entirely on observational data and in the cases of α Aql and α CMi (except for the ultraviolet flux at 1376 Å which is almost negligible) on data from observations of these stars, as opposed to averaged data for stars of the same spectral type. It follows that the effective temperatures for α Aql and α CMi are the best values that can be determined specifically for them.

TABLE II

Effective temperatures for 5 stars of spectral type A–F

BS	Star	Spectral type	BC	$T_e(K)^a$ (BC)	$T_e(K)^b$ (M)	$T_e(K)^c$ (HB)
7001	α Lyr	A0 V	-0.21	9330	9700	9500
2491	α CMa	A1 V	-0.21	9910	10750	10380
8728	α PsA	A3 V	-0.07	8970	9550	9300
7557	α Aql	A7 IV, V	$+0.01$	8110	8600	8250
2943	α CMi	F5 IV–V	-0.04	6470	–	–

[a] Effective temperature from empirical bolometric correction.
[b] Effective temperature from theoretical model bolometric correction.
[c] Effective temperature given by Hanbury Brown et al. (1967).

We note that the empirical effective temperatures could have been obtained directly from the absolute flux distributions by using

$$\sigma T_e^4 = \int_0^\infty \pi F_\nu \, d\nu = \frac{4}{\theta_{LD}^2} \int_0^\infty f_\nu \, d\nu, \tag{4}$$

where f_ν is the flux per unit bandwidth received outside the Earth's atmosphere and θ_{LD} is the angular diameter of the star. This is completely equivalent to using the bolometric correction since the solar data enter only because the Sun is used to fix the arbitrary constant in the definition of the bolometric correction (Section 3a).

Effective temperatures based on the theoretical bolometric corrections in Figure 3 will be systematically higher than those based entirely on empirical data as a consequence of the excess ultraviolet flux predicted by the Mihalas models. Although they have no real significance, effective temperatures from the theoretical bolometric corrections have been included in column 6 of Table II for comparison purposes.

Column 7 of Table II contains the effective temperatures assigned to the stars on the basis of a comparison of the empirical and theoretical model fluxes at 4425 Å by Hanbury Brown *et al.* (1967). It is to be expected that temperatures based on a comparison of observed fluxes in the visual region of the spectrum with those predicted by the Mihalas models should be higher than those based on empirical flux distributions and this is borne out by a comparison of the figures in Table II.

It is clear that effective temperatures based on the Mihalas models should be treated as upper limits to the interpretation of the observations. We believe that effective temperatures based on the empirical bolometric corrections are to be preferred but since additional angular diameter measurements should be available in the near future we have refrained from tabulating an empirically based effective temperature scale.

5. Conclusions

It has been demonstrated that empirical bolometric corrections and effective temperatures can be determined for main sequence stars in the spectral range B8 to F5 from the available data. These empirical results are preferred to results based on present model atmosphere predictions because of large differences between the predicted and observed ultraviolet fluxes.

The Mihalas (1966) Balmer line blanketed models predict ultraviolet fluxes significantly greater than are observed and this points to the need for new models for late B and A type stars which take into account additional sources of continuous and line absorption and which include the effects of convective energy transport for the cooler models.

Although it does not follow from the present discussion that the predictions of the hotter ultraviolet line blanketed models are incorrect, it is of great importance to have reliable far ultraviolet fluxes in order to test them. The availability of such measurements would allow empirical bolometric corrections and effective temperatures to be obtained for stars of spectral types earlier than B8.

The time is approaching when the combination of ultraviolet flux data and visual and infrared observations with angular diameter measurements will allow the effective temperature scale for early type stars to be put on a sound observational basis.

Acknowledgements

We are grateful to Drs. Daniel M. Popper and Donald C. Morton for their interest in this work and for valuable discussions. We are also indebted to Drs. O. Gingerich, D. C. Morton and T. P. Stecher for making available results in advance of publication, to Drs. D. Mihalas and A. M. Smith for supplementary data, and to Professor R. Hanbury Brown for his interest and support. This investigation formed part of the programme of the Stellar Intensity Interferometer group which is supported by the Air Force Office of Scientific Research (Office of Aerospace Research) of the United States Air Force under grant (AFOSR-68-1560), the Australian Research Grants Committee, the Research Grants Committee of the University of Sydney and the Science Foundation for Physics within the University of Sydney. One of us (R. J. W.) gratefully acknowledges the tenure of a Commonwealth Scholarship.

References

Bahner, K.: 1963, *Astrophys. J.* **138**, 1314.
Bless, R. C., Code, A. D., Houck, T. E., McNall, J. F., and Taylor, D. J.: 1968, *Astrophys. J.* **153**, 557.
Brown, R. Hanbury, Davis, J., Allen, L. R., and Rome, J. M.: 1967, *Monthly Notices Roy. Astron. Soc.* **137**, 393.
Chubb, T. A. and Byram, E. T.: 1963, *Astrophys. J.* **138**, 617.
Code, A. D.: 1960, in *Stars and Stellar Systems*, Vol. VI, *Stellar Atmospheres* (ed. by J. L. Greenstein), University of Chicago Press, Chicago, Chapter 2, p. 50.
Davis, J. and Webb, R. J.: 1970, *Astrophys. J.* **159** (in press).
Detwiler, C. R., Garrett, D. L., Purcell, J. D., and Tousey, R.: 1961, *Ann. Geophys.* **17**, 263.
Gingerich, O.: 1969, private communication.
Harris, D. L. 1963, in *Stars and Stellar Systems*, Vol. III, *Basic Astronomical Data* (ed. by K. Aa. Strand), University of Chicago Press, Chicago, Chap. 14, p. 263.
Hayes, D.: 1967, unpublished dissertation, University of California, Los Angeles.
Johnson, H. L.: 1965, *Comm. Lunar Planetary Lab.* **3**, 73.
Johnson, H. L.: 1966, *Ann. Rev. Astron. Astrophys.* **4**, 193.
Johnson, H. L., Mitchell, R. I., Iriarte, B., and Wiśniewski, W. Z.: 1966, *Comm. Lunar Planetary Lab.* **4**, 99.
Kuiper, G. P.: 1938, *Astrophys. J.* **88**, 429.
Labs, D. and Neckel, H.: 1968, *Z. Astrophys.* **69**, 1.
Matthews, T. A. and Sandage, A. R.: 1963, *Astrophys. J.* **138**, 30.
Mihalas, D.: 1965, *Astrophys. J.* **141**, 564.
Mihalas, D.: 1966, *Astrophys. J. Suppl. Ser.* **13**, 1.
Mihalas, D. and Morton, D. C.: 1965, *Astrophys. J.* **142**, 253.
Oke, J. B. and Conti, P. S. 1966, *Astrophys. J.* **143**, 134.
Popper, D. M.: 1959, *Astrophys. J.* **129**, 647.
Smith, A. M.: 1967, *Astrophys. J.* **147**, 158.
Smith, A. M. 1969, *Table of Photometric Observations* supplementary to *Astrophys. J.* **147**, 158 (private communication).

Stecher, T. P.: 1969, *Astron. J.* **74**, 98.
Strom, S. E. and Strom, K. M.: 1969, *Astrophys. J.* **155**, 17.
Van Citters, G. W. and Morton, D. C.: 1969, private communication.
Willstrop, R. V.: 1965, *Mem. Roy. Astron. Soc.* **69**, 83.
Wolff, S. C., Kuhi, L. V., and Hayes, D.: 1968, *Astrophys. J.* **152**, 871.

Discussion

Heintze: You are using the solar energy distribution as measured by Labs and Neckel to determine the constant in the formula for the bolometric correction. In spite of the fact that Labs and Neckel's measurements agree with the observed value of the solar constant I am more inclined to trust Peyturaux's (1968) measurements of the solar energy distribution. According to these observations the slope of the continuum from 7000 to 4500 Å is steeper than that according to Labs and Neckel and the difference between them is equal to the difference in slope between Hayes (1967) and the 1964 adoption of the energy distribution of α Lyr (Oke, 1964). According to me (Heintze, 1969; see also Aller *et al.*, 1966) Willstrop's (1965) observations are in agreement with the 1964 adoption of the energy distribution of α Lyr. Labs and Neckel show that their energy distribution of the Sun agrees very well with Willstrop's measurements of a G2 V star. Applying corrections Oke-Hayes on the energy distribution of this G2 V star it is in agreement with Peyturaux's observations. I wonder whether the discrepancies you mentioned will disappear when Peyturaux's measurements are used.

Aller, L. H., Faulkner, D. J., and Norton, R. H.: 1966, *Astrophys. J.* **144**, 1073.
Hayes, D. S.: 1967, Thesis Univ. California Los Angeles.
Heintze, J. R. W.: 1969, *Bull. Astron. Inst. Netherl.* **20**, 154.
Labs, D. and Neckel, H.: 1968, *Z. Astrophys.* **69**, 1.
Oke, J. B., 1964, *Astrophys. J.* **140**, 689.
Peyturaux, R.: 1968, *Ann. Astrophys.* **31**, 227.
Willstrop, R. V.: 1965, *Mem. Roy. Astron. Soc.* **69**, 83.

Davis: I disagree with you concerning the intercomparison of the relative photometry. I find that Willstrop's relative photometry is in good agreement with Hayes' calibration but is not in such good agreement with the calibration adopted by Oke in 1964. Hayes may care to comment?

Hayes: With respect to the comparison of my calibration and Willstrop's, the difficulty is that there are few stars in common. I have used my own and other reliable spectrophotometry reduced to my system to make this comparison, and I find that Willstrop's calibration agrees well with mine.

FAR-ULTRAVIOLET INTENSITIES OF ORION STARS

GEORGE R. CARRUTHERS

*E. O. Hulburt Center for Space Research, Naval Research Laboratory,
Washington, D.C., U.S.A.*

Abstract. Photometric data in the 1050–1180 Å and 1230–1350 Å wavelength ranges, and electronographic spectra in the 1000–1600 Å wavelength range, were obtained in an Aerobee rocket flight on January 30, 1969. The spectral intensities derived from these data for main-sequence stars are in good agreement with the model atmospheres of Morton and co-workers. Giant and supergiant stars, however, appear to be up to one magnitude weaker, at 1115 Å, than main-sequence stars of the same spectral class.

The correction for interstellar reddening appears to be not inconsistent with a $1/\lambda$ extrapolation of earlier determinations of Smith (1967) and Stecher (1965), except in the case of θ Ori, in which the predicted color excess appears to be much too great, confirming the existence of a peculiar reddening law in the Orion Nebula region.

This paper covers some of the results of an Aerobee rocket flight from White Sands Missile Range, New Mexico, on January 30, 1969, in which an electronographic objective spectrograph and ultraviolet photon-counter photometers were used to obtain spectra in the 1000–1600 Å wavelength range, and photometric data in the 1050–1180 Å and 1230–1350 Å wavelength ranges, for early-type stars in Orion. Details of the observations are published elsewhere (Carruthers, 1969b, c, d), as are details of the instrumentation (Carruthers 1969a, d); the present paper will only summarize some of the astrophysically significant results.

The regions of the sky which were covered in the photometer scans are shown in Figure 1. Two photometers for the 1050–1180 Å range (effective wavelength 1115 Å), differing by a factor of 5 in sensitivity, were used, in order to cover a wider range of stellar intensities in this previously little-explored wavelength range. Only one photometer for the 1230–1350 Å range (effective wavelength 1270 Å) was flown; its absolute sensitivity was intermediate between the two 1115 Å photometers. As this photometer saturated on the brightest stars, the number of stars covered in this wavelength range was somewhat less than in the 1050–1180 Å range. However, some stars observed by the 1270 Å photometer were in common with stars observed in an earlier flight in March 1967 (Carruthers, 1968); since we were much more confident of the detector calibrations for this flight than for the earlier one, the earlier results were 'corrected' using the present results. This was also done for the earlier 1115 Å results, which, on the basis of the present data, appear to have been about a factor of 4 too low.

Photometric data of useful quality were obtained for 20 stars or unresolved groupings of stars (see Table I). Tabulated are here the measured photon fluxes, in the ultraviolet and in the visible (the latter was obtained using the visual magnitudes of Iriarte *et al.* (1965) and taking the photon flux at 5560 Å to be 1065 photons cm^{-2} sec^{-1} Å^{-1} (Code, 1960)). Data obtained with the more sensitive of the two 1115 Å photometers are enclosed in brackets. The color excess E(B-V) was determined for

Houziaux and Butler (eds.), Ultraviolet Stellar Spectra and Ground-Based Observations, 100–108.

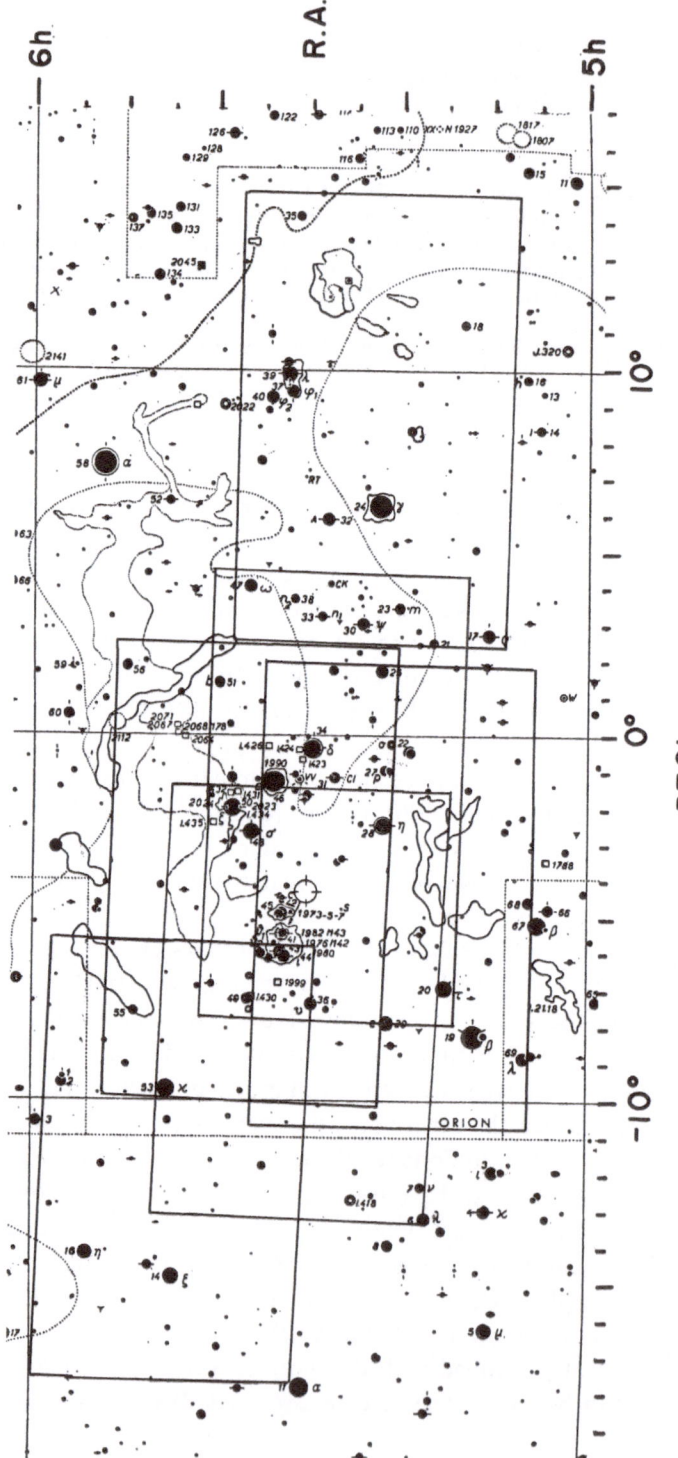

Fig. 1. Regions of the sky covered by the scanning photometers.

TABLE I

The photometric data

Star	SP.	# Obs.	V	N_{5560}	N_{1115}	N_{1270}	$\frac{N_{1115}}{N_{5560}}$ Obs.	$\frac{N_{1270}}{N_{5560}}$ Obs.	$E(B-V)$	$E(1115-V)$	$E(1270-V)$	$\frac{N_{1115}}{N_{5560}}$ C	$\frac{N_{1270}}{N_{5560}}$ C
κ Ori	B0.5 I$_a$	5	2.06	160	806	> 734	5.04	> 4.60	0.04	0.37	0.32	7.08	> 6.18
λ Eri	B2 IV	1 (B)	4.27	20.9	(74.3)	–	(3.56)	–	0.05	0.46	0.40	(5.43)	–
β Ori	B8 I$_a$	1	0.15	928	(92.8)	437.	(0.100)	0.464	0.00	0.00	0.00	(0.100)	0.464
υ Ori	B0 V	7	4.63	15	274(231)	215.	18.23(15.4)	14.4	0.04	0.37	0.32	25.63(21.6)	19.3
ι Ori	O9 III	5	2.76	84 } 96.8	1299	> 758	13.4	> 7.83	0.08	0.74	0.64	26.7	> 14.1
GC 6926(BS 1887)	B0 V		4.77	12.8 }									
θ¹C Ori	O6p	7	5.13	9.4					0.31	2.87	2.48		
θ¹D Ori	O9.5 V		6.70	2.22					0.38	3.52	3.05		
θ¹A Ori	B0.5 V		6.72	2.17 } 26.6					0.28	2.59	1.84		
θ² Ori A	O9.5 V$_p$		5.07	10.0					0.22	2.04	1.76		
θ² Ori B	B0.5 V$_p$		6.41	2.8					0.22	2.04	1.76		
42 Ori	B2 III	7	4.60	15.4	243(203)	197	15.8(13.2)	12.75	0.05	0.46	0.40	24.12(20.2)	18.4
σ Ori	O9.5 V	6	3.83	31.3	465	412	14.84	13.15	0.06	0.555	0.48	24.75	20.5
η Ori	B0.5 V	6	3.32	50.2	408	420	8.13	8.36	0.10	0.925	0.80	19.07	17.5
ζ Ori	O9.5 Ib	6	1.74	215	1813	> 987	8.43	> 4.6	0.06	0.555	0.48	14.06	> 7.15
ε Ori	B0 Ia	6	1.70	223	1548	> 925	6.94	> 4.15	0.05	0.46	0.40	10.60	> 6.00
δ Ori	O9.5 II	6	2.21	139	1214	> 860	8.73	> 6.20	0.06	0.555	0.48	14.56	> 9.65
25 Ori	B IV$_{pe}$	1	4.95	11.15 } 16.9	(63.1)	88.5	(3.73)	5.24	0.05	0.46	0.40	(5.76)	7.57
GC 6800	B2 IV		5.67	5.76									
ψ Ori	B2 IV	2	4.61	15.21 } 21.8	232(111)	173	10.6(5.10)	8.06	0.02	0.185	0.16	(6.05)	9.22
33 Ori	B3		5.52	6.57									
23 Ori	B1 V	1	5.00	10.65	(40.9)	67.6	(3.83)	6.34	0.11	1.02	0.88	(9.71)	14.28
ω Ori	B3 III	1	4.59	15.55	(32.8)	41.2	(2.11)	2.65	0.09	0.83	0.72	(4.54)	5.14
32 Ori	B5 IV	1	4.20	22.19	(32.8)	49.2	(1.48)	2.21	0.03	0.28	0.24	(1.90)	2.76
γ Ori	B2 III	2	1.63	238	1548	> 1032	6.50	> 4.34	0.03	0.2	0.24	8.40	> 5.40
φ¹ Ori	B0 IV	2	4.42	18.2	165(103)	135	9.12(5.67)	7.45	0.15	1.39	1.20	33.73(21.0)	22.5
λ Ori	O8	2	3.39	47	522	367	11.11	7.8	0.13	1.20	1.04	33.66	20.35

each star using the observed colors of Iriarte *et al.* (1965) and the intrinsic colors of Johnson (1963).

The measured ultraviolet fluxes were then corrected for interstellar extinction in the manner of Smith (1967), using extinction factors at 1270 Å and 1115 Å obtained by a $1/\lambda$ extrapolation of the extinction measurements of Stecher (1965) and Smith (1967) to shorter wavelengths. These extinction factors, combined with the observed $E(B-V)$, yield the tabulated ultraviolet color excesses and corrected photon flux ratios.

It was then found that, for the main-sequence stars σ, υ, and η Ori, there is good agreement between their corrected photon flux ratios and those of the blanketed B0 V model atmosphere of Hickok and Morton (see Figure 2), with the differences among the three stars essentially as expected from their different spectral classes (O9.5, B0, B0.5, respectively). Also, the combined flux of ι Ori (O9 III) and BS 1887 (B0V) is in good agreement with the main-sequence model. The later-type stars ω Ori (B3 III) and 32 Ori (B5 IV) are in good agreement with the blanketed B4 V model of Adams and Morton (1968). However, at 1115 Å, the giant and supergiant stars γ, δ, ε, ζ, and κ Ori are significantly less bright, relative to the visible, than main-sequence stars of similar spectral classes. In particular, κ Ori appears to be about one magnitude below the main-sequence brightness, as is confirmed by comparison with ι Ori in both this and the previous flight. This deficiency is in part expected, due to the low surface gravity of these stars, but in addition, there may be continuous absorption in the expanding shells of gas surrounding these luminous stars, which are made evident by P-Cygni-type profiles of the stronger ultraviolet lines (see, e.g., Morton, 1967a, b; Carruthers, 1968; Morton *et al.*, 1969).

Somewhat anomalous results are obtained for λ Ori (O8) and ϕ^1 Ori (B0 IV), in that λ Ori appears deficient at 1270 Å, particularly relative to 1115 Å, whereas ϕ^1 Ori appears too bright at 1270 Å. These stars are separated by only 0.5° in the sky, and are part of the same association, hence the reddening factor is presumably about the same for both. The relative deficiency at 1270 Å in λ Ori could be due to strong absorption in the NV 1239–1243 Å resonance line, which rocket observations have shown to be much stronger than expected in the O stars, such as ζ Pup (Carruthers, 1968; Smith, 1969; Morton *et al.*, 1969). The apparent excess at 1115 Å for λ Ori may be due to over-correcting for interstellar reddening at this wavelength (see discussion of θ Ori, below); this star is somewhat more reddened than most of the others

TABLE II

Corrected photon flux ratios for March 1967 flight
(includes reddening factors)

Star	N_{1115}/N_{5560}	N_{1270}/N_{5560}
γ Vel	25.0	16.8
ζ Pup	31.3	18.7
α CMa	0.07	0.10
κ Ori		8.24
γ Ori		6.63

that were observed. The excess brightness at 1270 Å for ϕ^1 Ori could be due to additional, fainter stars and/or emission nebulosity in the field of view of the photometer. A similar explanation may also apply to 42 Ori, which appears too bright in both wavelength ranges. Alternately, this latter star may be misclassified, particularly if it is responsible for exciting the emission nebulosity which surrounds it.

Recent ground-based spectroscopic studies of 42 Ori, by N. R. Walborn of Yerkes Observatory (private communication, 1970) indicate that, in fact, the spectral classification should be B1 V, instead of B2 III as originally listed.

As mentioned previously, the results of our March 1967 flight were corrected by comparison with the present results for stars in common. Table II gives corrected fluxes for three stars not observed in the present flight, and 1270 Å fluxes for two stars for which the 1270 Å photometer in this flight was saturated. Although this is a rather crude method for obtaining stellar fluxes, the results seem to indicate that the hot stars ζ Pup (O5f) and γ Vel (WC7+O7) are somewhat less bright than expected from the model atmosphere theory (Morton, 1969), as also indicated by Smith (1969, private communication) from satellite photometry, particularly in comparison with other O stars such as S Mon (O7). Both ζ Pup and γ Vel exhibit P-Cygni-type profiles in their stronger ultraviolet lines, indicating the presence of expanding shells of gas.

Perhaps the most interesting result of all is that for the group of stars θ Ori, which excites the Orion Nebula. If the observed flux is corrected for reddening as for the other stars, the resulting corrected flux would be far off scale in Figure 2, in fact,

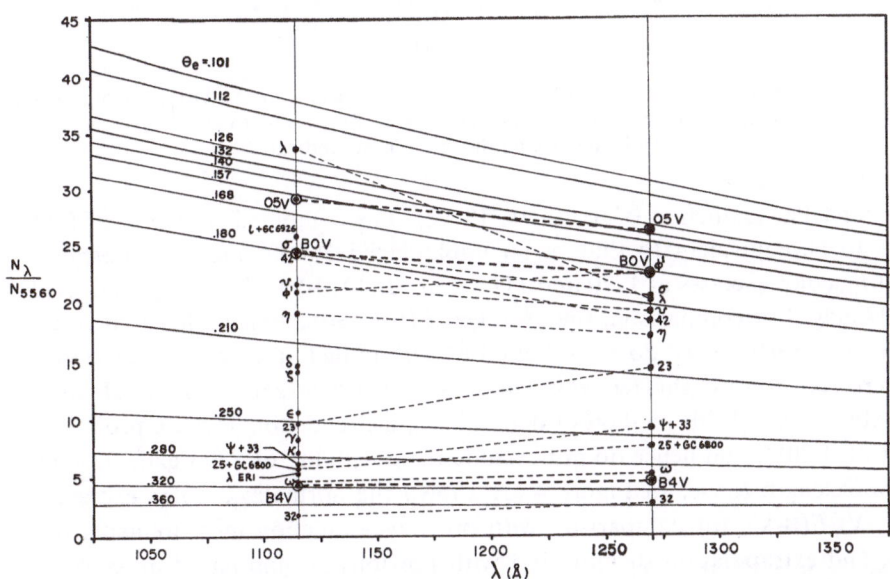

Fig. 2. Stellar photon flux ratios at 1115 Å and 1270 Å, (relative to 5560 Å), compared to model atmosphere predictions. The observed flux ratios have been corrected for interstellar extinction as per Smith (1967). Circled points are averages, over the photometer passbands, of the predicted flux ratios from O5 V and B0 V model atmospheres of Hickok and Morton (1968) and the B4 V model of Adams and Morton (1968). Also shown are curves for unblanketed models of Mihalas (1965).

it would correspond to a flux ratio greater than that of a black body of infinite temperature! Therefore, it is apparent that the reddening correction must be in error. The relative brightnesses of 42, θ, and ι Ori, as measured by the photometers, are confirmed by densitometer traces of the electronographic spectra, as shown in Figure 3. The appearance of the spectrum of θ Orionis does not indicate any substantial contribution of discrete nebular emission lines in the 1000–1350 Å range, and the two subgroups θ^1 and θ^2 Ori appear to contribute comparably (Carruthers, 1969b), though the spatial resolution was not adequate to allow a definitive evaluation of the separate contributions.

Therefore, instead of correcting for the interstellar extinction to obtain intrinsic fluxes, the opposite procedure was used in that intrinsic fluxes were assumed to derive

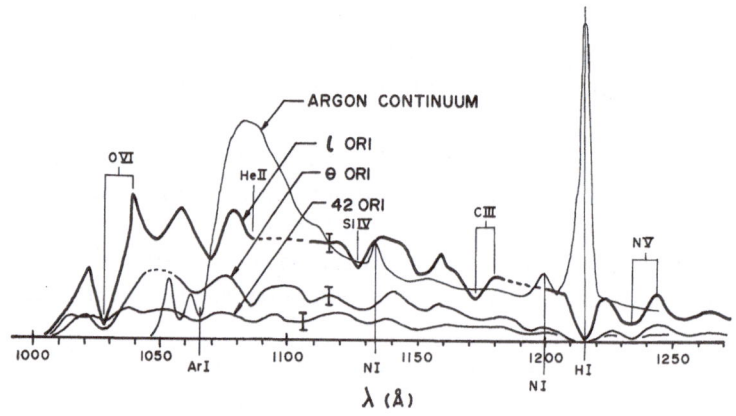

Fig. 3.   Microdensitometer tracings of ultraviolet stellar spectra in the Orion Sword region and preflight calibration spectrum. The limits of error due to grain noise and background fluctuations are indicated.

the interstellar extinction (Carruthers, 1969c). The combined spectrum of θ Ori was taken to be equivalent to that of an unblanketed 30000° model atmosphere. The resulting color excesses at 1270 and 1115 Å are $E(1270\text{-V})=0.53$ mag, and $E(1115\text{-V}) =0.37$ mag. For comparison, the averaged $E(\text{B-V})=0.28$ mag. Hence, it is apparent that a $1/\lambda$ extinction law does not hold for θ Ori, in fact, it appears that the extinction is *decreasing* toward shorter wavelengths in the 1050–1350 Å range. However, the probable errors of this method of determining the reddening law are probably of the order of $\pm 20\%$, and hence do not exclude the possibility of a flat extinction curve in this wavelength range. In Figure 4 are shown the normalized color excesses, $\Delta E= E(M_\lambda\text{-V})/E(\text{B-V})$, for comparison with previous extinction measurements for other stars, and extrapolation thereof, and with previous ground-based measurements of the extinction law for θ Ori.

The present ultraviolet measurements confirm the anomalous nature of the reddening law for the Orion Nebula region. Although no results are available for the intermediate wavelength range 1350–3000 Å, the present results seem to imply that the

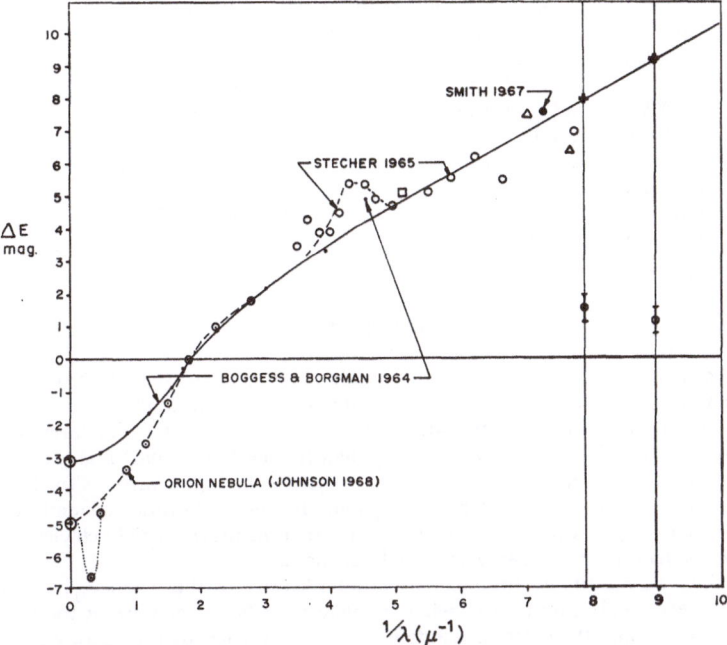

Fig. 4.   Comparison of the observed interstellar extinction in the Orion Nebula region with that observed in other regions of the sky by Boggess and Borgman (1964), small dots; Stecher (1965), unfilled symbols; and Smith (1967). Extrapolation of these results to 1270 Å and 1115 Å assuming a 1/λ law gives values shown by crosses. The dotted portion of the extinction curve for the Orion Nebula indicates the possible effect of circumstellar infrared emission (Johnson, 1967). The present results are indicated by circles with error bars, where the error limits include both measurement errors and uncertainties in the intrinsic stellar fluxes.

absorption is essentially neutral in the far ultraviolet, i.e., that the attenuating particles are large compared to the wavelengths of interest. A relative lack of small particles had already been inferred by the anomalous reddening curve in the wavelength range accessible from the ground, and was attributed to their having been driven out of the nebula by the radiation pressure of the exciting stars (Johnson and Morgan, 1955).

References

Adams, T. F. and Morton, D. C.: 1968, *Astrophys. J.* **152**, 195.
Boggess, A. and Borgman, J.: 1964, *Astrophys. J.* **140**, 1636.
Carruthers, G.: 1968, *Astrophys. J.* **151**, 269.
Carruthers, G.: 1969a, *Appl. Opt.* **8**, 633.
Carruthers, G.: 1969b, *Astrophys. J.* **156**, L97.
Carruthers, G.: 1969c, *Astrophys. J.* **157**, L113.
Carruthers, G.: 1969d, *Astrophys. Space Sci.* **5**, 387.
Code, A. D.: 1960, *Stellar Atmospheres*, University of Chicago Press, Chicago, p. 50.
Hickok, F. R. and Morton, D. C.: 1968, *Astrophys. J.* **152**, 203.
Iriarte, B., Johnson, H. L., Mitchell, R. I., and Wisniewski, W. K.: 1965, *Sky and Telescope* **30**, p. 21.
Johnson, H. L.: 1963, *Basic Astronomical Data*, University of Chicago Press, Chicago, p. 204.
Johnson, H. L.: 1967, *Astrophys. J.* **150**, L39.

Johnson, H. L.: 1968, *Nebulae and Interstellar Matter*, University of Chicago Press, Chicago, p. 197.
Johnson, H. L. and Morgan, W. W.: 1955, *Astrophys. J.* **122**, 142.
Mihalas, D.: 1965, *Astrophys. J. Suppl.* **9**, 321.
Morton, D. C.: 1967a, *Astrophys. J.* **147**, 1017.
Morton, D. C.: 1967b, *Astrophys. J.* **150**, 535.
Morton, D. C.: 1969, to be published. (See also the present volume, p. 59.)
Morton, D. C., Jenkins, E. B., and Brooks, N. H.: 1969, *Astrophys. J.* **155**, 875.
Smith, A. M.: 1967, *Astrophys. J.* **147**, 158.
Smith, A. M.: 1969, to be published. (See also the present volume, p. 164.)
Stecher, T. P.: 1965, *Astrophys. J.* **142**, 1683.

Discussion

Campbell: What is the accuracy of your absolute photometry in the region 1230–1350 Å?
Do you recover your instruments in a form suitable for re-calibration?

Carruthers: The accuracy of our calibrations, I feel, is about $\pm 10\%$. However, we have much more confidence in the calibrations for this flight than for the March 1967 flight, because this time we were able to recalibrate in the field, up to 2 days before the flight, whereas for the March 1967 flight the last calibration was one month before the flight. In view of the observed tendency of detectors to decrease in efficiency with time, it was not surprising that the previous flight efficiencies were found to have been considerably lower than was thought at the time.

When possible, we recalibrate the detectors after recovery of the payload, but the results are generally not a very good indication of what the detector efficiencies were at the time of the observations. This is because the detector windows are contaminated by the heating effects of atmospheric re-entry, and by dirt and moisture after ground impact.

Greenberg: In connection with your observation that the far ultraviolet extinction for θ Ori appears to have levelled off (or even dropped) relative to the ground based measurements I should like to point out that this is consistent with the extinction curve of θ Ori in the 1–3 μ^{-1} range deviating from the average extinction by being more concave downward. This behavior can be attributed to a larger mean interstellar grain size. The effect of the larger particles would be to bring about extinction saturation at longer wavelengths with a subsequent gradual drop toward shorter wavelengths.

Carruthers: Johnson's curve for the reddening in the visible also hints at a downward curvature of the reddening law even in the ground-accessible ultraviolet above 3000 Å, as indicated on the slide.

Heintze: In the past the shape of the extinction curve is also determined by comparing measured energy distributions with calculated ones. The use of the 1964 adoption of the energy distribution of α Lyr and the adoption of too high effective temperatures of the stars considered causes a steeper slope of the extinction curve towards the UV than by using Hayes' energy distribution of α Lyr and the lower effective temperatures generally adopted now.

Davis: Did your observation of θ Ori include the contribution from the Orion Nebula? Our Celescope measurements of the Orion sword region also show θ Ori and the Orion Nebula to be brighter than would result from stellar models of the appropriate spectral type for θ by itself.

Carruthers: We do not feel that the nebula made an appreciable contribution to the measured fluxes, unless it is from a part of the nebula very close to the exciting stars, because our spectra of θ Ori do not show any emission lines or other indications of the nebula, and the widths of the spectra are no greater than would be expected for the two components, θ^1 and θ^2 Ori, separated by 2 arc min.

Bless: We have obtained spectral scans of θ^1 and θ^2 Ori with the OAO. The field of view of the instrument also includes much of the Orion nebula. The integrated spectral type and color of $\theta^1 + \theta^2$ is 0.1 spectral type earlier than ζ Oph and about 0m.05 bluer in (B-V). However, after correction for the relative magnitude, the flux observed from the Orion region is about 10 times greater at 1200 Å than that from ζ Oph, slowly decreasing to 4 times greater at 1800 Å. To the extent that a composite spectrum and color is meaningful, this suggests that either these stars are very strange, that the nebula contributes a large flux, or that the interstellar reddening in the nebula is abnormal, as has been observed in the visual. At the moment the last explanation seems the most plausible, in agreement with Carruthers' interpretation of his observations.

Underhill: Is the ultraviolet brightness of $(\theta^1 + \theta^2)$ Ori (which I take to be chiefly due to θ^1C Ori, O6) the same as that of ζ Pup, O5f?

Carruthers: Even after accounting for reddening, ζ Pup appears to be considerably fainter than expected for an O5 star; it appears more comparable to an O9 star such as ι Ori. In the case of θ Ori, the intrinsic brightness could not be determined directly because of the anomalous reddening; we had to assume the former to derive the latter. However, it appears that θ Ori is considerably brighter in the UV, relative to the visible, than ζ Pup, unless there is even less extinction for θ Ori in the UV than in the visible, which seems unlikely.

Stecher: The ratio of selective to total extinction for θ Ori was found to be 5.6 by Sharpless instead of the normal value of ~ 3. It is reasonable to expect a deviation from the usual curve but quite exciting to find that it is the case.

Haupt: Can you exclude completely some residual atmospheric extinction?

Carruthers: Yes, we feel that residual atmospheric extinction was negligible, except for the last few scans of γ, ϕ^1, and λ Ori at the end of the flight. These, which were observed at the last pointing position showed decreasing intensities related to earlier scans of these stars. In general, in repeated scans of the same stars we got very good agreement except for this one case in which atmospheric attenuation was apparent just before re-entry.

ULTRAVIOLET PHOTOMETRY OF STARS
OBTAINED WITH THE CELESCOPE EXPERIMENT IN THE
ORBITING ASTRONOMICAL OBSERVATORY

ROBERT J. DAVIS

Smithsonian Astrophysical Observatory, Cambridge, Mass., U.S.A.

Abstract. We have used the television photometers in the Celescope OAO experiment to measure the far ultraviolet brightnesses of several thousand stars, including parts of the constellations Draco, Lyra, Puppis, Vela, Taurus, and Orion; and the Moon. As of this date (22 July 1969), three of our four cameras continue to operate satisfactorily, and we are making measurements in additional star fields distributed throughout the sky. Our shortest wavelength band, which includes the Lyman α line of atomic hydrogen, provides information on the Earth's outer atmosphere, as well as on star brightnesses. The intensity of the Lyman α radiation from the geocorona is a maximum when the Sun is near the horizon as seen by the OAO, and a minimum when the Sun is in the nadir. The direction that the telescope points does not much affect the measured intensities.

Because of the heavy logistic problems of identification, calibration, and measurement for so many stars and because of the survey character of the program, the scientific interpretation of the results is, as expected, lagging the measurement program. However, one consistent picture beginning to emerge from our data is that our observed stars are about equally divided between those that fall within 0.5 magnitude of the predicted ultraviolet brightnesses and those that are significantly fainter than predicted. Most of the giant stars we observe exhibit these ultraviolet deficiencies. Since some of these giants are stars for which little or no interstellar reddening is predicted, we attribute the observed deficiencies to the stars themselves.

Many of the objects we observe do not have accurate ground-based published data regarding magnitude, color, and spectral type; new ground-based observations of these objects are required to ensure satisfactory interpretation of our results.

1. Introduction

The Celescope experiment consists of four ultraviolet television photometers, each with an aperture of 31 cm, mounted in the Orbiting Astronomical Observatory (OAO). These cameras produce pictures, 3° square, in four spectral regions: U1, extending from 2200 to 3200 Å; U2, 1600 to 3200 Å; U3, 1350 to 2000 Å; and U4, 1050 to 2000 Å. A complete description of the instrumentation appears in Smithsonian Astrophysical Observatory Special Report No. 282, 18 July 1968.

Since launch on 7 December 1968, the Celescope experiment has taken 3652 pictures, in 1276 different directions, covering approximately 6% of the sky. Most of these observations are within 20° of the galactic plane; however, we have taken enough pictures at high galactic latitude to confirm our expectation that very few objects at high latitudes are brighter than our limit of 9th magnitude at spectral type A0. Near the north galactic pole, we detected about 1 star for every 3 sets of pictures (1 star per 12 square degrees), whereas near the Orion nebula, we measured 20 stars per picture, and in the galactic plane in Vela, we measured 50.

Each Celescope picture consists of 256 lines, each of which contains 251 picture elements. The Celescope instrumentation encodes the signal amplitude in each ele-

Houziaux and Butler (eds.), Ultraviolet Stellar Spectra and Ground-Based Observations, 109–119.

ment to 7-bit accuracy. Although Celescope is capable of other modes of picture transmission, these modes have not yet been used for the collection of scientific data. These pictures, recorded on magnetic tape at both the data-acquisition station and the central control center, are re-formatted by NASA's data-analysis center and processed on Smithsonian Astrophysical Observatory's CDC-6400 computer to provide identification and ultraviolet magnitudes for the stars observed. A detailed description of these analysis techniques will appear in Smithsonian Astrophysical Observatory Special Report No. 310, scheduled for publication later this year.

We will publish these Celescope observational results in a series of Celescope Observational Data Reports, which will be regularly distributed to all institutions and individual scientists who request them from the Publications Division, Distribution Section, Smithsonian Astrophysical Observatory, Cambridge, Mass. 02138. Our preliminary review of the pictures and of the resultant data indicates that for nearly every object we have observed, the ultraviolet brightness can be reasonably explained on the basis of atmospheric theory and interstellar reddening. We have, therefore, concentrated our attention on compiling and refining our observational results, postponing the scientific interpretation to a later date. The remainder of this paper is an example of one set of Celescope pictures, typical of the more than one thousand such sets obtained to date.

2. Sample Celescope Observational Data Report

Celescope observations are obtained only by live video transmission from the OAO to one of five ground stations: Madagascar (M); Orroral, Australia (O); Quito, Ecuador (Q); Rosman, North Carolina (R); and Santiago, Chile (S). Such transmission can occur only during a 'contact' or 'pass' of the satellite over the station. Contacts are identified by a letter designating the station and a number indicating the number of times the satellite has crossed the ascending node of its orbit. During a contact, the Celescope experiment may transmit video data of one or more exposures. The first of these exposures may precede arrival of the OAO over the station.

The reports are divided into sections, one for each contact. The first page of each section is a 'data sheet' listing the conditions under which the exposures were made. If there are more than three exposures, subsequent exposures are listed on a second page. On this data sheet, right ascension and declination are given for the Celescope optic axis in current-epoch coordinates. Magnetic coordinates are based on an approximate magnetic coordinate system whose north pole is at geographic latitude $78°.3$ N and longitude $69°.0$ W; magnetic hour angle is measured eastward from the sun; beta angle is the distance of the target from the Sun. Roll (direction to the north celestial pole) is measured clockwise from the left end of the filter split line. The directions to zenith, magnetic pole, and Sun are measured counterclockwise from the left end of the filter split line, as indicated by the horizontal lines in the plots and pictures.

The most important parameters of Celescope operation are exposure time and operating mode, which control sensitivity and noise level, respectively. Normally, the exposure time is set at 60 sec and the operating mode is set at pulse code modulation (PCM). Blank entries indicate which, if any, cameras were not used.

Status at estimated time of arrival (ETA) lists the less important parameters of operation. Special Report 310 describes these entries in full.

Following the data sheet for each contact are charts, pictures, and numerical data for each exposure. The picture presents the video data exactly as recorded on the magnetic data tape, free from any processing other than that necessary to convert the magnetic data into printed picture. It can thus be used both to assess image quality and to serve as orientation while the numerical data are studied. The small plot, to the same scale as the picture, indicates rough brightnesses and positions for all stars found in the picture and the position of the filter split line. Star-like objects evident in the picture but missing from the small plot include noise pulses, the signal from the calibrator lamp, permanent target blemishes, and stars too near the edge of the field to allow proper analysis of the data. These objects are indicated on the large plot, as described in the paragraph below.

The large plot identifies the star-like images in the picture and cross references them to the data listing. Numbers refer to stars; letters refer to other star-like objects in the picture: C for calibrator lamp; G for ghosts of the calibrator lamp; F for target blemishes ('false stars'); N for noise pulses; and X for objects more than $1°4$ from the optical axis. Asterisks refer to merged images for double stars that are seen as separate objects on some, but not all, cameras. These reports list numerical data only for the numbered objects and for those represented by asterisks.

Ultraviolet brightnesses and identifications are presented in the observational data listing following the pictures and identification charts for each exposure. Objects are listed and assigned identification numbers in order of increasing right ascension. For each object, the first line summarizes ground-based data, and subsequent lines list the Celescope observational data, one line per camera. If the Celescope identification catalog contains more than 10 literature references for an object, a second line is used for printing additional reference numbers.

The information is listed as follows:

ID	Identification number.
HD	*Henry Draper Catalogue* number, or variable star name.
DM	Durchmusterung number or IC/NGC number.
R.A.	Right ascension (1950.0).
DEC.	Declination (1950. 0).
MAGNITUDES AND MCODE	Magnitudes (M1, M2, M3), and magnitude code describing same. The preferred set of magnitudes (V, $B-V$, $U-B$) is assigned a magnitude code of 1. Refer to CDL-100 for other code assignments.
SP	Spectral type.
L	Luminosity class.

Station ID Ororal   Orbit No. 692   Date   24 January 1969

		First Exposure	Second Exposure	Third Exposure	
Target	Universal time	12 : 60	0 : 0	0 : 0 : 0	H, M
Target	Right ascension	5 : 30 : 25	0 : 0 : 0	0 : 0 : 0	H, M, S
	Declination	−5 : 16 : 45	0 : 0 : 0	0 : 0 : 0	DEG, M, S
	Roll	283	0	0	DEG
Target	Zenith distance	33	0	0	DEG
Target	Azimuth	325	0	0	DEG
	Direction to zenith	255	0	0	DEG
Target	Magnetic polar distance	106	0	0	DEG
Target	Magnetic hour angle	9 : 17	0 : 0	0 : 0	H, M
	Direction to magnetic pole	106	0	0	DEG
	Beta angle	130	0	0	DEG
Solar	Zenith distance	122	0	0	DEG
Solar	Azimuth	209	0	0	DEG
	Direction to sun	343	0	0	DEG
Satellite	Latitude	−33	0	0	DEG
Satellite	Longitude	142	0	0	DEG
Satellite	Magnetic latitude	−43	0	0	DEG
Satellite	Magnetic hour angle	10 : 48	0 : 0	0 : 0	H, M
	Distance to illumination	528	0	0	KM
	Height of illumination	1223	0	0	KM
Camera 1	Exposure time	5			SEC
	OP mode	PCM			
Camera 2	Exposure time				SEC
	OP mode				
Camera 3	Exposure time	5			SEC
	OP mode	PCM			
Camera 4	Exposure time	5			SEC
	OP mode	PCM			

Status at E.T.A.

Bay E-4	OP Mode	Thresh	Temp	+6V	+18V	-12V	+28V	ADC	HBC	VBC	HO	VO
	OFF	6	+28.6	A +6.10	A +18.0	A -12.2	A +28.0	A	A	A	A	A

	Temperature										
	CAM	UVICON	CAL	SWP	LAMP	FIL	EXP	MODE	FOCUS	BEAM	ASTIG
Camera 1				A	OF	ON	OF	HV	2	2	4
Camera 2	-35.1	-17.3	-43.0	A	OF	ON	OF	HV	2	1	4
Camera 3				A	OF	ON	OF	HV	2	3	4
Camera 4	-31.3	-16.6	-42.1	A	OF	ON	OF	HV	2	1	4

Comments

(1) On camera 4, the over-exposed image of object no. 11 is merged with the images of object no. 10 and object no. 15, and exceeds the brightness range of accurate calibration. This image appears through the U4 filter. The combined light of these objects exceeds a U_4 magnitude of -2.5.

(2) Object no. 18 appears in the extreme right-hand edges on cameras no. 1 and 3, but its image is truncated by the scanning raster so that no accurate brightness could be determined.

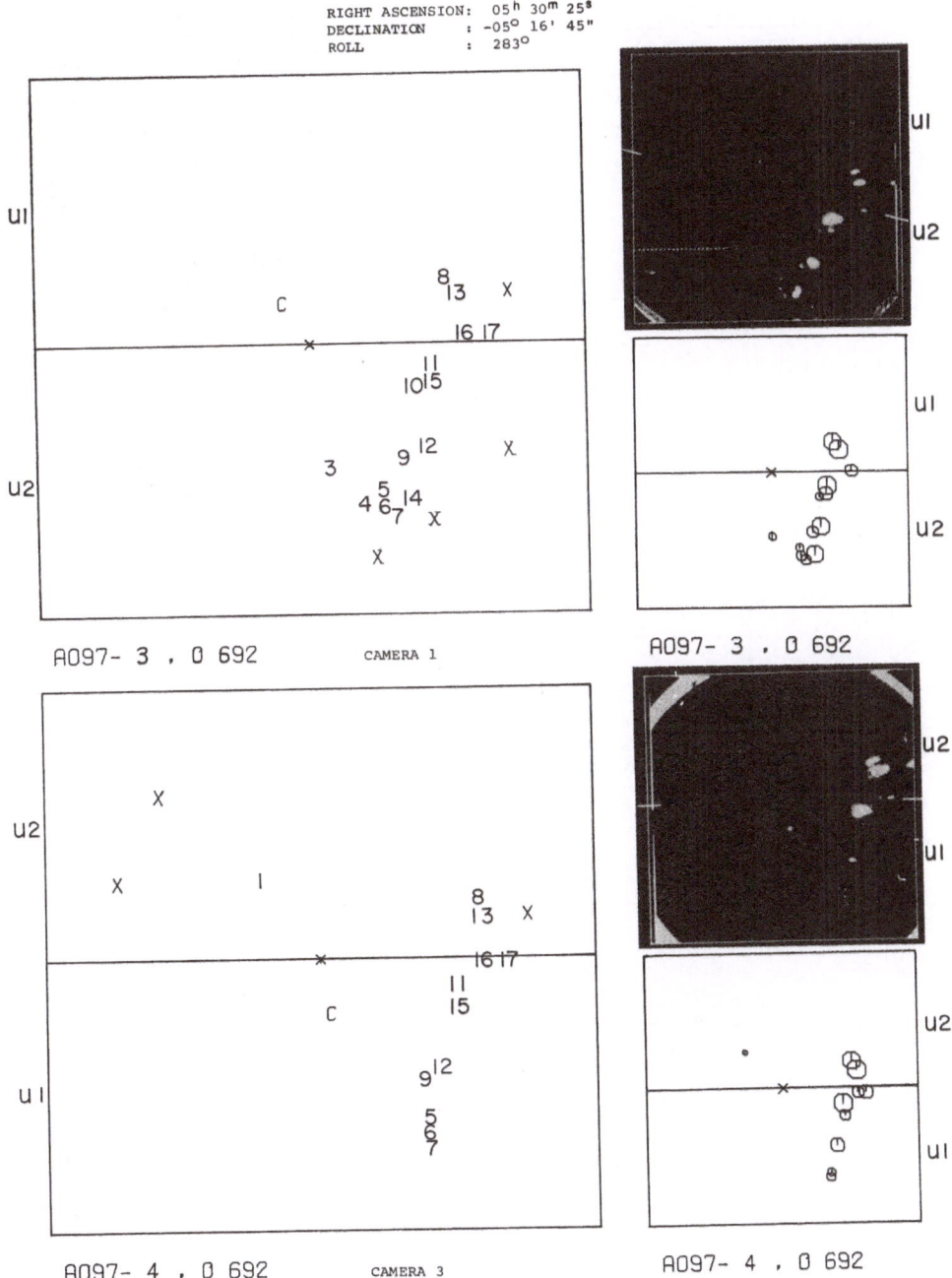

Fig. 1.

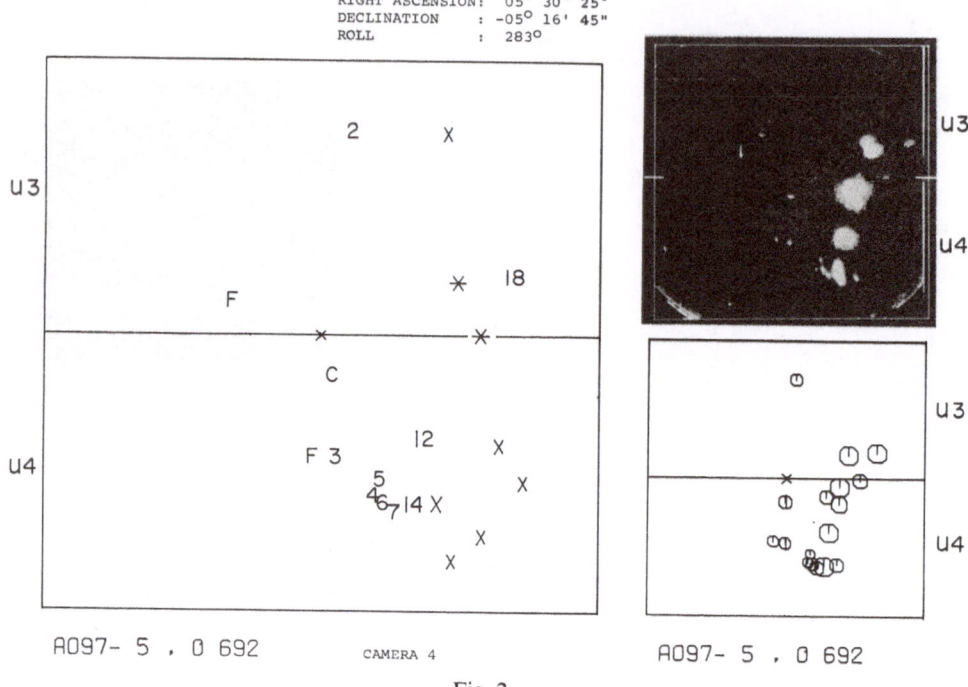

RIGHT ASCENSION: 05^h 30^m 25^s
DECLINATION : -05° 16' 45"
ROLL : 283°

A097- 5 . 0 692   CAMERA 4   A097- 5 . 0 692

Fig. 2.

ROBERT J. DAVIS

1950.0 ID HD	DM	R.A.	DEC.	MAGNITUDES AND MCODE	SP	L	6——	7——.——8	REFERENCES/ CAM	U	FIL	NUM	CODES	TAPE	FM	CON
1 36120	−5 1269	5 26 42	−5 49.8	7.96 −.03 −.36 1	B8		A		R A19	897	898					
1	5	5 26 42	−5 49.8						3P	11.94	2	3	00 0A	A097	4	o 692
2 36430	−6 1207	5 28 55	−6 44.7	6.22 −.17 −.74 1	B2v	s		P	R A19	840	898	20	36	259	901	884 2 13
2		5 28 55	−6 44.7						897 4P	10.56	3	19	00 0A	A097	5	o 692
3 36629	−4 1164	5 30 29	−4 36.0	7.65 .02 −.66 1	B2v			P	R A19	A 7	2	898	36	340	897	
3	5	5 30 29	−4 36.0						1P	9.73	2	9	00 0A	A097	3	o 692
3	5	5 30 29	−4 36.0						4P	10.88	4	8	00 0A	A097	5	o 692
4 36842	−4 1168	5 31 57	−4 23.6	8.09 −.11 −.51 1	B9		A		R A19	404	898	897	756			
4	5	5 31 57	−4 23.6						1P	10.62	2	5	00 0A	A097	3	o 692
4	5	5 31 57	−4 23.6						4P	11.58	4	22	01 0A	A097	5	o 692
5 36865	−4 1171	5 32 3	−4 31.2	7.40 −.07 −.43 1	B9		A		R A19	404	898	897				
5	5	5 32 3	−4 31.2						3P	10.32	1	13	00 0A	A097	4	o 692
5	5	5 32 3	−4 31.2						1P	9.49	2	7	00 0A	A097	3	o 692
5	5	5 32 3	−4 31.2						4P	11.90	4	23	01 0A	A097	5	o 692
6 36883	−4 1172	5 32 15	−4 25.5	7.22 −.08 −.47 1	B8				A19	897	897					
6	5	5 32 15	−4 25.5						3P	10.02	1	14	00 0A	A097	4	o 692
6	5	5 32 15	−4 25.5						1P	8.61	2	18	01 0A	A097	3	o 692
6	5	5 32 15	−4 25.5						4P	10.49	4	21	01 0A	A097	5	o 692
7 36936	−4 1176	5 32 30	−4 23.2	7.52 −.11 −.58 1	B8				A19	897	897					
7	5	5 32 30	−4 23.2						3P	9.94	1	15	00 0A	A097	4	o 692
7	5	5 32 30	−4 23.2						1P	8.52	2	4	01 0A	A097	3	o 692
7	5	5 32 30	−4 23.2						4P	10.06	4	20	01 0A	A097	5	o 692
8 36959	−6 1233	5 32 34	−6 2.5	5.67 −.22 −.90 1	B1v		v	YP	R A19	12	897	884	898	259	629	20 756 158
8									699	783	36	921	2			

1950.0 ID HD	DM	R.A.	DEC.	MAGNITUDES AND MCODE	SP	L 6	7	8	REFERENCES/ CAM	U	FIL	NUM CODES	TAPE	FM	CON
8 36960	−6 1234	5 32 36	−6 2.0	4.78 −.25 −1.01 1	B0v	P S	V	P	R A19	12	897 884	898 259	629	158 699	20
8		5 32 36	−6 2.1						36	756	921 13	2			
8		5 32 36	−6 2.1						1P	5.17	1 16	00 0A	A097	3	o 692
8 also 13		5 32 36	−6 2.1						3P	5.96	2 4	01 0A	A097	4	o 692
									4P	4.80	3 15	00 0A	A097	5	o 692
9 KX OR I	−4 1179	5 32 37	−4 45.8	7.40 3	B3v			P	2	A 7	898 259	340 B97		4	o 692
9		5 32 37	−4 45.8						3P	10.54	1 12	00 0A	A097	4	o 692
9		5 32 37	−4 45.8						1P	7.87	2 19	01 0A	A097	3	o 692
10 36981	−5 1311	5 32 38	−5 14.3	7.82 −.11 −.59 1	B8		A		897	314	898 A19	36	A097	3	o 692
10		5 32 38	−5 14.3						1P	9.62	2 11	00 0A			
11 37020	−5 1315	5 32 49	−5 25.3	6.73 .02 −.88 1	B0v		V	P	A19	2	884 901				
11 37021	−5 1315	5 32 49	−5 25.3	8.10 6	B3			Y	884	901					
11 37022	−5 1315	5 32 49	−5 25.3	5.13 .02 −.95 1	06	P	V	A R	A19	901	897 884	898 350	474	12 13	816
11 37023	−5 1315	5 32 49	−5 25.3	6.70 .09 −.71 1	B0v		V	P	883	2	895 314	883 756	895	314	
11 Theta −15		5 32 49	−5 25.3	4.60 .02 −.90 1	08		V		A19	901	884 2				
11 IC/NGC1976		5 32 52	−5 25.0	4.00 6			E		A19						
11 37041	−5 1319	5 32 56	−5 26.8	5.07 −.09 −.94 1	O9v	P	V	YP	R A09	975	695 940				
									R A19	895	699 921	756 884	897	898 340	350
									629	901	764 2	12 13	336	377 883	
11 37042	−5 1319	5 32 56	−5 26.8	6.38 −.09 −.93 1	B1v	P	V	YP	R A19	895	699 921	756 884	897	898 340	350
11		5 32 52	−5 25.1						629	901	764 2	12 13	336	377 883	
11		5 32 52	−5 25.1						3P	2.50	1 8	00 0A	A097	4	o 692
									1P	2.40	2 20	01 0A	A097	3	o 692
12 37018	−4 1185	5 32 55	−4 52.2	4.59 −.20 −.93 1	B2III	s	V	YP	R A19	783	897 756	921 898	36	377 785	2
12		5 32 55	−4 52.2						884	12	901 13	212 158			
12		5 32 55	−4 52.2						3P	7.55	1 11	00 0A	A097	4	o 692
									1P	4.96	2 8	01 0A	A097	3	o 692

1950.0 ID HD	DM	R.A.	DEC.	MAGNITUDES AND MCODE	SP	6—	7—	8	REFERENCES/ CAM	U	FIL	NUM	CODES	TAPE	FM	CON
12		5 32 55	−4 52.2						4p	7.07	4	7 10	0A	A097	5	o 692
13 37043	−6 1241	5 32 59	−5 56.5	2.76 −.25 −1.07 1	o9III	ES	USV	P	A19 9	699	158	793	7 921	891	934	2 10
									12		13	336	377 765	766	785	895 882
13		5 32 59	−5 56.5						1p	4.19	1	15	00 0A	A097	3	o 692
13		5 32 59	−5 56.5						3p	4.38	2	16	01 0A	A097	4	o 692
13 also	8	5 32 36	−6 2.1						4p	4.80	3	15	00 0A	A097	5	o 692
14 37017	−4 1183	5 32 53	−4 31.5	6.54 −.14 −.78 1	B1V			P	R A19	12	897	756	898 36	901	2	884
14 37016	−4 1184	5 32 54	−4 27.3	6.23 −.16 −.68 1	B3V	s	v	P	A19	897	901	2	13 884	839	756	898
14 37040	−4 1186	5 33 2	−4 23.7	6.30 −.15 −.68 1	B3		v	P	R A19	13	897	36	2 901	840	884	898
14		5 33 2	−4 23.8						1p	5.31	2	3	00 0A	A097	3	o 692
14		5 33 2	−4 23.7						4p	6.15	4	3	01 0A	A097	5	o 692
15 NU ORI		5 33 4	−5 11.0	6.90 5	B3V	P		J	969							
15		5 33 4	−5 11.4						3p	9.22	1	9	00 0A	A097	4	o 692
15		5 33 4	−5 11.4						1p	6.90	2	10	01 0A	A097	3	o 692
16 37115	−5 1330	5 33 26	−5 38.9	7.06 −.08 −.57 1	B6V	EN			314	342	898	260	A 7 A19			
16		5 33 26	−5 38.9						1p	9.36	1	13	00 0A	A097	3	o 692
16		5 33 26	−5 38.9						3p	8.79	2	6	00 0A	A097	4	o 692
16 ALSO	17	5 33 48	−5 40.8						4p	9.59	3	14	00 0A	A097	5	o 692
17 37150	−5 1334	5 33 48	−5 40.7	6.54 −.19 −.80 1	B3V			P	A19	13	898	756	897 901	835	884	2
17		5 33 48	−5 40.7						1p	7.45	1	12	00 0A	A097	3	o 692
17		5 33 48	−5 40.7						3p	7.67	2	7	00 0A	A097	4	o 629
17 ALSO	16	5 33 48	−5 40.8						4p	9.59	3	14	00 0A	A097	5	o 692
18 37209	−6 1255	5 34 9	−6 5.7	5.70 −.24 −.91 1	B1V	s	v	P	R A19 839	897	756	898	259 36	901	884	13
18		5 34 9	−6 5.7						4p	7.62	3	17	00 0A	A097	5	o 692

6...7...8	Special characteristics and peculiarities. Refer to CDL-100 for code assignments.
REFERENCES/	References to original publications of the data. CDL-100 contains the bibliography. References are listed in the same columns used to report the Celescope observational results.
CAM	Camera number and operation mode; P is for direct PCM mode, A for analog mode, and S for store mode.
U	Ultraviolet magnitude: $U = -2.5\log I$, where I is the measured intensity in MKS units (watts m^{-3}).
FIL	Filter number.
NUM	An identification number by which the computation for each object can be traced and verified.
CODES	Five columns describing the quality of the data. The first column describes the background in the vicinity of the object: codes 0 and 2 are normal background; 1 and 3 indicate unreliable data because of high background surrounding the object. The second column describes the relationship of the object to the filter split line; codes greater than or equal to 2 indicate unreliable data because of proximity to the filter split line. The third and fourth columns indicate general picture quality, larger numbers indicating poorer quality. The fifth column is a letter indicating which calibration table was used for data reduction. Special Report 310 includes a complete description of these quality codes.
TAPE, FM	Tape number and frame number by which the supporting data can be located in the Celescope files.
CON	Contact designation.

Discussion

Henize: I understand that you find the excess UV radiation of θ Ori to originate, at least partially, in the Orion nebula. Could you say whether your observation of η Car relates to the star itself or to its surrounding nebula?

Davis: The observation of η Car does not have sufficient resolution to determine definitely whether the observed UV comes only from the star, or partly from the nebula. However, I believe it is best interpreted as including an important contribution from the nebula.

Morton: Does the Orion nebula appear bright in all your wavelength bands?

Davis: Yes.

Malaise: When will Celescope data be released and how will it be made available?

Davis: Celescope data will be released in regular reports (Celescope Observational Data Reports), beginning in about two months. Any scientist wishing to receive these reports should write to the Publications Division, Distribution Section, Smithsonian Astrophysical Observatory, Cambridge, Mass. 02138, USA, requesting that he be placed on the mailing list.

PHOTOGRAPHIC MAGNITUDES OF 201 STARS AT 2600 Å

JEAN-PIERRE SIVAN and MAURICE VITON

Laboratoire d'Astronomie Spatiale du CNRS et Observatoire de Marseille, France

Résumé. Les magnitudes de 201 étoiles à 2600 Å (longueur de la bande passante: 1000 Å) ont été obtenues grâce à deux photographies de la Voie Lactée d'hiver données par une caméra à grand champ. Une estimation préliminaire du rougissement interstellaire a permis de tracer un diagramme couleur-type spectral. Il semble que les étoiles O sont plus brillantes que prévu à 2600 Å.

Abstract. The magnitudes of 201 stars at 2600 Å (1000 Å passband) were derived from two plates of the winter Milky Way obtained with a large field camera. A preliminary investigation of the interstellar reddening allowed us to plot a color-spectral type diagram. Stars of type O seem to be brighter than predicted.

1. Introduction

The preliminary results, presented here, are from the second flight of a sounding rocket programme.

These experiments were proposed by G. Courtès to the Centre National d'Etudes Spatiales (CNES). They are designed to photograph the sky in an ultraviolet passband, by night, at a high altitude (200–300 km), with different cameras having large fields (actually 5700 sq deg), a high luminosity ($f/1$) and a low angular resolution (about 10'). These parameters allow the use of poor pointing and guiding systems.

Because of the large field of these cameras, it is theoretically possible to cover the entire sky in 10 flights.

2. Description of the Experiment

The experiment was launched on April 4, 1967. Despite the difficult recovery, it gave the two expected photographs.

The optical system (Figure 1) is as follows: a hyperbolic convex mirror forms a large field image of the sky, which is refocussed by a Maksutoff-Brouwers camera. This very simple design is free of astigmatism and the convex curvature of the sky image given by the hyperbolic mirror has been calculated so that the Maksutoff-Brouwers camera gives a flat field. This field was 82° radius and was limited to a sector of about 120°.

The ultraviolet passband, which we call U', was produced by a multilayer coating on the spherical mirror of the Maksutoff-Brouwers camera. The transmission curve is shown in Figure 2.

The exposure times were successively 20 sec and 210 sec, with Kodak 103 a0-UV film.

3. Description of the Photographs

The shorter exposure (Figure 3) shows only about 50 stars but during this time the

Houziaux and Butler (eds.), Ultraviolet Stellar Spectra and Ground-Based Observations, 120–129.

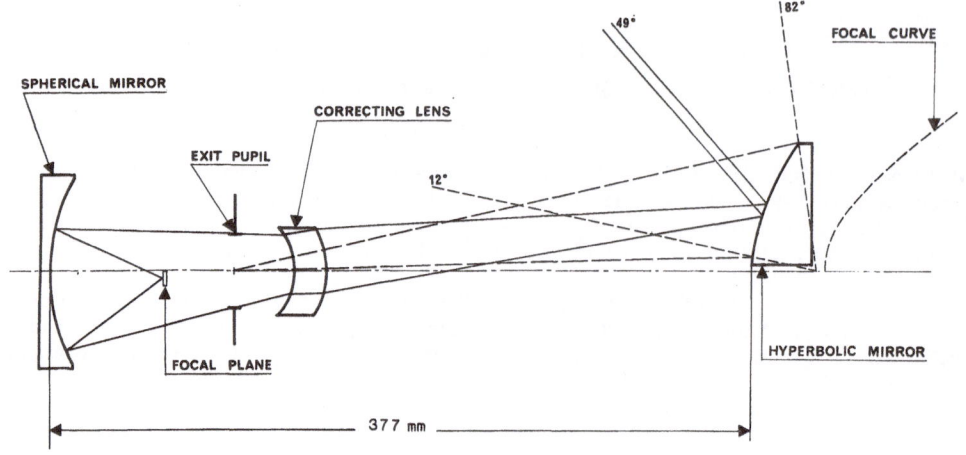

Fig. 1.   The optical layout.

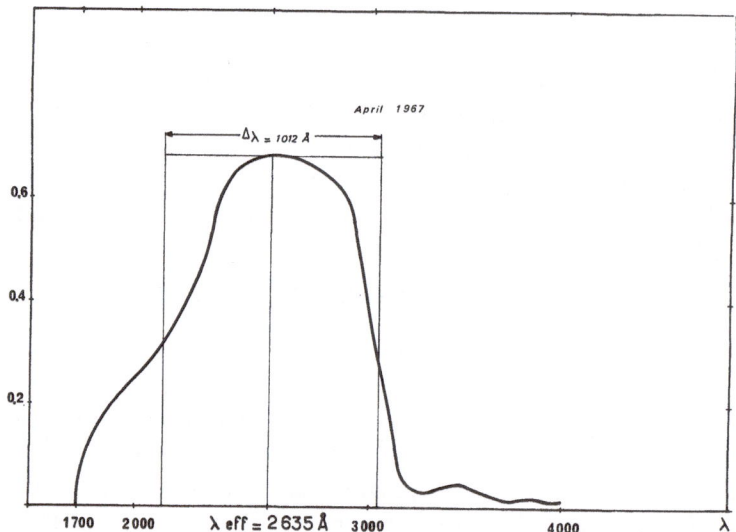

Fig. 2.   Camera transmission for a flat energy spectrum.

guiding was very good and the optical resolution is reached. The brightest stars, particularly the Orion's Belt stars, are available for measurements. The individual stars of Pleiades are almost separated.

On the longer exposure (Figure 4), more than 700 stars were identified. The visual limit V-magnitude is about 8 for type O5 stars and about 6 for type F0 stars.

The thin distribution of population I stars in the Milky Way is to be seen and the Zodiacal Light (bottom right) is detectable up to the Milky Way (58° from the sun).

The high contrast of these two phenomena shows that the night UV brightness of the very high atmosphere (more than 200 km high) is not detectable in our instrument

JEAN-PIERRE SIVAN AND MAURICE VITON

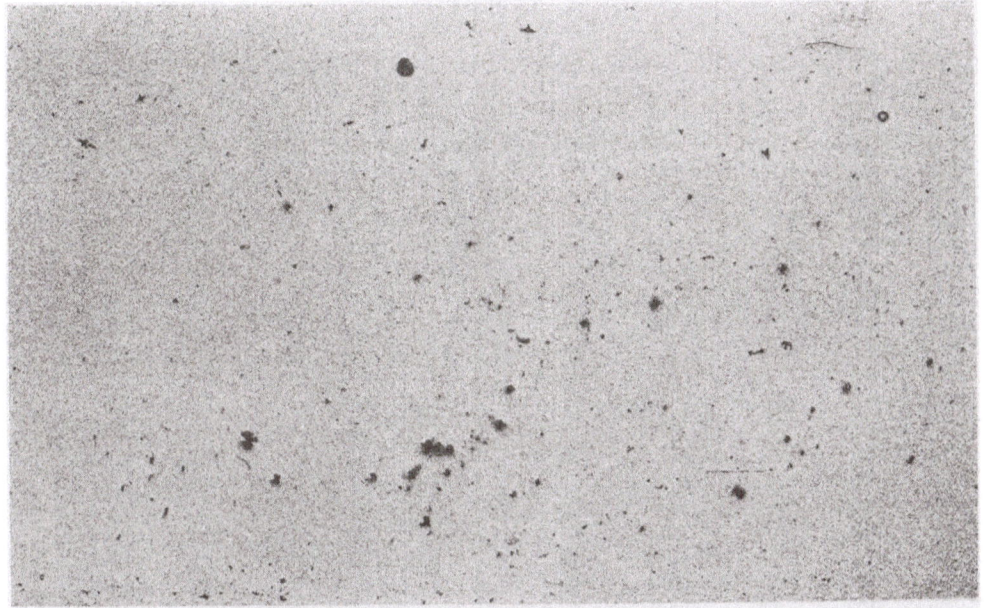

Fig. 3.   20 sec exposure photograph.

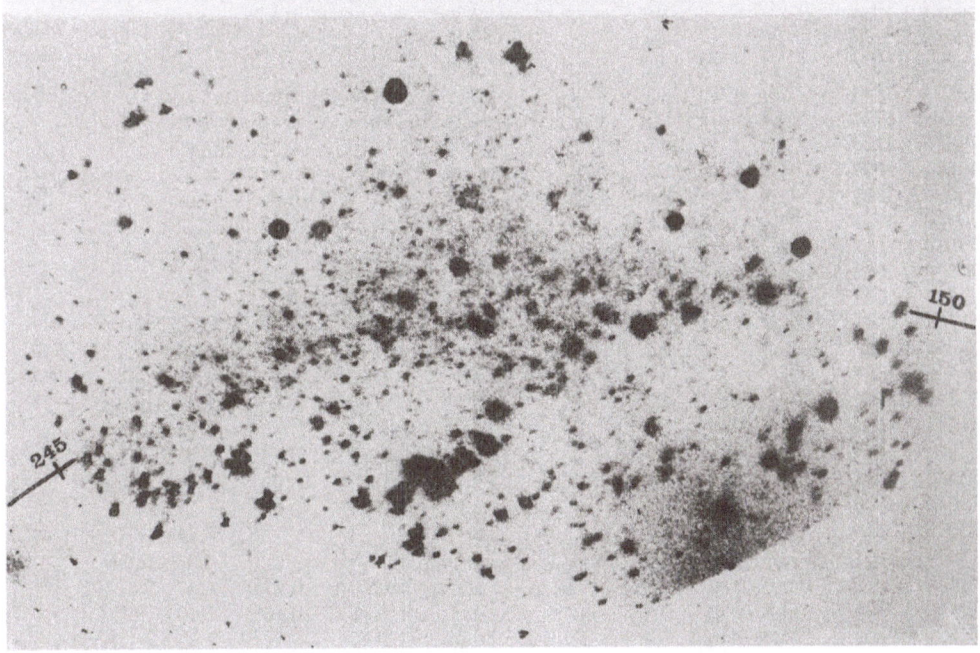

Fig. 4.   210 sec exposure photograph. Extreme galactic longitudes (l^{II} system) are given. Constellations of Orion (bottom center), Auriga (top right) and Canis Major (bottom left) are easy to identify. Among the Zodiacal Light (bottom right) are the Pleiades, much over-exposed, as for Jupiter (top center).

TABLE I

HD	Chart No.	Name		Spectral Type		V	B-V	U-B	U'	Remarks
20995	3420			B9.5	V	5.62	−0.03	−0.16	5.31	
21856	3414			B1	V	5.88	−0.06	−0.86	3.69	
22091	2804	7	Tau	A3	V	5.90	+0.13		6.19	doub.
22951	3416	40	Per	B0,5	V	4.99	−0.02	−0.84	2.76	
23016	2806	13	Tau	B8	Ve	5.56	−0.01		5.07	
23793	2113	30	Tau	B3	V	6.00			2.99	doub. var.?
24155	2104			B9	II-III	6.16	−0.06	−0.48	5.66	
24398	3422	44 ζ	Per	B1	Ib	2.84	+0.12	−0.77	1.32	mult. var.?
24640	3413			B2	V	5.48	−0.03		3.19	var.?
24760	3408	45 ε	Per	B0.5	III-V	2.89	−0.18	−0.99	+0.53	
24912	3412	46 ξ	Per	O7	I	4.03	+0.01		2.05	var.?
25330	2112			B8		5.67	+0.02	−0.41	4.67	doub.
25204	2105	35 λ	Tau	B3	V	3.8			1.61	3.8–4.1 var.
25823	2701	41	Tau	Asi		5.21	−0.13	−0.48	4.10	
25940	3904	48	Per	B3	Vpe	4.03	−0.03	−0.55	2.79	var.?
27026	3302			B8	V	6.10	−0.08	−0.30	5.81	
27396	3808	53	Per	B6	III	4.86	−0.02	−0.54	3.62	var.?
27638	2704	59 χ	Tau	B9.5	V	5.38	−0.02		4.92	doub.
27742	2712			B9	V	5.90	+0.03	−0.26	4.86	
28217	2106			B7	III	5.83	+0.05		5.55	doub.
28929	3315			Ap		5.70	−0.05		5.00	doub.
29140	2107	88	Tau	Am		4.25	+0.18	+0.09	5.10	doub. var.?
29365	2710			B8	V	5.72	−0.05	−0.34	4.80	
29499	2108			dA9		5.39	+0.25	+0.12	6.16	
29646	3316			A2	V	5.58	−0.02		5.79	doub.
29722	3814	59	Per	A1	V	5.26	+0.01	+0.02	5.80	
29763	2706	94 τ	Tau	B3	V	4.31	−0.13	−0.56	2.25	
29866	3305			B7e		6.07	+0.10	−0.29	5.93	
30652	2009	1 π³	Ori	F6	V	3.19	+0.45	−0.01	4.99	doub.
30739	2008	2 π²	Ori	A0	V	4.34	+0.01	−0.01	4.08	
30780	2711	97	Tau	dA5		5.11	+0.21	+0.12	6.11	
30836	2017	3 π⁴	Ori	B2	IV-III	3.69	−0.17	−0.81	1.42	var.?
30870	2004			A0-B5n		6.09	+0.08	−0.45	5.14	
31237	2025	8 π⁵	Ori	B2	III	3.71	−0.19	−0.82	1.40	var. 0.05
31295	2003	7 π¹	Ori	A0p		4.68	+0.08	+0.09	4.66	
31331	2027			B5		5.92	−0.13	−0.55	4.83	
31373	2643			B8	III	5.71	−0.08	−0.46	4.54	
31592	2606	98	Tau	B9.5	V	5.54	0.00		5.79	doub.
31647	3308	4	Aur	A0	V	4.93	+0.02		5.28	doub.
32301	2619	102 ι	Tau	A7	V	4.65	+0.15	+0.14	6.05	
32549	2641	11	Ori	A0si		4.66	−0.07	−0.09	4.46	
32630	3816	10 η	Aur	B3	V	3.17	−0.18	−0.67	1.17	var.?
32977	2620	106	Tau	A3		5.17			6.44	
32990	2607	103	Tau	B2	V	5.41			4.06	doub.
32991	2618	105	Tau	B2	Vp	5.87	+0.20	−0.55	4.60	
33641	3306	11 μ	Aur	Am		4.80	+0.18	+0.10	5.10	
34029	3806	13 α	Aur	G8	III + F	0.09	+0.80		2.62	doub var.?
34203	2644	18	Ori	A0	III	5.48	−0.02	+0.05	5.60	
34656	3206			O7		6.71	+0.01		4.74	
34759	3711	20 ϱ	Aur	B5	V	5.09	−0.18		3.24	
34989	2005			B1	V	5.78	−0.13	−0.88	3.39	

(Table I, continued)

HD	Chart No.	Name		Spectral Type		V	B-V	U-B	U′	Remarks
35 149	2016	23	Ori	B1	V	4.99	−0.16	−0.86	2.53	
35 239	3224			B9	III	5.92	+0.04	−0.12	5.88	
35 439	2020	25	Ori	B1.5	Vpe	4.94	−0.21	−0.91	2.33	var.?
35 468	2010	24 γ	Ori	B2	III	1.64	−0.24	−0.87	−0.26	var.?
35 497	3234	112 β	Tau	B7	III	1.66	−0.13	−0.49	0.50	
35 671	2628	115	Tau	B5	V	5.30			3.67	doub.
35 708	2608	114	Tau	B3	V	4.83			2.61	doub.
36 351	2014	33	Ori	B1.5	V	5.44	−0.19	−0.81	2.85	doub.
36 408	2627			B7	IV	5.42	−0.04		4.73	doub.
36 486	2029[1]	34 δ	Ori	O9.5	V-II	2.21	−0.21	−1.06	0.02	trip. var.
36 576	2622	120	Tau	Bp		5.52			3.80	
36 653	2636	35	Ori	B3		5.56			3.76	
36 741	2022			B2	V	6.58	−0.20	−0.77	4.97	
36 819	2603	121	Tau	B3	V	5.25	−0.06?		3.57	
37 098	3232			B8	III	5.69	−0.05		4·92	doub.
37 128	2029[2]	46 ε	Ori	B0	Ia	1.69	−0.19	−1.04	−0.40	
37 202	2609	123 ζ	Tau	B2	IVp	2.99	−0.15	−0.68	0.91	var.?
37 320	2006			B8		5.88	−0.08	−0.37	4.63	
37 339	3202			B9?		6.89?			5.37	
37 438	3231	125	Tau	B2	V	5.07	−0.16	−0.69	3.04	
37 490	2012	47 ω	Ori	B3	IIIe	4.52	−0.09	−0.78	2.80	
37 519	3215			B7	V	6.01	+0.03	−0.20	5.57	
37 711	2625	126	Tau	B3	IV	4.85			2.77	doub.
37 742 ⎱ 37 743 ⎰	2029[3]	50 ζ	Ori	O9.5 ⎱ B3 ⎰	Ib	1.75	−0.21	−1.06	−0.48	trip. var.?
38 478	2624	129	Tau	B7	IIIp	5.90	−0.06	−0.44	5.10	
38 622	2635	133	Tau	B2	V	5.15	−0.18		3.10	doub.
38 670	2610			B7	V	5.92	−0.09		4.52	doub.
39 317	2531	137	Tau	Ap		5.54	−0.04		5,20	
39 357	3222	136	Tau	A0	III	4.52	−0.02		4,74	doub.
39 698	2612	57	Ori	B2	V	5.86			3.51	
39 777	1945			B2	V	6.55	−0.19	−0.80	4.80	
39 970	2601			A0	Ia	6.02	+0.39		5.52	
39 985	1906			B9		5.98	−0.06	−0.14	5.64	
40 005	2524			B3?		6.91?			5.07	
40 111	3229	139	Tau	B1	Ib	4.80	−0.07	−0.93	2.61	
40 183	3706	34 β	Aur	A2	V	1.90	+0.03		1.40	doub. var.
40 312	3201	37 θ	Aur	B9.5pv		2.69	−0.08		2.18	doub. var.?
40 446	1936	60	Ori	A1		5.22	+0.01	+0.01	5.18	
40 932	1905	61 μ	Ori	Am		4.12	+0.15	+0.10	5.00	doub. var.?
40 978	3704			B3		7.12	−0.06	−0.70	6.32	
41 076	2539			B9.5	V	5.94	−0.04		6.37	
41 335	1302			B2	IV-Vne	5.22	−0.08	−0.84	3.37	var.?
41 692	1946			B5	IV	5.37	−0.15	−0.53	4.03	
41 753	2530[2]	67 ν	Ori	B3	V	4.42	−0.27		2.30	doub. spectro.
42 509	2514	68	Ori	B9.5	V	5.67	−0.09		5.21	
42 545	2525	69	Ori	B5	V	4.92	−0.15	−0.60	3.13	
42 560	2530[1]	70 ξ	Ori	B3	V	4.38	−0.20		2.25	
42 657	1947			B9		6.17	−0.09	−0.36	5.22	
42 690	1301			B2	V	5.06	−0.22	−0.77	2.86	
43 112	2530[3]			B1	V	5.91	−0.24	−0.96	3.27	doub.

(Table I, continued)

HD	Chart No.	Name		Spectral Type		V	B-V	U-B	U'	Remarks
43 153	2526	72	Ori	B7	V	5.24	−0.14	−0.46	3.86	
43 247	2536	73	Ori	B9	II-III	5.34	−0.03		5.37	
43 285	1916			B5e-B6	V	6.00	−0.12	−0.53	4.60	
43 362	1306			B9		6.10	−0.08	−0.30	5.07	doub.
43 819	2521			Ap		6.16	−0.08	−0.34	5.50	
44 092	3107			A1	V	6.27	+0.06	+0.01	5.74	
44 112	1305	7	Mon	B2	V	5.24	−0.20	−0.74	3.08	
44 173	2541			B5n		6.40			5.48	
44 700	1921			B3	IV	6.32	−0.16	−0.62	4.55	
44 701	1943			B5?		6.58?			4.73	
44 769	1920	8	Mon	A5	IV	4.48	+0.21	+0.09	5.20	doub.
44 783	1902			A0		6.25	−0.08	−0.30	5.24	doub.
45 542	2509	18 ν	Gem	B7	IVe	4.15	−0.13		2.79	doub.
46 052	3104	WW	Aur	Am-A7	V	5.80			5.98	var.
46 300	1907	13	Mon	A0	Ib	4.48	+0.01	−0.25	4.37	
46 487	1939			B6	V	5.07	−0.14	−0.56	3.65	
46 553	3110	49	Aur	B9.5	V	5·07	−0.03	−0.08	5.16	
46 769	1934			B8	Ib	5.72	0.00	−0.46	4.81	
47 054	1950			B8ne		5.51	−0.10	−0.39	4.57	
47 100	3601	52 ψ³	Aur	B8	III	5.25	−0.07	−0.40	4.05	
47 105	2522	24 γ	Gem	A0	IV	1.93	0.00	+0.04	2.07	
47 129	1914			O8	O9	6.04	+0.05	−0.90	3.02	var.?
47 152	3111	53	Aur	A0p		5.53	−0.01	−0.08	5.81	
47 395	3112	54	Aur	B6	III	5.86	−0.09		4.67	doub.
47 432	1929			O9.5	II	6.18	+0.15	−0.85	4.42	
47 839	2544	15	Mon	O7		4.65	−0.25	−1.06	1.66	doub. var.
47 887				B2	III	7.02				
47 964	1933			B8	III	5.78	−0.10	−0.35	4·85	
48 099	1913			O6-O7		6.36	−0.05	−0.96	3.34	
48 434	1919			B0	III	5.83	−0.02	−0.90	3.84	
48 977	1901	16	Mon	B3	V	5.91	−0.18	−0.68	3.76	
49 147	1218			A0	IV	5.65	−0.06	−0.10	5.68	
49 567	1832			B3	II-III	6.14	−0.14	−0.67	4.32	
49 606	2523	33	Gem	B8	III	5.71	−0.13	−0.52	4.75	
49 643	1848			B8	V	5.70	−0.10	−0.46	4.65	doub.
49 908	2501	36	Gem	A2	V	5.18	−0.02		5.72	doub.
50 019	3101	34 θ	Gem	A3 III-A2 I		3.59	+0.10	+0.13	3.98	doub.
50 635	2417	38	Gem	F0	Vp	4.63			6.23	doub. var.?
50 820	1847			B3 Ve + K2 II		6.22	+0.56	−0.36	4.70	
51 104	2422			B7	V	5.88	−0.08	−0.35	4.88	
52 266	1850			O9	V	7.23	−0.01	−0.90	5.47	
52 312	1207			B9	III	5.84			4.97	doub.
52 559	1817			B2s		6.52	−0.02	−0.64	5.35	
52 721	1215			B3?		6.52?			4.66	doub.
52 918	1846	19	Mon	B1	V	4.93	−0.21	−0.93	2.77	var.?
53 205	1828			B9		6.52	+0.02	−0.05	6.61	
53 244	1233	23 γ	CMa	B8	II	4.10	−0.12	−0.48	3.08	
53 257	3017	44	Gem	B9.5	V	5.89	−0.03	−0.08	6.05	
53 744	3009			B9	V	6.22	−0.10	−0.26	5.77	
53 755	1214			B0	V	6.48	−0.05		4.28	triple
53 929	1815			B8		6.05	−0.14	−0.47	4.77	

(Table I, continued)

HD	Chart No.	Name		Spectral Type		V	B-V	U-B	U'	Remarks
53974	1216			B0.5	IV	5.38	+0.05		3.31	mult.
54662	1213			O6		6.21	+0.03	−0.94	4.50	
54801	3010	47	Gem	A4	V	5.58	+0.12		6.40	
55879	1212			B0	IV	6.00	−0.18		3.55	
56310	1231			B1	V	6.79			4.99	
56386	3004			B9.5	V	6.01	−0.04	−0.11	6.33	
56446	1804			B9		6.57	−0.12	−0.40	5.57	
56537	2407	45 λ	Gem	A3	V	3.58	+0.11	+0.09	3.94	doub. var.?
56986	3016	55 δ	Gem	F0	IV	3.52	+0.34		4.75	doub.
57539	1201					6.53	−0.10		5.29?	
57682	1210			O9	V	6.42	−0.20		3.59	
57744	3014	58	Gem	A1	V	5.96	−0.01		6.23	
58050	2406			B3	III	6.35	−0.13	−0.93	3.95	var.?
58187	2419	1	CMi	A4	III	5.30	+0.10	+0.13	6.27	
58343	1229			B3	Ve	5.29	−0.05	−0.60	4.10	
58580	1839			B9		6.75	−0.01	−0.11	6.45	
58599	2420			B6	IV	6.30	−0.13	−0.47	4.37	
58715	1801	3 β	CMi	B8	V	2.87	−0.10	−0.29	1.99	var.?
58923	1803	5 η	CMi	gF0		5.28	+0.22	+015	6.40	doub.
59037	3007	64	Gem	A6	V	5.01	+0.11	+0.12	5.91	
59059	2405			B9	V	6.05	−0.05	−0.11	5.88	
59211	1221					6.62?			5.63?	
60107	2404	68	Gem	A1	V	5.08	+0.05	+0.06	5.63	
60325	1226			B1	V	6.21	−0.04		4.05	
60357	1821	9	CMi	A0n		5.80	−0.02	−0.09	5.29	
61421	1810	10 α	CMi	F5	IV-V	0.35	+0.41	−0.01	1.46	doub var.?
61887	1820			A0n		5.92	−0.04	−0.08	6.03	var.?
62832	2309	11	CMi	A1	V	5.26	+0.01	−0.02	4.94	
63655	1108			B9		6.12	−0.09	−0.48	5.35	
63975	1711	13 ζ	CMi	B8		5.14	−0.13	−0.47	3.77	
64648	2902	85	Gem	B9.5	V	5.34	−0.04	−0.06	5.29	
65241	1703			B9		6.35	−0.04	−0.06	5.77	
65396	1706			B9?		6.78?			6.17?	doub.
65810	0606			A3	V	4.61	+0,08	+0.08	5.46	
65873	2302	5	Cnc	B9	V	5.89	−0,02	−0.02	5.73	
65875	1713			B3	Vp	6.48	−0,08	−0.83	4.55	
65900	1707			A0		5.64	0.00	+0.01	5.69	
66664	2303	8	Cnc	A0	IV	5.10	0.00	0.00	4.86	
66834	0605	14	Pup	B3	V	6.12	−0.17		4.58	
67159	1104			A0		6.00	−0.04		5.78	doub.
67797	0604	16	Pup	B5	V	4.40	−0.17	−0.59	2.73	
67880	1115			B3s		5.67	−0.18		3.58	doub.
68099	2306			B7	III	6.07	−0.11	−0.42	4.56	
69686	2305			B8?		7.02?			5.50	
72310	0601			A0		5.41	−0.06		5.00	doub.
72660	1608			A1		5.69	0.00	0.00	5.88	
73262	1604	4 δ	Hya	A0	V	4.15	0.00	0.00	3.22	
74280	1605	7 η	Hya	B3	V	4.30	−0.20	−0.74	1.42	
74988	1607			A2-A0n		5.26	+0.04	+0.08	5.87	
75333	1001	14	Hya	Ap		5.25	−0.09	−0.34	3.80	

between 2000 and 3000 Å. Neither is the sky brightness to be seen far away from the Milky Way (galactic latitude up than 40°). Isophotes and quantitative data will be published later.

4. Calibrations

Using a deuterium-lamp source, we made three series of calibrations with different backgrounds for the longer exposure and a single calibration with no background for the shorter one. Each series covered a range of five magnitudes for each of 13 field angles.

When comparing the different series, we found errors generally less than 0.1 magnitude.

For the flight photographs, we determined the accuracy of the magnitude measurements by comparing the stars common to both the exposures. We found a mean error of about 0.2 magnitude.

5. First Results

Among the 700 stars detected, only 201 were suitable for measurement. The remaining stars were either too faint or too badly defined or the background was too irregular or the part of the field was too vignetted for accurate correction.

Measurements were made with a Becker type iris photometer. The information in Table I is as follows:

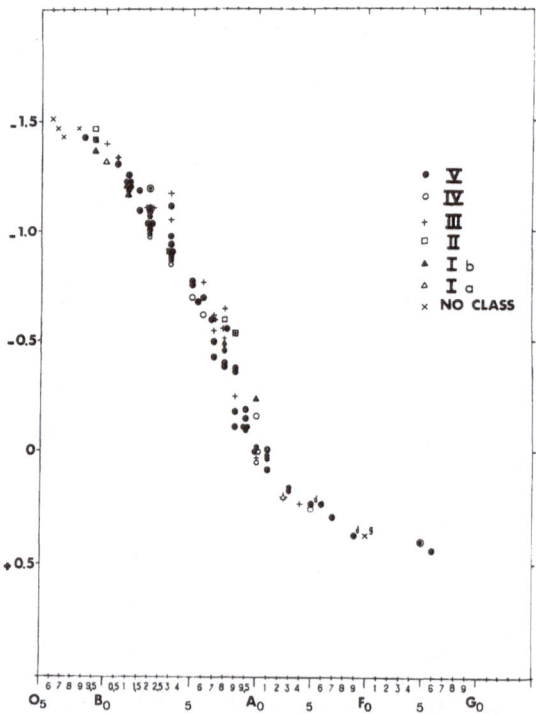

Fig. 5. $(U—V)_0$ color diagram for 100 stars.

Column 1 – HD number
       2 – Our code or 'Chart' number
       3 – Star names or numbers
       4–7 – Spectral type, luminosity classification of UBV data from the BSC
          or from Jaschek (1968) or from other sources
       8 – Our U'
       9 – Remarks, generally from the BSC.

The zero adjustment for U' magnitudes was defined such that $(U'-V)_0=0(\pm0.1)$ for A0 V stars.

We tried first to determine the mean interstellar reddening by comparing stars of identical type. The preliminary results thus obtained seem to show that the color excess ratio $(E(U'-V))/(E(B-V))$ is slightly higher than those found by Stecher (1965), Boggess and Borgman (1964) and others at 2600 Å, e.g. about 4. But because of the large scatter for this ratio, it is probably not a real effect and we think that it is due to the actual lack of accuracy of UBV and spectral type data. Nevertheless, it is probable that there is some scatter in the reddening law itself.

For this first paper, we have adopted a color excess ratio of 4 and knowing the intrinsic colors, we were able to plot two color-spectral type diagrams.

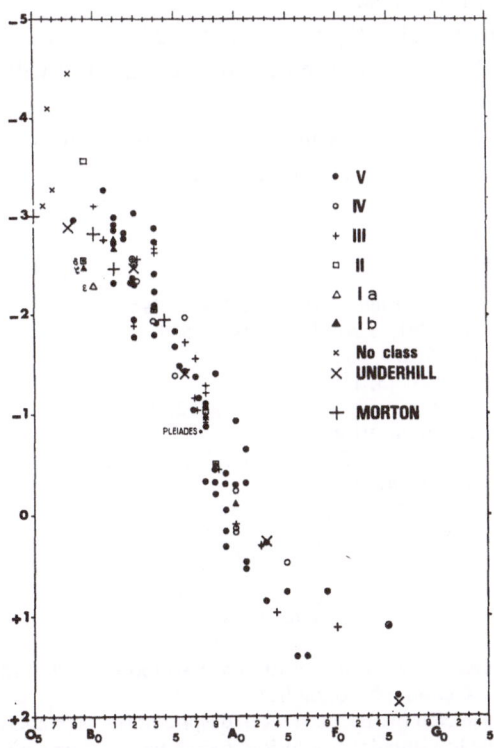

Fig. 6. $(U'-V)_0$ color diagram for 100 stars.

The first one (Figure 5), a $(U-V)_0$ diagram, shows that the 100 stars for which U, B, V, and spectral types are well known, are quite normal.

The second one (Figure 6), a $(U'-V)_0$ diagram, shows the large scatter of O type stars and that they are brighter than theoretical models, the A type stars being fainter.

It can be seen that there is no significant difference between giant and main sequence stars. A recent rough study showed us that giant stars seem to be fainter by 0.2 or 0.3 magnitude. The Pleiades appear to be quite normal (integration in the V band was made for 13 component stars). The Orion's Belt stars (ζ, ε and δ) seem to be fainter than other O9–B0 stars, which is in agreement with a paper presented by Carruthers during the Symposium.

6. Conclusions

It is to be noted that we have arbitrarily adjusted the theoretical models in the linear and well-defined part of our diagram, e.g. for B4 (Morton's models) (Mihalas and Morton, 1965; Adams and Morton, 1968; Hickok and Morton, 1968), and B6 (Underhill's models) (Underhill, 1963), spectral types, because we have no absolute calibrations.

We are now trying to integrate Stecher's spectral energy measurements for ζ and ε Persei, given in a paper presented during the session, which will give us a better adjustment of theoretical models.

More complete reductions for the interstellar reddening, intrinsic colors and Milky Way and Zodiacal Light isophotes are now in progress and will be given in a further publication.

Our next experiment, 'JANUS' (which will be launched in one year's time) will give directly and simultaneously an accurate color index between two ultraviolet bands.

References

Adams, T. F. and Morton, D. C.: 1968, *Astrophys. J.* **152**, 195.
Boggess, A. and Borgman, B.: 1964, *Astrophys. J.* **140**, 1638.
Courtès, G.: *Astronomie*, Juin 1952.
Courtès, G.: 1960, *Ann. Astrophys.* **23**, 115.
Hickok, F. R. and Morton, D. C.: 1968, *Astrophys. J.* **152**, 203.
Jaschek, M.: 1968, private communication.
Mihalas, D. M. and Morton, D. C.: 1965, *Astrophys. J.* **142**, 253.
Stecher, T. P.: 1965, *Astrophys. J.* **142**, 1683.
Underhill, A. B.: 1963, *Space Sci. Rev.* **1**, 749.
Viton, M.: 1967, *C. R. Acad. Sci. Paris* **264**, 1761.

Discussion

Morton: Why do you not see the nebulosity of the Barnard loop found in the ultraviolet by Henize and his colleagues from the Gemini photographs?

Viton: No, we have not detected the Barnard loop, despite of the high aperture ratio of our camera (f/1) probably because of the low angular resolution, too short exposure time, and wavelength range.

SPECTROPHOTOMÉTRIE INTÉGRÉE DES GALAXIES PROCHES DANS L'ULTRAVIOLET (EXPÉRIENCE PERSÉE)

P. CRUVELLIER, A. ROUSSIN et Y. VALERIO

Laboratoire d'Astronomie Spatiale du CNRS, et Observatoire de Marseille, France

Abstract. Preliminary results obtained from a sounding rocket launched in December 1968 show a strong ultraviolet excess in the central part of the Andromeda nebula, as compared to the color of the sky background.

The color index of M 31 in this spectral range may be explained by an abnormally high proportion of B stars.

1. Introduction

'Persée' est avant tout un spectrophotomètre photoélectrique de précision à 4 bandes passantes simultanées qui a été réalisé grâce à la collaboration du Centre National d'Etudes Spatiales.

La définition de l'instrument est liée essentiellement à la stabilisation; elle est également liée à la précision probable d'acquisition ($\pm 1°5$). Cela conduit à la nécessité, pour être sûr d'acquérir un astre donné, de concevoir un instrument avec au moins 3° de champ. Nous avons choisi 4°.

Un spectromètre de 4° de champ risquait d'avoir bien peu d'intérêt. On a préféré concevoir l'appareil, dont la pupille d'entrée a 15 cm de diamètre, pour qu'il donne un champ de 1°, mais en effectuant un balayage bidimensionnel sur 16° carré ($4° \times 4°$). Ainsi, l'acquisition est certaine, et la définition angulaire sur le ciel, permet d'étudier plusieurs astres isolés situés dans le champ, ou bien un astre étendu de grande dimension dont la spectrophotométrie peut être étudiée en 16 régions différentes de 1° chacunes.

'Persée' a été tirée en décembre 1968 sur la nébuleuse d'Andromède (M 31). Cette expérience avait pour but de préciser le taux des populations d'étoiles de M 31 (région centrale) et de mieux définir les limites de cette galaxie par rapport au fond du ciel qui, au sol, sature rapidement à cause de la luminance du ciel nocturne et de la répartition générale d'étoiles galactiques.

2. Principe de l'Optique

Le principe du monochromateur du type filtre B. P. M. a été décrit en détail (Bonnet et Courtès, 1962; Courtès, 1962); nous n'y reviendrons pas ici. Rappelons simplement que le monochromateur qui disperse les pupilles au lieu de disperser l'image de l'astre a, grâce à ce dispositif, des bandes passantes indépendantes des défauts de guidage de la plateforme de stabilisation de la fusée (Figure 1) et présente l'avantage d'illuminer d'un éclairement uniforme la photocathode des photomultiplicateurs, sans qu'il soit nécessaire d'adjoindre une lentille de Fabry.

Houziaux and Butler (eds.), Ultraviolet Stellar Spectra and Ground-Based Observations, 130–133.

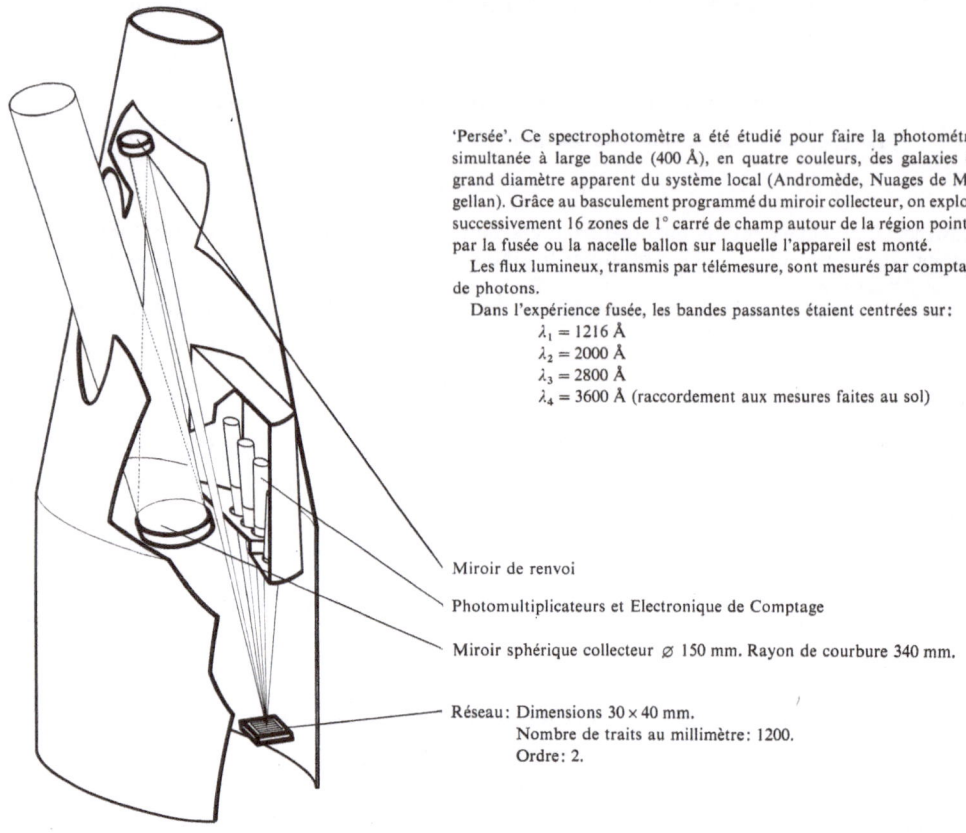

'Persée'. Ce spectrophotomètre a été étudié pour faire la photométrie simultanée à large bande (400 Å), en quatre couleurs, des galaxies de grand diamètre apparent du système local (Andromède, Nuages de Magellan). Grâce au basculement programmé du miroir collecteur, on explore successivement 16 zones de 1° carré de champ autour de la région pointée par la fusée ou la nacelle ballon sur laquelle l'appareil est monté.

Les flux lumineux, transmis par télémesure, sont mesurés par comptage de photons.

Dans l'expérience fusée, les bandes passantes étaient centrées sur:

$\lambda_1 = 1216$ Å
$\lambda_2 = 2000$ Å
$\lambda_3 = 2800$ Å
$\lambda_4 = 3600$ Å (raccordement aux mesures faites au sol)

Miroir de renvoi

Photomultiplicateurs et Electronique de Comptage

Miroir sphérique collecteur ⌀ 150 mm. Rayon de courbure 340 mm.

Réseau: Dimensions 30 × 40 mm.
Nombre de traits au millimètre: 1200.
Ordre: 2.

Fig. 1.

Le flux monochromatique est recueilli par le photomultiplicateur travaillant en comptage d'impulsions. L'information est transmise au sol par télémesure.

Les 4 bandes passantes de 400 Å de large, étaient centrées sur: 1216, 2000, 2750, et 3450 Å.

La Figure 2 donne la forme de ces bandes passantes.

Avant le tir, des calibrations avaient permis, non seulement de connaître la forme exacte des bandes passantes, mais encore de déterminer la fonction de transfert de l'appareil permettant ainsi l'établissement des rapports absolus des différents éléments spectraux.

3. Résultats

Par suite d'une saturation due à la télémesure, seules les deux bandes passantes (3450 Å et 2750 Å) ont donné des résultats exploitables.

Dans le domaine 2750 Å, les étoiles de la région marginale de la Voie Lactée superposable à la nébuleuse d'Andromède jouent un rôle de moins en moins grand dans la luminance moyenne du fond du ciel (chute de la répartition spectrale énergétique des étoiles galactiques). C'est ce que l'on constate sur les résultats de 14 des champs de 1°.

Ainsi, à l'exception des deux carrés correspondant au signal de M 31, le fond du ciel dans la bande 3450 Å présente, d'un degré carré à l'autre, les fluctuations dues aux irrégularités de la répartition des étoiles galactiques. Ces fluctuations disparaissent complètement sur les résultats du fond du ciel 2750 Å ce qui confirme le bien fondé de la recherche des limites extrêmes de M 31 (Figure 3) ou de tout autre système extragalactique.

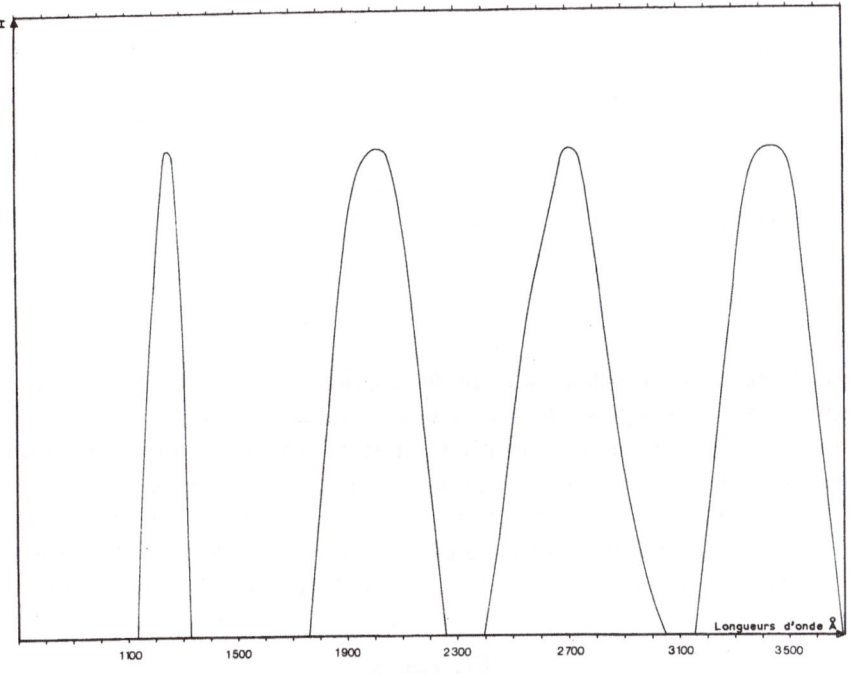

Fig. 2.

On peut donc considérer que la luminance du fond du ciel à 2750 Å est pratiquement indépendante du fond d'étoiles et est essentiellement représentative de la diffusion des poussières interplanétaires. Le fond du ciel peut donc être assimilé à une diffusion en λ^{-1} du spectre solaire. Ce résultat est en bon accord avec ceux obtenus par l'OAO II Expérience 'Wisconsin' (Bless, 1969).

L'indice de couleur $U'_{2750} - U_{3450}$ a pu être déterminé et l'on trouve que les régions centrales de la nébuleuse d'Andromède sont 3 fois plus brillantes en ultraviolet que l'extrapolation à 2750 Å du spectre visible global du noyau de la nébuleuse.

Ce résultat très intéressant est à rapprocher d'un autre résultat obtenu sur le grand nuage de Magellan (Cruvellier, 1967), puis sur M 33 en 1967 (Carranza *et al.*, 1967, 1968), et depuis, sur plusieurs autres galaxies du système local.

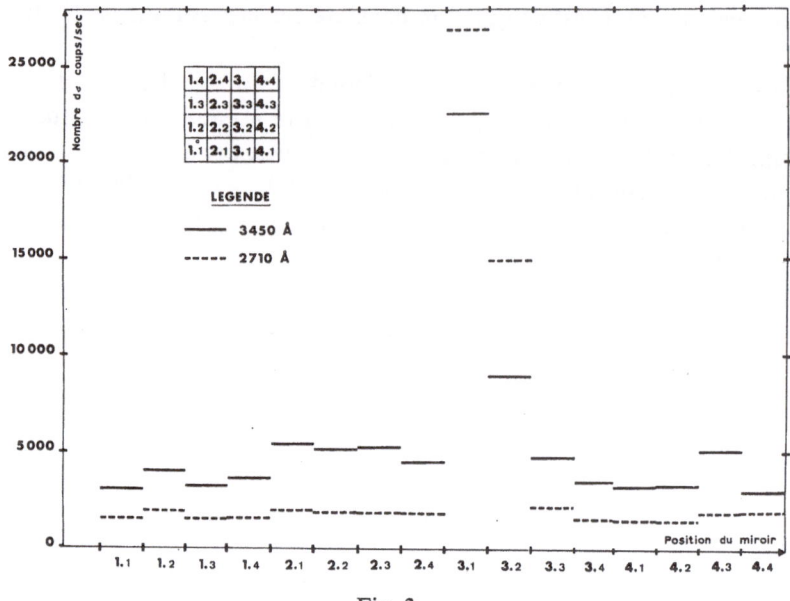

Fig. 3.

Dans les régions centrales de ces nébuleuses, il a été mis en évidence pour la première fois l'émission Hα générale qui indique la présence d'hydrogène ionisé. L'interprétation la plus simple du mécanisme d'excitation est vraisemblablement la présence d'étoiles B dont le taux dans la population des régions centrales de certaines galaxies (le plus souvent des Sb) a certainement été sousestimé. C'est l'interprétation que nous avons retenue pour expliquer l'intensité ultraviolette anormale de M 31 et c'est également cette interprétation que donne dans un article récent Goldberg (1969).

Références

Bless, R. C.: 1969, Communication privée.
Bonnet, R. M. et Courtès, G.: 1962, *Ann. Astrophys.* **25**, 367.
Carranza, G., Courtès, G., Georgelin, Y., et Monnet, G.: 1967, *C. r. hebd. Acad. Sci. Paris* **264**, 191.
Carranza, G., Courtès, G., Georgelin, Y., Monnet, G., et Pourcelot, A.: 1968, *Ann. Astrophys.* **31**, 63.
Courtès, G.: 1962, *C. r. hebd. Acad. Sci. Paris* **254**, 1738.
Cruvellier, P.: 1967, *Ann. Astrophys.* **30**, 1059.
Goldberg, L.: 1969, *Scient. Amer.*, juin.

LOW RESOLUTION STELLAR SPECTROPHOTOMETRIC OBSERVATIONS IN THE REGION 1500 Å–3000 Å

G. C. SUDBURY

Royal Observatory, Edinburgh, Great Britain

The success of Stecher and Milligan (1962) in obtaining UV photoelectric objective dispersion spectra of stars in 1962 inspired similar work at the Royal Observatory, Edinburgh. For a number of reasons we preferred to build a system using grating dispersion in the convergent beam from a 21-cm diameter paraboloidal mirror (Sudbury, 1969). The same principle of random sky scan from a spinning rocket was employed.

The first flight of this instrument in August 1965 on an ESRO Skylark rocket from Sardinia obtained what can now be identified as stellar spectra but for a number of reasons (partial telemetry failure, lunar attitude sensor failure, and pitch-yaw lock-in producing a large precession and very slow roll) it was not possible to resolve the attitude solution. A subsequent payload fired on 3rd December 1968, this time fitted with roll-rate control, was entirely satisfactory. There were two parallel instruments, similar except that one employed wide-range, photon-pulse counting from an EMI photomultiplier (spectral range 1650 Å–3000 Å) while the other continued to employ the system of current measurement from an Ascop tube over the range 1500 Å–3000 Å. The slit-width resolution was 190 Å.

At the present moment of the reduction some 75 stars have been identified, all with spectral types between O7 and A5, and extending to 6th magnitude. Consistency checks of stars observed more than once show that good multicolour photometric data should be derivable for the majority of the stellar spectra.

References

Stecher, T. P. and Milligan, J. E.: 1962, *Astrophys. J.* **136**, 1.
Sudbury, G. C.: 1969, *Appl. Opt.* **8**, 2013.

Houziaux and Butler (eds.), Ultraviolet Stellar Spectra and Ground-Based Observations, 134.

ABSOLUTE STELLAR PHOTOMETRY IN THE REGION
1200–3000 Å

J. W. CAMPBELL

Royal Observatory, Edinburgh, Great Britain

1. General

Since 1962 the Space Research Division of the Royal Observatory Edinburgh have been planning and launching a number of rocket experiments aimed at observing the absolute stellar fluxes of early-type stars in the region 1000 Å–3000 Å. Although some earlier attempts were unsuccessful due to rocket failures, good photometric data have been obtained from four successful flights in 1967 and 1968, and it is the purpose of this short report to indicate the nature of the data rather than give a complete scientific analysis.

Because of the non-availability of stabilised rockets it was necessary to use the roll and precessional motion of the rocket to scan the telescope axes across the sky, the scan rate being optimized by means of a single axis gas jet system operating around the roll axis.

In addition to the ultraviolet photometers, a number of stellar photometers sensitive in the region 2500 Å–5000 Å was also flown in order to facilitate the determination of the rocket attitude, each photometer having a one degree field of view.

2. Stellar Observations in the Region 1200 Å–2000 Å

The photometer for this region consisted of an $F/3$, Cassegrain telescope of aperture 23 cm. The mirrors were aluminized and given a protective coating of magnesium fluoride, typical reflectances being of the order of 85% at 1216 Å. The light from the primary mirror was imaged on to the photo-cathode of an Ascop 542G photomultiplier by means of a lithium fluoride Fabry lens and the field of view defined by means of a circular diaphragm ($1\frac{1}{2}°$ diameter) in the focal plane. The waveband was 250 Å wide and centred on 1600 Å. The detector was operated with the cathode at -3000 V with respect to the rocket body potential and the output from the anode measured by means of a dual range solid state d.c. electrometer operating in the range 3×10^{-10} A–3×10^{-7} A.

Each photometer was calibrated in an absolute manner using microwave excited gas discharges as well as xenon and low pressure mercury lamps. The exact procedure will be described elsewhere.

Approximately 120 observations of identified stars of spectral types between O8 and A2 down to a limiting magnitude of 6.5 were obtained in a flight in December 1968.

Houziaux and Butler (eds.), Ultraviolet Stellar Spectra and Ground-Based Observations, 135–137.

3. Stellar Observations in the Region 2000 Å–3000 Å

In this wavelength region the stellar photometer consisted of two parallel identical Newtonian telescopes of aperture 18 cm × 8 cm and field of view 1° also mounted perpendicular to the thrust axis of the rocket.

The light from the telescopes was imaged on to the photo-cathodes of two solar blind photomultipliers by means of quartz Fabry lenses. Again the detectors were operated at a high negative cathode potential and the output measured by d.c. electrometers.

The wavelength isolation in the May 1967 flight was achieved by means of interference filters of passband 250 Å, centred on 2150 Å and 2550 Å. In two later flights in 1968, this isolation was achieved by selectively reflecting surfaces, with an increase in efficiency of the order of 4. The absolute calibration was achieved by means of a d.c. discharge of the Hinteregger type as well as by low pressure mercury lamps.

Preliminary analysis of some thirty observations from the May 1967 flights indicates good agreement with the other observational data available and is in agreement with stellar models exhibiting no blanketing, such as proposed by Mihalas (1965). The photometers of the 1968 flight produced some 150–200 stellar observations which are at present under reduction.

4. Sky Background Observations in the Region 1300 Å–5000 Å

In addition to the stellar photometry experiments, a separate experiment was flown in 1968, aimed at obtaining sky background measurements in the region 1650 Å–3200 Å. The photometer optical system consisted of a combination of a 6.5 cm objective lens and a 2.5 cm Fabry lens, both made of quartz. Periodically inserted filters at the focal plane of the photometer isolated the following wavelength regions: 1650 Å–3200 Å, 2500 Å–3200 Å, and 2800 Å–3200 Å. Eleven sky scans were achieved in a period of 400 sec of flight above a height of 110 km.

Measurements of the sky background were also obtained from the stellar attitude sensors in the region 1650 Å–5000 Å and from the stellar photometer sensitive in the region 1300 Å–2000 Å. The data are in process of being analysed.

5. Comment

Although the first European results are now beginning to appear, there is still considerable difficulty in intercomparing observations made by different experimental groups. It is a matter of extreme urgency that a system of cross calibrations between European and, of course, American instruments be established if the maximum use of the data is to be achieved. Such a system at its simplest could consist of the exchange of a suitably calibrated source (or detector) at reasonable intervals between the various scientific groups.

Reference

Mihalas, D.: 1965, *Astrophys. J. Suppl. Ser.* **9**, 321.

Discussion

Carruthers: We have an interlaboratory calibration comparison program going on in the Washington area in which we pass around from one laboratory to another, a nitric oxide ion chamber with magnesium fluoride window, which each lab calibrates at 1216 Å by its own technique, and we compare the results.

I would be interested in getting together with anyone interested in interlaboratory calibration comparisons, so that we could come up with a set of standard detectors for various wavelength ranges for passing around among laboratories.

Campbell: I am very glad that there has been an immediate response to my plea for an inter-comparison of ultraviolet standards between co-workers in the field of ultraviolet astronomy. This is particularly important in order to compare the European absolute photometry with that of the U.S.A. groups. We would be very interested in joining the U.S.A. calibartion network with a particular view to comparing measurements in the region 1600 Å–2500 Å. As a preliminary step however the transfer of your nitric oxide chamber to interested European experimenters could be of great value and should be undertaken as soon as possible. May I mention in passing that I have already mentioned a similar procedure to Dr. Boldt who intends to inter-compare various European calibrations procedures with a view to selecting the best system of standardization.

PRELIMINARY NOTE ON THE ASTRONOMICAL
SATELLITE KOSMOS 215

N. DIMOV

Crimean Astrophysical Observatory, U.S.S.R.

The astronomical satellite Kosmos 215 was launched in April 1968 to examine ultra-violet and X-ray stellar radiation. During a period of 40 days the spectral region from 1250 Å to 2700 Å was recorded by 7 photon-counters and photomultipliers mounted in the focal planes of similar telescopes with apertures of 70 mm, fields of view 1° and focal ratios about $\frac{1}{3}$. The optical axes of all the telescopes were parallel. The satellite was not stabilized but special damping was used to reduce the rotational velocity to $0°.12 \sec^{-1}$. The optical telescopes scanned the sky and recorded all stars which crossed their fields of view.

The stellar identifications were determined by means of a magnetometer and two telescopes which gave information about visual B and V magnitudes.

An X-ray counter had a $4°.5$ half-width of field of view and recorded in the region $1-54°$.

From 10 to 25 stars can be identified for each revolution and the reductions are now in progress.

The work has been carried out by several institutions, namely, Dr. V. Kurt at the Sternberg State Institute, Dr. V. Tiyt at the Estonian Astronomical Institute and Dr. V. Prokofiev and myself at the Crimean Astrophysical Observatory.

Houziaux and Butler (eds.), Ultraviolet Stellar Spectra and Ground-Based Observations, 138.

ULTRAVIOLET PHOTOMETRY OF STARS FROM OSO II

KENNETH L. HALLAM

Goddard Space Flight Center, Greenbelt, Md., U.S.A.

Discussion

Abstract. Ultraviolet stellar fluxes from 1500 to 3200 Å were from February through August 1965 on OSO II. A 15-cm diameter Gregorian telescope with a stepped grating spectrophotometer provided flux measurements in ten adjacent 180 Å wide band passes.

By comparing fluxes of stars showing B-U color excesses smaller than 0.19 and larger than 0.26, an ultraviolet extinction curve has been derived, which agrees with others which have been published as far as the magnitude effect is concerned, but the shape differs somewhat, the slope at $1/\lambda = 3.8\ \mu^{-1}$ being somewhat greater than at $1/\lambda = 5.0\ \mu^{-1}$.

If the stars' fluxes are corrected for reddening, it is found that there is a good agreement between observation and models for stars earlier than B3, but that many of the later type stars have a residual apparently intrinsic reddening.

Greenberg: Has Hallam investigated the possibility that apparently anomalous properties of the later type stars may be due to the fact that the extinction curve derived from the early type stars is not applicable to the later types? The early-type stars could modify the neighbouring interstellar medium and since the stars which have been studied so far for extinction are relatively nearby (total extinction rather small) a significant proportion of the overall extinction could be of local origin.

Roman: I doubt it. However, I would be very surprised if these stars are sufficiently reddened to explain the problem in this way.

Houziaux and Butler (eds.), Ultraviolet Stellar Spectra and Ground-Based Observations, 139.
All Rights Reserved. Copyright © 1970 by the IAU.

THE ULTRAVIOLET SOLAR OPACITY

OWEN GINGERICH

Smithsonian Astrophysical Observatory and
Harvard College Observatory, Cambridge, Mass., U.S.A.

Abstract. Ultraviolet solar observations are compared with predictions from a new solar model. From 3600 to 1700 Å there is heavy line blanketing; probably one or more major sources of opacity are missing from the theoretical calculation in this region.

The Sun provides a particularly well-observed example of the ultraviolet spectrum of a typical G2 V star with a chromosphere. Because of the large increase in opacity toward shorter wavelengths, the ultraviolet continuum radiation arises at progressively higher levels. Hence, the construction of an empirical model becomes possible using continuum intensities alone. Because the opacity also rises with longer wavelengths, such a model can be checked from observations in the infrared.

Ultraviolet intensities predicted from a model newly constructed along these principles are shown in Figure 1. This model, which will be described in more detail in *Solar Physics*, differs from the Bilderberg Continuum Atmosphere (Gingerich and De Jager, 1968) in having a somewhat deeper temperature minimum and an earlier chromospheric temperature rise.

The solar spectrum beyond the Balmer continuum (911 to 3643 Å) can be divided into three general regions. From the Balmer limit down to 1683 Å (the edge produced by the first excited state of silicon), the spectrum is characterized by heavy absorption line blanketing and limb darkening. Between 1683 and 1525 Å (the emission edge from the ground state of silicon), the spectrum is at first glance deceptively similar to that at the longer wavelengths, but a closer inspection shows a dense pattern of *emission* lines. No atomic absorption lines have been identified in this interval, although certain strong CO lines have been found (Porter *et al.*, 1967). The center-to-limb variation is almost flat (Tousey, 1963). Radiation in this interval must arise from near the solar temperature minimum (Gingerich and Rich, 1968). Finally, between 1525 Å and the Lyman limit, the spectrum arises from the low chromosphere; it is presumably limb brightened, the bound-free opacity edges appear in emission, and the strong emission lines become sufficiently separated so that the continuum is readily isolated.

These characteristics of the spectral regions are important for understanding the mixed success of the model predictions shown in the figure. Toward the shorter wavelengths, where the continuum is well defined, the model represents the observations quite well – in fact, better than the observational accuracy warrants. Above 1683 Å, the heavily line-blanketed spectrum falls well below the predicted continuum level.

Nevertheless, the difficulty of finding the continuum cannot mask the fact that one or more opacity sources must still be missing from the theoretical calculations, especially in the 1700 to 1900 Å region. Through this part of the spectrum, the model assumes an opacity we formerly identified with the distant resonance broadening wing

Houziaux and Butler (eds.), Ultraviolet Stellar Spectra and Ground-Based Observations, 140–142.

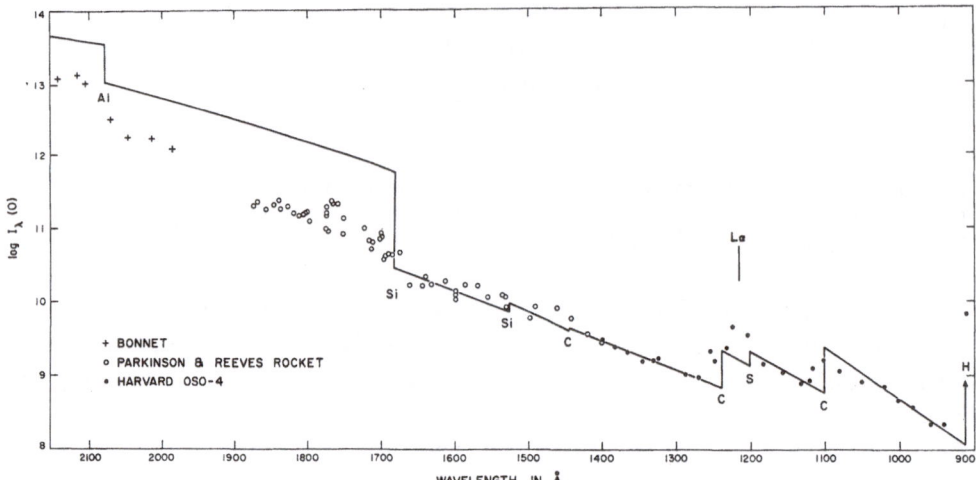

Fig. 1. The predicted solar ultraviolet continuum compared with observations by Bonnet (1968) and Parkinson and Reeves. (The observations at 1460 and 1490 Å are still provisional.)

of Lyman α; but this is incorrect, so the situation is actually worse than graphed here. Carbon monoxide, the quasi-H_2 molecule, and the proper (molecular) form of the Lyman α wing all absorb in the 1700 to 1800 Å region, but not very strongly. One good possibility for the missing opacity would be the several lower levels of bound-free iron, especially if the solar iron abundance is higher than that recently accepted.

Perhaps I should point out explicitly that the discrepancies between the predictions and the observations at 1700 to 2000 Å cannot be reconciled by adjusting the temperature structure of the new model. This would wreak havoc with the agreement at longer wavelengths. Therefore, the search for additional absorbers must continue.

Acknowledgements

I should like to thank E. M. Reeves and W. H. Parkinson for giving me the results of their September 1968 rocket flight, as well as the OSO-4 observations and calibration curve. The new model has been established with the help of R. Noyes and Y. Cuny.

References

Bonnet, R.: 1968, *Ann. Astrophys.* **31**, 597.
Gingerich, O. and De Jager, C.: 1968, *Solar Phys.* **3**, 5.
Gingerich, O. and Rich, J. C.: 1968, *Solar Phys.* **3**, 82.
Porter, J. R., Tilford, S. G. and Widing, K. G.: 1967, *Astrophys. J.* **147**, 172.
Tousey, R.: 1963, *Space Sci. Rev.* **2**, 3.

Discussion

Morton: Does your new solar UV flux distribution change the effective temperature of 5180 K derived by Labs and Neckel?

Gingerich: No. There is so little flux in the solar UV that a change of this size will have no appreciable effect.

Müller: I am referring to the discrepancy between the observed and predicted opacity near 1700 Å. You mentioned to me yesterday that a higher photospheric iron abundance might account for the strong increase in the observed opacity near 1700 Å. Since, apparently, there are some evidences for a higher photospheric iron abundance (i.e. a factor of 10 to 20 higher than the GMA value), I wonder whether the 1700 Å region is the only one in which a large photospheric iron abundance would affect the opacity distribution and whether the increase of the photospheric iron abundance by a factor 10 to 20 would be sufficient to produce the observed opacity.

Gingerich: Unfortunately there are no laboratory or quantum mechanical values for the iron cross sections, but if the hydrogenic approximation has any validity, then with a 10 × GMA iron abundance, about five bound-free levels will be important around 1700–1800 Å, where we find now a severe discrepancy between the models and observations. I hope that laboratory cross sections will be determined soon in the Harvard Shock Tube Laboratory. At how much longer wavelengths the iron might contribute noticeably, I can't guess, but we do have definite evidence for a comparatively smooth unidentified absorber diminishing at longer wavelengths up into the visual spectrum. An increased iron abundance should also give a desirable increase in electron density in the solar chromosphere.

Conti: From the reports by Davis and by Hallam, we understand there is an ultraviolet deficiency for A, F, and G stars. This is also true for the Sun. Would Gingerich care to comment on the possible explanations for the discrepancy?

Gingerich: With the possible exception of iron, I believe we have the major bound-free opacity sources. I have looked for neutral and ionized atoms without finding other candidates for these temperatures. Most molecules are out – carbon monoxide, for example, is overwhelmed by silicon in the Sun. At 1700 Å the resonance wing of Lyman α, which acts like a molecule here, may be significant, and there is always the possibility of something strange like HHe. Consequently it appears that line blocking may be the most likely cause of the ultraviolet deficiency both in the Sun and in earlier stars. However, in the Sun the major discrepancy is ∼ 2500 Å, whereas in Sirius the disagreement also includes the resonance line region down to 1200 Å.

PART II

STELLAR LINE SPECTRA

A. ROCKET AND SATELLITE OBSERVATIONS OF ULTRAVIOLET SPECTRA

OBSERVATIONS OF ULTRAVIOLET STELLAR SPECTRA

R. WILSON

Science Research Council, Astrophysics Research Unit,
Culham Laboratory, Abingdon, Berksh., England

1. Introduction

The dividing line between photometry and spectrometry is not always obvious and for the purpose of this review, I will define ultraviolet stellar spectroscopy as observations with sufficient spectral resolution to allow the detection of individual spectral lines and their measurement in terms of wavelength and strength. From an examination of the existing observations this results in a resolution requirement of $\delta\lambda < 10$ Å. Since the best spectral resolution so far obtained is about 1 Å then this places the results to be discussed within the range 1–10 Å. In terms of $\lambda/\delta\lambda$ this corresponds to a range of about 2000–200 and it is important to bear in mind that these represent low resolution spectra. In fact the limit of 200 that I have imposed would rarely be used for spectroscopic studies in ground based observatories where it corresponds, in the notation of the optical astronomer, to a dispersion of about 1000 Å/mm, the resolution limit being set by the photographic plate, typically taken as 20 μ. Hence, even the faintest objects like quasars are usually studied with a dispersion of a few hundred Å mm^{-1}. The fact that such a resolution can be included here is an indication of the exceptionally strong resonance lines which occur in the ultraviolet and which can be detected with such a resolution. On the other hand, the richness of the ultraviolet spectrum is making and will continue to make, demands on improved resolution in order to separate the many features. The best achieved resolution of about 1 Å goes only part-way to solving this problem.

In such a rapidly advancing field, this review cannot be completely up-to-date. However, I believe it is comprehensive as far as published information is concerned and therefore a reasonable starting point for the subsequent papers of the Conference which will present some of the very recent data.

2. The Observations

The first UV stellar spectra were obtained by Morton and Spitzer [1] of δ and π Scorpii during a rocket flight in 1965 and were made possible by the development of a star-pointing control system for the Aerobee rocket. This pointing control, the *Inertial Attitude Control System* (*IACS*) of Space General Corporation (U.S.A.) is a 3-axis stabilized, cold gas reaction jet system used to orient the entire rocket at any given time during free fall. More than five pre-programmed positions can be acquired during a single flight by changing the inertial attitudes of two free gyro sensors. In each orientation, the control system continuously aligns the rocket with the null positions of

Houziaux and Butler (eds.), Ultraviolet Stellar Spectra and Ground-Based Observations, 147–162.

the sensors. A 3-axis rate gyro is used for system damping. In basic form, the IACS is capable of pointing a payload within 3° of a desired direction and has a limit cycle of ±15 arcmin. The performance of this system has been considerably improved in the *Stellar Tracking Rocket Attitude Positioning* (*STRAP*) system which has been developed at the NASA Goddard Space Flight Center from the basic IACS with the addition of a star sensor and an auxiliary low-thrust jet system. This achieves a limit cycle performance of about ±30 arcsec for 3rd magnitude objects, improving to about ±10 arcsec for brighter objects.

All published UV stellar spectra have been obtained from Aerobee rockets using either the IACS or STRAP pointing systems. After their pioneering flights, the Princeton group have obtained further observations and have been joined by Carruthers of the Naval Research Laboratory and Stecher and Smith, both at Goddard. Although this review is not concerned with techniques as such, it is necessary to consider them to a degree sufficient to appreciate the data. This will be done by briefly describing the instrumental concepts of the four groups involved.

Morton and his colleagues (Princeton) employed a fine stabilisation system in conjunction with the IACS to give an improved pointing stability in the direction of dispersion. This was done passively by pivoting the instrument platform and attaching a large gyro. Instrument stability of better than ± 20 arcsec and a spectral resolution of about 1 Å have been achieved in this way. The components of the system used in the first few flights [1, 2] are indicated in Figure 1. The platform was mounted to the rocket bay with a single degree of freedom. Two spectrographic cameras (only one is shown) were rigidly attached to one side of the platform and the gyro to the other. The dispersion direction was placed perpendicular to the platform axis. The spectro-

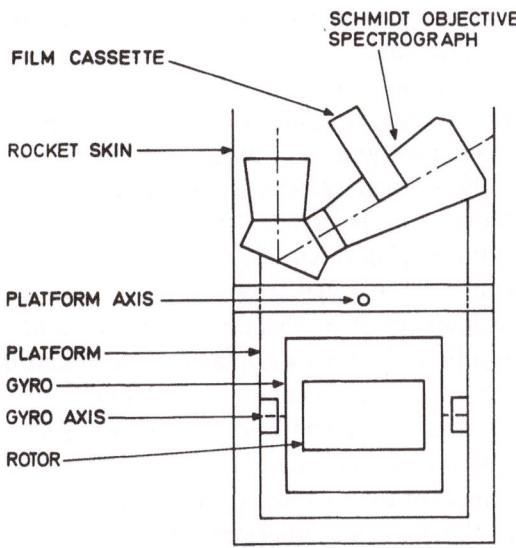

Fig. 1.   Schematic of a Princeton objective spectrograph payload. For example of UV data recorded with such an instrument, see Figure 5.

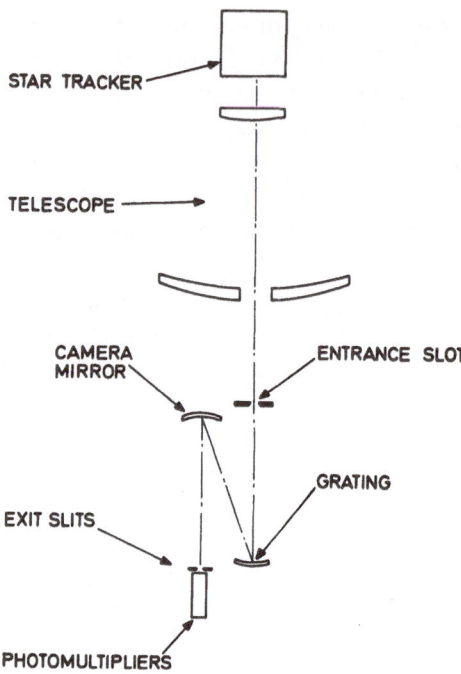

Fig. 2. Schematic of a scanning photoelectric spectrometer system with telescope as flown by Stecher (see text). For example of UV data recorded with such an instrument, see Figure 6.

graphic cameras each consisted of a plane objective grating followed by an $f/2$ Schmidt camera with a 10° diameter field. Correctors of CaF_2, LiF and fused silica were used in various flights. A solenoid actuated magazine gave 6 exposures on Kodak Pathé SC5 film for each camera. On later flights, an all-reflective camera with LiF overcoating on all surfaces was used for improved short wavelength performance [3–5].

Stecher [6, 7] (Goddard) uses the STRAP system with a payload consisting of a 33 cm $f/10$ telescope feeding a scanning photoelectric spectrometer. This instrument is indicated schematically in Figure 2. The spectrometer section contains a concave grating, rotated about an axis in the plane of the diagram, a folding mirror and three exit slits with photomultipliers, here seen side-on. The latter were used simultaneously and a continuous pointing-error signal from the star sensor, telemetered with the spectrum data, allowed subsequent wavelength corrections to be made at each point in the spectrum. A resolution of 10 Å was obtained, the limitation in this case being due to the exit slit widths rather than to pointing instability.

Smith (Goddard) has used the STRAP system with an objective grating camera (a concave grating in Wadsworth mount) to obtain spectra below 1200 Å [8]. Because of the low reflection efficiencies of the coatings available in these spectral regions, such a single element system, although of small aperture, may still be comparable in speed to a large telescope with spectrograph for which at least two additional reflections are required. Furthermore, the system has the advantage of a relaxed stability requirement compared to a telescope with spectrograph as can be seen from the following expression

$$\delta w = (G/T) D \, \delta\lambda,$$

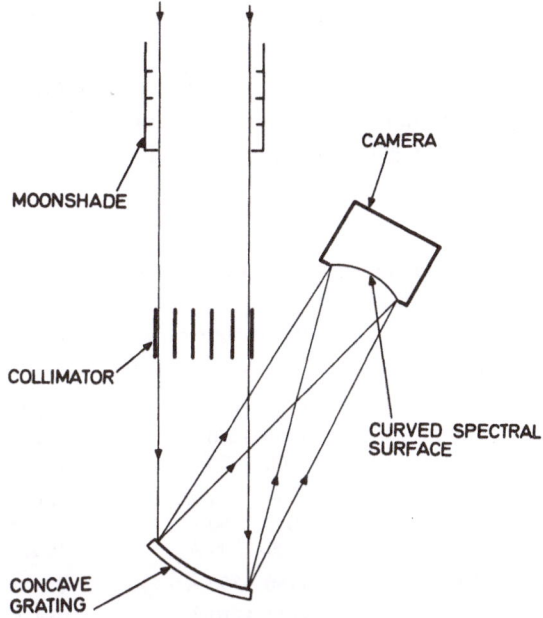

Fig. 3. Schematic of an objective grating camera in a Wadsworth mounting using the same basic principle as the payload launched by Smith [8]. For example of UV data, recorded with such an instrument, see Figure 7.

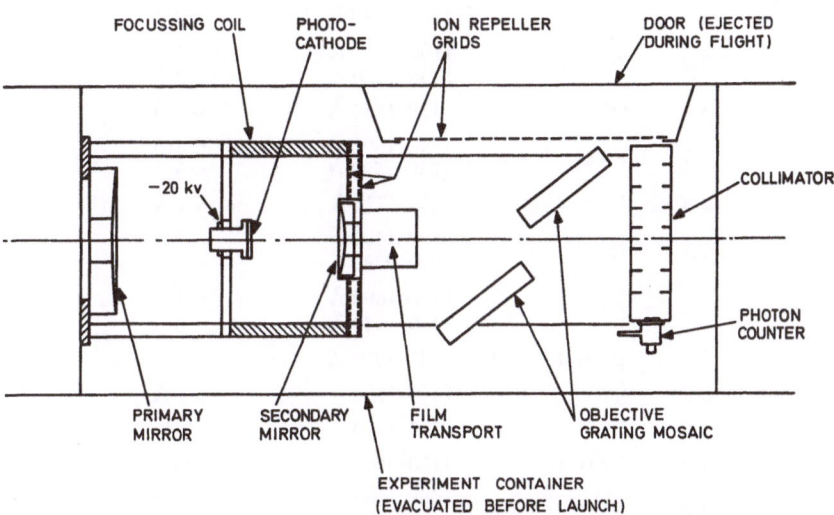

Fig. 4. Diagram of the objective grating camera incorporating an electronographic detector flown by Carruthers [9]. (By permission of the University of Chicago Press.) For example of UV data, recorded with such an instrument, see Figure 8.

which relates the degradation in resolution $\delta\lambda$ produced only by an angular pointing noise δw. G is the aperture of the grating and D its angular dispersion; T is the aperture of the telescope. In this particular configuration, the grating is also the telescope and G/T is unity compared with typical values of ~ 0.1 for telescope/spectrometer systems. With this system Smith has obtained UV spectra with the best spectral resolution (0.8 Å) yet achieved.

A representation of a Wadsworth camera is shown in Figure 3 which includes a baffled hood to reduce instrumentally scattered moonlight, if applicable, and a collimator system, to limit geometrically the background fogging due to Lyman α night glow.

TABLE I

Observations of UV stellar spectra

Star	m_v	MK	Wavelength range	Resolution	Reference source
γ^2 Vel	1.9	WC7 + O7	1127–1193 Å 1050–1250 Å 1100–3000 Å	1.6 Å 2–3 Å 10 Å	[5] [9] [10]
ζ Pup	2.2	O5f	1100–1965 Å 1050–1250 Å 1100–3000 Å	1.6 Å 2–3 Å 10 Å	[5] [9] [10]
λ Ori	3.7	O8	1150–1350 Å	2–3 Å	[9]
ι Ori	2.8	O9 III	1130–1310 Å 1100–1400 Å 1230–2100 Å	~ 1 Å 2–3 Å ~ 3 Å	[4] [9] [2]
ζ Ori	1.8	O9.5 Ib	1100–1670 Å 1070–1400 Å 1200–1730 Å	~ 1 Å 2–3 Å ~ 3 Å	[4] [9] [2]
δ Ori	2.2	O9.5 II	1100–1951 Å 1200–1420 Å	~ 1 Å ~ 3 Å	[4] [2]
σ Ori	3.8	O9.5 V	1138–1634 Å	~ 1 Å	[4]
ε Ori	1.7	B0 Ia	1100–1806 Å 1230–1590 Å	~ 1 Å ~ 3 Å	[4] [2]
δ Sco	2.3	B0 V	1260–1720 Å	~ 1 Å	[1]
κ Ori	2.1	B0.5 Ia	1120–1400 Å 1620–2780 Å	2–3 Å ~ 3 Å	[9] [2]
η Ori	3.3	B0.5 V	1178–1800 Å 1230–1440 Å	~ 1 Å ~ 3 Å	[4] [2]
β CMa	2.0	B1 II–III	1050–1300 Å	2–3 Å	[9]
π Sco	2.9	B1 V	1260–2180 Å	~ 1 Å	[1]
α Vir	1.0	B1 V	928–1350 Å	0.8 Å	[8]
ε CMa	1.5	B2 II–III	1100–3000 Å	10 Å	[10]
γ Ori	1.6	B2 III	1668–2747 Å 1150–1370 Å	~ 1 Å 2–3 Å	[4] [9]
α CMa	−1.4	A1 V	1100–3000 Å	10 Å	[10]

Carruthers (N.R.L.) has obtained objective grating spectra using an electrono-graphic detector system [9] which is displayed schematically in Figure 4. The system incorporated a KBr photocathode and Schwarzschild optical system and was preceded by an objective mosaic of 4 plane gratings operated in the second order. LiF overcoated Al was used for all reflecting surfaces. The effective aperture and field of view were 15 cm and 7° respectively and the inherent resolution capability, about 1 Å. Care was taken to reduce the background fog level resulting from positive ions either of ionospheric origin or from ionizing collisions in the residual gas. The IACS was used with an additional feedback system to reduce the jitter rate within its limit cycle to less than 1 arcmin sec^{-1}; the resulting spectral resolution was typically 3 Å for a 10 sec exposure.

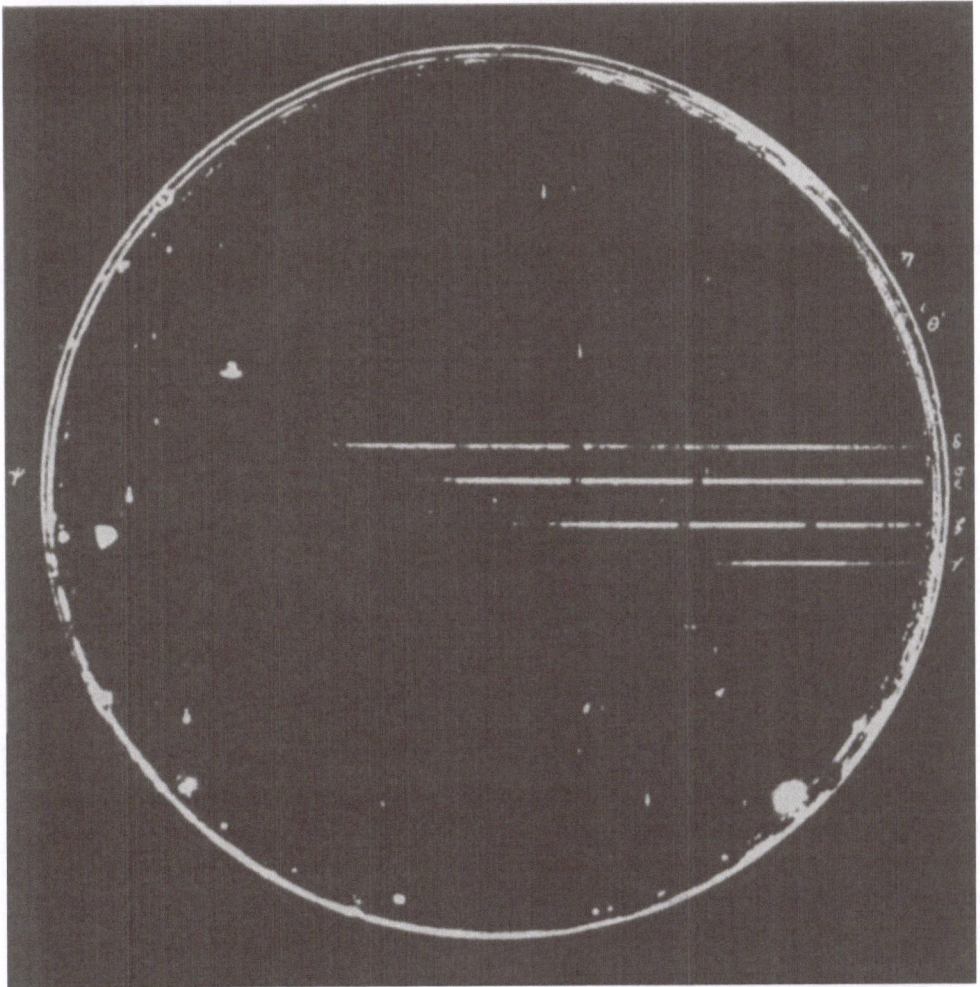

Fig. 5.   UV objective spectra of the Orion region obtained by Morton *et al.* [4]. (Reproduced from a print kindly supplied by Morton.)

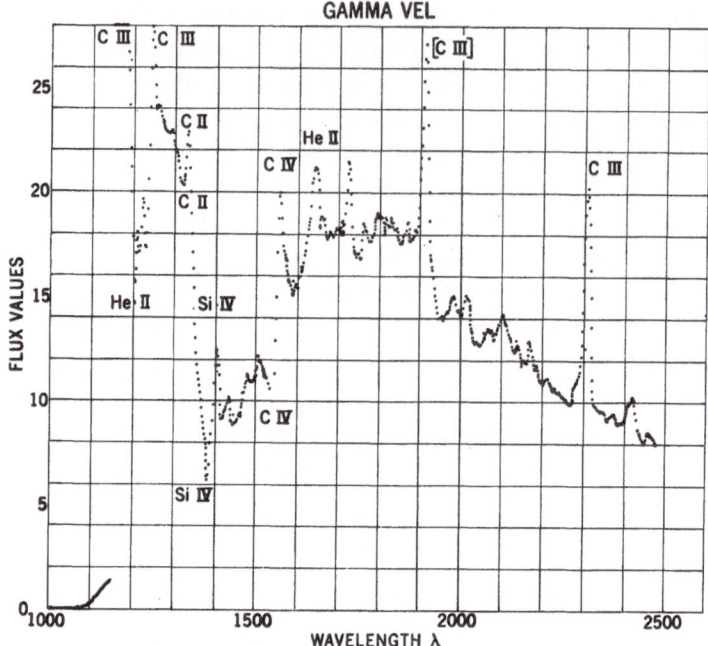

Fig. 6.   A spectral scan of γ Vel from 1200–2500 Å obtained by Stecher [10]. The flux is in units of 10^9 erg cm^{-2} sec^{-1} Å^{-1} (from Goddard preprint [10] by courtesy of Stecher).

3. Results

The observed data on UV stellar spectra that are available in the literature at the time of writing are summarized in Table I, which gives for each star the observed wavelength range and spectral resolution, together with the reference source. The table is selective to the extent that observations which are labelled 'very weak' by the authors or which involve superimposed spectra have been excluded. Not surprisingly, the observations are limited to bright early-type stars.

An example of the type of data obtained by each of the four groups who have contributed to Table I is given in Figures 5–8. These correspond to the instrumental concepts given in Figures 1–4. Figure 5 displays an objective spectrogram of the Orion stars obtained by Morton and his colleagues [4]; Figure 6 gives a spectrum scan of γ Vel obtained by Stecher [10]; Figure 7 gives a microdensitometer tracing of the spectrum of α Vir obtained by Smith [8] and Figure 8 gives microdensitometer tracings of a selection of spectra obtained by Carruthers [9].

It is indicative of the rapid expansion of UV astronomical spectroscopy that Table I is already out of date, as will be made clear by subsequent papers. Each of the four groups has carried out further successful rocket flights and the first UV stellar spectra from a satellite have been obtained by Code and his colleagues at Wisconsin using a scanning spectrometer with resolution ~ 10 Å on the *Orbiting Astronomical Observatory* (OAO-A2). This work is being described by Bless in a number of papers in this conference.

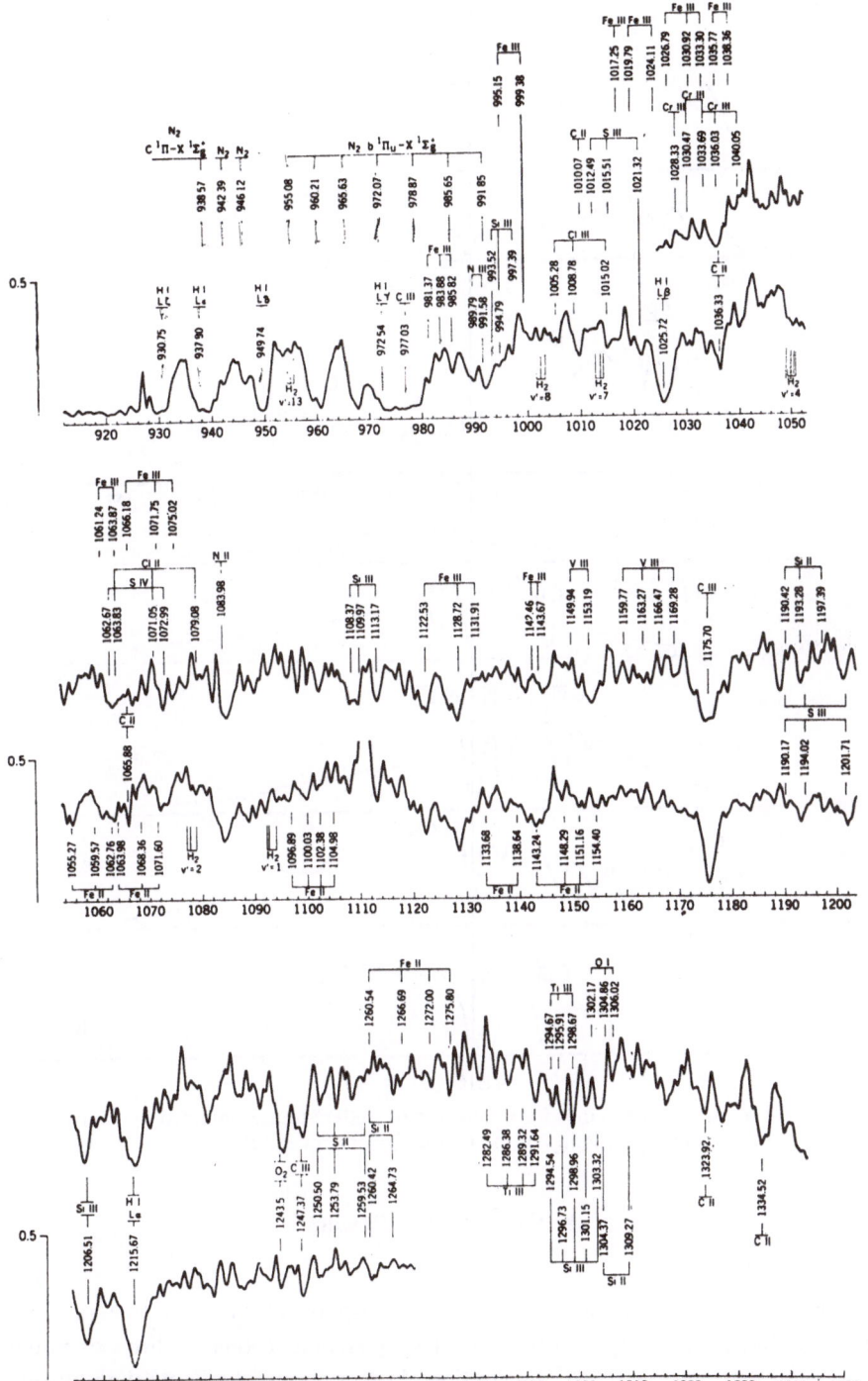

Fig. 7.   Microdensitometer traces of the spectrograms of α Vir obtained by Smith [8] over the range 920–1350 Å. The upper and lower tracings correspond to short and long exposures respectively with the ordinate density scale reflecting the lower trace only.

(By permission of the University of Chicago Press.)

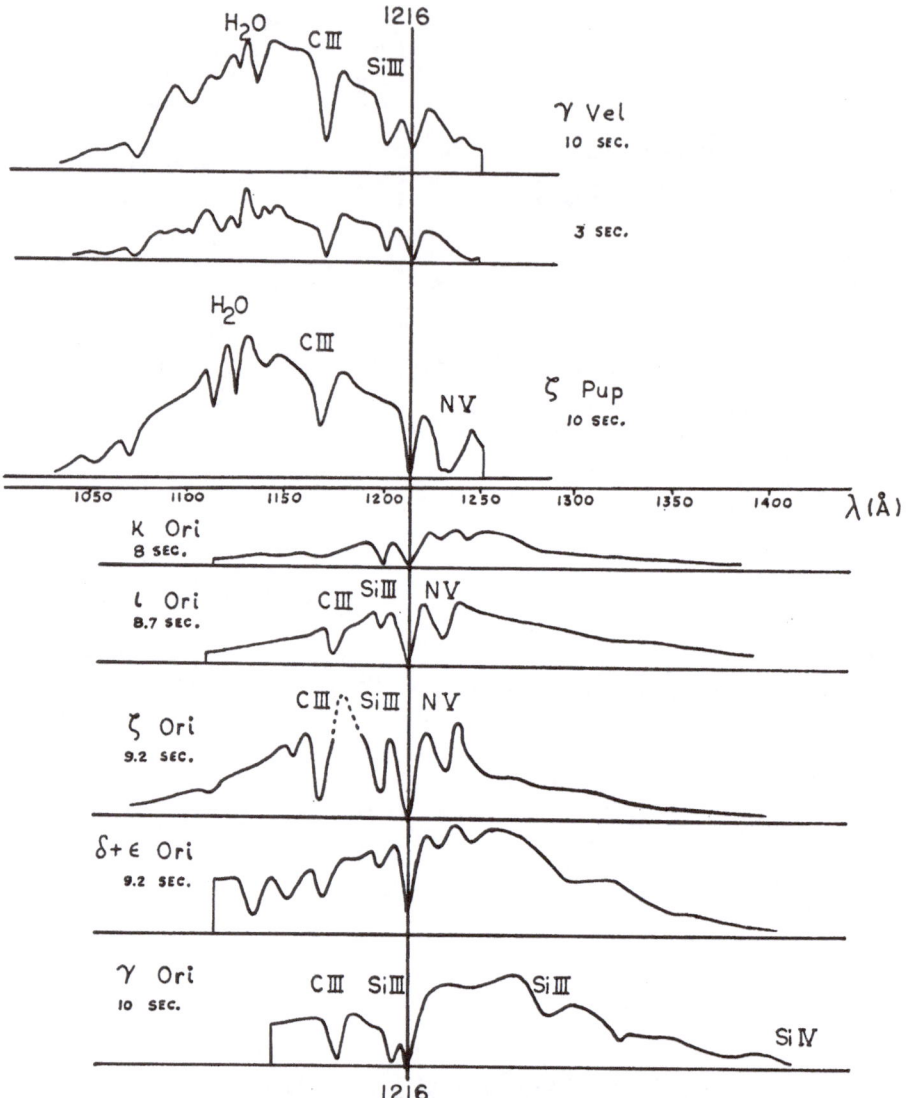

Fig. 8. Microdensitometer tracings for a selection of stellar spectra obtained by Carruthers [9]. (By permission of the University of Chicago Press.)

4. Discussion of Results

A. STELLAR FEATURES

The observed absorption line spectra give general support to the theoretical predictions, in the sense that they show the presence of very strong resonance lines of abundant ions and demonstrate directly the severe line blanketing which affects the photometric data. The analysis of photospheric lines has concentrated on the identification of ob-served features. This is made necessary by the richness of the ultraviolet spectra and is best illustrated by the observations of Morton *et al.* [4] and Smith [8] which combined the best spectral resolution yet obtained (~ 1 Å) with well exposed spectra. In the spectra of δ, ε and ζ Ori, Morton *et al.* have observed nearly 200 absorption lines in the region 1100–2000 Å of which more than 100 remain unidentified. Smith's observation of α Vir (B1 V) in the range 928–1350 Å gave about 90 lines for which classifications are proposed, leaving many more features unidentified (see Figure 7).

Calculations of emergent flux are now available for a number of theoretical line-blanketed model atmospheres of early-type stars against which the observed UV stellar spectra can be compared. The first estimate of the strengths of ultraviolet spectral lines was made by Gaustad and Spitzer [11] for a B2 star using a curve of growth analysis based on a Schuster-Schwarzschild model. Since then, more refined calculations have been made of the emergent ultraviolet flux from early-type stars by introducing line absorption into model atmospheres of the type developed by Underhill [12, 13]. The models are based on a plane-parallel atmosphere in hydrostatic equilibrium, radiative equilibrium and local thermodynamic equilibrium with a chemical composition of He/H=0.15 by number plus heavy elements as determined from the solar photosphere. Calculations are available for a number of early-type main-sequence models with surface gravity = 10^4 cm sec^{-2} as follows: O5 V [14], B0 V [14], B1 V [15], B2 V [16, 17, 18], and B4 V [19].

In addition to these, a series of models for B8–F2 is available [20] based on the same assumptions but with line blanketing calculated for the hydrogen Balmer lines only.

The most extensive comparison of model atmosphere calculations has been made in respect of photometric observations by Bless *et al.* [21] who examined the UV data of a number of observers [22–25] which covered the range 1115 Å–2800 Å and included 35 main-sequence stars of spectral types B and A. The models were identified with the observations through the effective temperature scale of Morton and Adams [26]. It was concluded that the existing model atmospheres could adequately represent the UV photometric observations of main sequence stars within their possible errors ($\pm$ 0.5 mag.) at wavelengths above 1500 Å but that gross discrepancies could occur at shorter wavelengths, particularly 1150 Å.

A more detailed check on model atmospheres is possible with the UV spectroscopic data but an extensive comparison has yet to be made. However, discrepancies are apparent between the theoretical and observed strengths of some spectral lines. Further, Smith's [8] comparison of his observations of α Vir with the B1 V model atmosphere calculations of Mihalas and Morton [15] shows that many more lines are present than are included in the calculations; he concludes that the net effect of these additional lines will be significant in terms of line blanketing. It is therefore apparent that more sophisticated model atmospheres are required, not only to include all significant line absorption, but also to examine departures from the fundamental hypotheses of the 'classical' model atmosphere. Convective energy transport has been considered in model atmospheres by Mihalas who shows that the ultraviolet flux can be appreciably smaller than in radiative models. Also, Guillaume *et al.* have studied the effect of microturbulence on spectral lines and show that lines can be increased substantially in strength by such effects. A review and discussion of the problems of model atmospheres in predicting UV stellar spectra is to be given by Miss Underhill [27].

All the main-sequence stars so far observed have been characterized by an absorption line spectrum. However, many other stars listed in Table I show emission as well as absorption lines, indicating gross departures from the classical model atmospheres and the presence of extended or circumstellar atmospheres. This is not surprising for γ^2 Vel (WC7 + O7) and ζ Pup (O5 f) which are known to be emission line objects from observations in the visible but in addition to these objects all super-giants listed in Table I include some emission features in their spectra. Thus, emission lines have been reported [2, 4, 9] in ε Ori (B0 Ia), κ Ori (B0.5 Ia), ζ Ori (O9.5 Ib), δ Ori (O9.5 II) and ι Ori (O9 III). The lines show a P Cygni type profile with violet shifted absorption components indicating expansion velocities of the order of 1500 km/sec. This effect occurs in the strong resonance lines of Si III (1207 Å), Si IV (1394–1403 Å), C IV (1548–51 Å) and N V (1239–43 Å), together with the low lying transition of C III (1157 Å). The emission component appears in varying degrees of strength and is sometimes totally absent, but the effect is recognized by the large shift in the absorption line produced in the expanding shell. The several other lines observed in the spectra show no shift, indicating their formation in lower photospheric layers. The observed velocities are comparable with those associated with Wolf-Rayet stars and their presence in early type supergiants is somewhat surprising although the ground-based observations of Wilson [28, 29] and Underhill [30] had indicated velocities of this order associated with emission lines of He II, C III and N III.

Since the observed velocities are considerably in excess of the escape velocities, it is clear that the early-type supergiants are losing mass. The loss rate has been calculated by Morton [31, 32] for δ, ε and ζ Ori using the Si IV and N V lines. Other shifted lines were excluded because they were saturated. The broad and shallow nature of the N V line eliminates radiation damping as the broadening mechanism since core saturation would be needed to explain the observed width. The broadening is therefore due to Doppler motions and the observed profile indicated turbulence rather than differential expansion. Assuming microturbulence rather than macroturbulence, Morton calculated column densities of N V and Si IV from the observed strengths of the absorption lines. A simple model was then adopted in which the lines are formed in a shell of constant expansion velocity where the ionization balance is imposed by the dilute radiation from the star. This requires that the electron temperature in the shell be less than 10^5 K and the electron density less than 10^{10} cm^{-3}. An electron temperature of 10^4 K was adopted and a colour temperature of 26 000 K for the star was derived from the analysis. Since this refers to radiation below the Lyman limit, it appears a reasonable value when compared with the effective temperatures of about 32 000 K for these objects [26].

Morton's resulting estimates of mass loss lie between $1-2 \times 10^{-6}$ $M_\odot$ yr^{-1}. This is a factor of 10 higher than Lucy and Solomon [33] deduced from a theoretical model in which the radiation pressure in the strong resonance absorptions produced expansion.

Since the stars have a mass of about 30 $M_\odot$ and spend about 10^6 yrs as hot super-giants, a few percent of their mass will be lost in this time. However, the simplifying assumptions in the model must render this estimate a very approximate one.

Observations of ζ Pup and γ^2 Vel have also shown emission features with shifted absorption lines indicating large expansion velocities. In ζ Pup, Morton et al. [5] derived an expansion velocity of about 1800 km sec^{-1} from the resonance lines of Si III, Si IV, C IV and N V. However, from the lines of He II (1640 Å) and N IV (1718 Å) formed from well excited levels they derived much lower velocities of 350 and 780 km/sec respectively. They deduced that these lines are formed at lower levels, in the region where the acceleration process occurs. The observations of γ^2 Vel which cover the most extensive wavelength range are those of Stecher [10] which are reproduced in Figure 6 and show a number of carbon lines due to C II, C III and C IV. Since then Stecher [34] has also reported N IV 1718 Å in emission (apparent in Figure 6) and N V (1239–43 Å) in absorption. Since γ^2 Vel is a WC star this questions the Beals hypothesis [35, 36] of a chemical separation in Wolf-Rayet objects and supports Miss Underhill's [37] arguments that a physical rather than a chemical interpretation may be possible. However, a proper analysis of the data will be needed to settle the point.

B. INTERSTELLAR LINES

Ultraviolet spectroscopic studies of early-type stars will undoubtedly provide the most powerful techniques for studying the physics and chemistry of the interstellar gas. In reviewing the present position, I should preface my remarks with the opinion that the only unambiguous detection of an interstellar line in the ultraviolet is the Lyman α line of neutral hydrogen. Many of the observations listed in Table I embrace this line which is seen in absorption and attributed to an interstellar origin, measurements of its strength then being used to determine the density of interstellar atomic hydrogen. The observations fall into two regions in Orion and the Gum nebula. For Orion, 6 stars (δ, ε, ζ, η, ι and σ Ori) have been measured by Morton et al. [4] and 4 (κ, ζ, ι and γ Ori) by Carruthers [9]. In the Gum nebula, measurements for ζ Pup and γ^2 Vel have been made by Morton et al. [5] and Stecher [10]. An intercomparison of this data shows that column densities of atomic hydrogen derived from the measurements of Lyman α by the different observers all lie within a factor of 2.

The broadening of the strong interstellar Lyman α line is due entirely to radiation damping, hence the column density N_H is given in terms of equivalent width W_λ by the expression

$$N_\mathrm{H} = 1.9 \times 10^{18}\ W_\lambda^2\ \mathrm{cm}^{-2}.$$

Using the Princeton observations [4] for 6 Orion stars, Jenkins and Morton [3] derived column densities N_H in the range 1–3×10^{20} cm^{-2}. Using a distance of 450 pc to Orion obtained from Blaauw and Borgmann [38] they obtained an average density of 0.1 cm^{-3} for interstellar atomic hydrogen. On the other hand, observations of the 21 cm line [39] suggest a column density $N_\mathrm{H} = 1.5 \times 10^{21}$ cm^{-2} in front of Orion, and hence an average density of about 1 cm^{-3}, a factor of 10 higher than that obtained from Lyman α.

The analysis of ζ Pup by Morton *et al.* [5] gives a column density $N_H = 5 \times 10^{19}$ cm^{-2}. They estimate a distance of 450 pc but allowing for the extent of the Gum H II region they deduce a column length for neutral hydrogen of 390 pc, giving an average interstellar density of 0.04 cm^{-3}. This compares with the value of 1.5 cm^{-3} obtained by McGee [40] from 21 cm observations. In this case, the discrepancy is greater than a factor of 10.

The large difference in the estimates of the concentration of interstellar atomic hydrogen from the radio and ultraviolet data seems to be outside the experimental errors and is a considerable puzzle. The estimate from Lyman α assumes that the observed line is entirely interstellar, which seems a reasonable assumption in view of the early spectral type of the objects being studied. Any stellar absorption line would increase the discrepancy and therefore a strong stellar emission line would be needed in order to bring the two sets of data into agreement. However, such an emission line would have to have a profile which matched the observed absorption wings which resemble those expected by radiation damping. This seems highly unlikely, particularly since such an emission line would have to be present in all the observed stars. If we accept the discrepancy as real then a possible explanation is that the 21 cm emission is produced by matter lying beyond the stars in question. This will require the interstellar medium to be locally deficient (by a factor of about 10) since Orion and the Gum nebula are separated by 65°. Failing this, the mechanism for the production of the 21 cm line would have to be re-examined. It will be interesting to see if this discrepancy is repeated in new observations covering a wider range of galactic position.

The analysis of the observed spectra have included a search for interstellar molecular hydrogen by means of the resonance lines of the Lyman band $^1\Sigma_g^+ - ^1\Sigma_u^+$ which lie between 1040 Å and 1120 Å. The spectra which covered this wavelength range with a reasonable spectral resolution for the purpose are those of Carruthers [9] and Smith [8]. In neither case were the transitions detected and from his observations of ζ Pup Carruthers [41] estimated an upper limit to the density of interstellar molecular hydrogen of 10% of atomic hydrogen. Smith's observations of α Vir were made with a spectral resolution of 0.8 Å and resulted in an upper limit of 3×10^{-4} cm^{-3} for the interstellar density of the molecule. A direct comparison with atomic hydrogen was not made because the Lyman α line was expected to have a strong stellar component, but if the atomic hydrogen densities derived from this line are adopted then an upper limit to the molecular hydrogen density is about 1% of atomic hydrogen.

The observed spectra have also been examined for the presence of interstellar lines of other abundant elements. Although a number of identifications are suggested, no reference claims unambiguously that the observed features are totally interstellar. The only analysis is that of Stone and Morton [42] using the observations of δ and π Sco [1]. Spectral lines of OI, CII, SiIII and AlII are assumed to be interstellar and their measured equivalent widths analysed to yield densities of the ions. The results were modified by Gaillard and Hesser [43] using improved oscillator strengths for the OI lines, to give abundance ratios for these elements in reasonable agreement with those in the Sun except for Al, which appears overabundant. A direct comparison with interstellar atomic hydrogen was not possible because the Lyman α line was not covered by the observations, but the density of atomic hydrogen was estimated by assuming solar abundances and gave a value of about 35 cm^{-3}. This is nearly 2 orders of magnitude greater than that derived directly from Lyman α in Orion and Gum nebula stars. The observations used were the first stellar spectra obtained and were stated by the authors [1] to be of poor photometric quality. In addition to this, the spectral types B0 and B1 are such that a stellar contribution to the observed line strengths is a distinct probability. Because of these factors, the conclusions must be suspect and any analysis of the chemistry of the interstellar medium must await new observations, particularly of the hottest stars using improved spectral resolution.

The nearest present approach to this requirement is the observation of ζ Pup by Morton *et al.* [5] in which interstellar lines of C II, N I, Al II and Fe II are suspected but in which O I, Si I, and Si II seem to be absent. The presence of N I and the absence of O I, if real, may be explained by the oxygen being concentrated in the interstellar grains.

5. Conclusion

It is normal to end a review with some kind of 'look ahead'. It is apparent that ultraviolet astronomical spectroscopy is still in its early stages and that the future looks bright and exciting. The published results described above were obtained entirely with pointing rockets and are confined to bright early-type stars. They have already made a considerable impact on the study of these objects and the intervening interstellar medium. The need to extend the observations to other objects is abundantly clear. Deserving of mention are the late-type stars, for studies of stellar chromospheres and coronae as well as their Fraunhofer spectra; a wide variety of variable stars, including magnetic variables, novae, eclipsing binaries and cepheids; galactic nebulae, including planetaries, diffuse nebulae and supernova remnants; and extra-galactic nebulae including both normal and abnormal galaxies. Further the rich character of the ultraviolet spectrum revealed by the pioneering rocket observations has demonstrated the need for improvements in spectral resolution, at least for the brighter objects. It is clear that the full exploitation of such a wide field will need satellite systems with a full pointing capability in addition to stabilised rockets. This is demonstrated by the important new results obtained by OAO-A2 and presented at this conference.

References

[1] Morton, D. C. and Spitzer, L.: 1966, *Astrophys. J.* **144**, 1.
[2] Morton, D. C.: 1967, *Astrophys. J.* **147**, 1017.
[3] Jenkins, E. B. and Morton, D. C.: 1967, *Nature* **215**, 1257.
[4] Morton, D. C., Jenkins, E. B., and Bohlin, R. C.: 1967, *Astrophys. J.* **154**, 661.
[5] Morton, D. C., Jenkins, E. B., and Brooks, N. H.: 1968, Princeton University Observatory Report.

[6] Stecher, T. P.: 1967, *Astron. J.* **72**, 831.
[7] Stecher, T. P.: 1967, Report on Commission 44, IAU 13th General Assembly, Prague.
[8] Smith, A. M.: 1969, *Astrophys. J.* **156**, 93.
[9] Carruthers, G. R.: 1968, *Astrophys. J.* **151**, 269.
[10] Stecher, T. P.: 1968, Goddard Preprint: Stellar Spectrophotometry from a Pointed Rocket.
[11] Gaustad, J. E. and Spitzer, L.: 1961, *Astrophys. J.* **134**, 771.
[12] Underhill, A. B.: 1963, *Pub. Dom. Astrophys. Obs. Victoria* **11**, 433.
[13] Underhill, A. B.: 1962, *Pub. Dom. Astrophys. Obs. Victoria* **11**, 467.
[14] Hickok, F. R. and Morton, D. C.: 1968, *Astrophys. J.* **152**, 203.
[15] Mihalas, D. M. and Morton, D. C.: 1965, *Astrophys. J.* **142**, 253.
[16] Morton, D. C.: 1964, *IAU Symposium* **23**, p. 163.
[17] Morton, D. C.: 1965, *Astrophys. J.* **141**, 73.
[18] Guillaume, C., Van Rensbergen, W., and Underhill, A. B.: 1965, *Bull. Astron. Inst. Netherl.* **18**, 106.
[19] Adams, T. F. and Morton, D. C.: 1968, *Astrophys. J.* **152**, 195.
[20] Mihalas, D.: 1966, *Astrophys. J. Suppl. Ser.* **13**, 1.
[21] Bless, R. C., Code, A. D., and Houck, T. E.: 1968, *Astrophys. J.* **153**, 561.
[22] Byram, E. T., Chubb, T. A., and Werner, M. W.: 1965, *Ann. Astrophys.* **28**, 594.
[23] Chubb, T. A. and Byram, E. T.: 1963, *Astrophys. J.* **138**, 617.
[24] Smith, A. M.: 1967, *Astrophys. J.* **147**, 158.
[25] Bless, R. C., Code, A. D., Houck, T. E., McNall, J. F., and Taylor, D. J.: 1968, *Astrophys. J.* **153**, 557.
[26] Morton, D. C. and Adams, T. F.: 1968, *Astrophys. J.* **151**, 611.
[27] Underhill, A. B.: 1969, this volume, p. 215.
[28] Wilson, R.: 1955, *Observatory* **75**, 222.
[29] Wilson, R.: 1957, *Mem. Soc. Roy. Sci. Liège*, Série 4 **20**, 85.
[30] Underhill, A. B.: 1957, *Mem. Soc. Roy. Sci. Liège*, Série 4 **20**, 91.
[31] Morton, D. C.: 1966, *Astron. J.* **71**, 172.
[32] Morton, D. C.: 1967, *Astrophys. J.* **150**, 535.
[33] Lucy, L. B. and Solomon, P. M.: 1967, *Astron. J.* **72**, 310.
[34] Stecher, T.: 1969, *Wolf-Rayet Stars* (ed. by K. B. Gebbie and R. N. Thomas), USDC-NBS Special Pub. 307.
[35] Beals, C. S.: 1930, *Monthly Notices Roy. Astron. Soc.* **90**, 202.
[36] Beals, C. S.: 1930, *Publ. Dom. Astrophys. Obs. Victoria* **4**, 271.
[37] Underhill, A. B.: 1957, *Mem. Soc. Roy. Sci. Liège*, Série 4, **20**, 17.
[38] Borgmann, J. and Blaauw, A.: 1964, *Bull. Astron. Inst. Netherl.* **17**, 358.
[39] Clark, B. G.: 1965, *Astrophys. J.* **142**, 1398.
[40] McGee, R. X.: 1968, private communication (quoted by Morton *et al.* [5]).
[41] Carruthers, G.: 1967, *Astrophys. J.* **148**, 2141.
[42] Stone, N. E. and Morton, D. C.: 1967, *Astrophys. J.* **149**, 29.
[43] Gaillard, M. and Hesser, J. E.: 1968, *Astrophys. J.* **152**, 695.

Discussion

Underhill: That the mass of supergiants of types near B0 is about 30 solar masses is a conclusion not supported by the evidence from radii (interferometric measures), brightness temperatures in the V band, M_v, and g estimated from the spectrum *via* N_e. Model atmospheres quite clearly show that g relates uniquely to N_e with very little dependence on T_{eff} over the range $8000\,\mathrm{K} < T_{eff} < 33\,000\,\mathrm{K}$. In the case of ε Ori, B0 Ia, the mass may well lie in the range 3 to 8 $M_\odot$.

Wilson: The estimate that is obtained from the UV data is, of course, the rate of mass loss, the value of 30 $M_\odot$ for the mass of the objects being quoted in the subsequent discussion by the author (Morton) of the particular paper I was referring to. Since he is here, I can let him explain its source. If the masses are the lower values you have deduced, then the mass loss to be expected in the lifetimes of these objects will be a substantial fraction of their total mass.

Solomon: (Regarding the mass loss rate of $10^{-6}\ M_\odot\ \mathrm{yr}^{-1}$ which was quoted from Morton's analysis of the observations.) It should be pointed out that there is no self-consistent analysis of the ionization equilibrium which reproduces at a single electron density, in the high velocity flow, the simultaneous existence of CIII, CIV, SiIII, SiIV and Nv. For example, if the Nv data are fitted one gets extremely small mass loss rates (since this requires very small electron densities) with $\dot{M} < 10^{-9}$ $M_\odot$ yr^{-1}. This difficulty makes any observational estimate of the mass loss rate highly questionable.

Morton: To estimate the masses of the Ori supergiants I assumed that they have evolved from the main sequence. Reasonable estimates of the absolute visual magnitudes and bolometric corrections correspond to the theoretical luminosity of an evolved 30 solar mass star calculated by Stothers.

Bless: Isn't it dangerous to produce arguments about early-type supergiants on the basis of model atmospheres? Both the models and observations indicate that these objects are not in hydrostatic equilibrium.

Underhill: From the present admittedly inadequate model atmospheres and the line and continuous spectra predicted from them we do have an idea what the electron density is in a supergiant atmosphere (from the break-off of the Balmer and Paschen series and from the Stark wings of high series members), we can estimate the brightness temperature in the V band from the few absolute flux measurements available and compare this with predicted values to confirm that our temperature scale is not seriously in error and we can estimate the radius of the photosphere from the known M_v and the brightness temperature. The radii found for one or two stars by the Narrabri group confirm that these estimates are at least self-consistent.

Hydrostatic equilibrium is assumed in order to obtain a depth-pressure relationship. The known stellar winds are not sufficiently strong to make one suspect that the hydrostatic equilibrium pressure-depth relationship is wrong by a large factor.

PHOTOELECTRIC ROCKET SPECTRA AT 10 Å RESOLUTION

THEODORE P. STECHER

NASA, Goddard Space Flight Center, Greenbelt, Md., U.S.A.

Abstract. Spectral scans of a number of early-type stars have been made with a three-channel spectro-photometer attached to a 32 cm telescope. The telescope was pointed at the individual program stars with an accuracy of at least 20 sec of arc. The spectral resolution was 10 Å over the range 1150 Å to 4000 Å. The data were recorded in digital form and processed in a computer.

The profile of hydrogen Ly-α, as smeared by the exit slit, was obtained for α Lyr, α CMA, β Ori, and β Tau. For the earlier-type stars observed the Ly-α line of the star is masked by interstellar hydrogen absorption. For ζ Oph and ζ Per the Ly-α absorption is considerably larger than for the other early-type stars. The identification of lines is relatively simple for the light elements but becomes more difficult with the heavier ions since there are many more weaker lines and 10 Å is then insufficient resolution. Extreme P Cygni profiles are present in the earliest stars and are interpreted as mass loss at high velocity.

A full description appears in *Astrophys. J.* **159**, Feb. 1970.

Discussion

Jenkins: At this time would you care to quote measurements of the Lyman-absorption equivalent widths in your spectra?

Stecher: I have not measured them in detail because of the difficulty in assigning the position of the continuum. Qualitatively only ζ Oph and ζ Per have large equivalent widths that are interstellar. The rest are in agreement with your measures. One indicates a low hydrogen abundance.

Jenkins: I had the impression that the continuum was reasonably well defined on either side of Lyman-α in some of your tracings. Since a few of these stars have not been observed at Lyman-α by other investigations, it might be useful to add your information to the collection of H I column densities in various directions.

Stecher: Yes, I plan to in the near future.

Morton: I believe you also have a scan of α Lyr. Does it show the extensive line blanketing in the 2000 to 3000 Å region found in α CMa?

Stecher: α Lyr appears to have less blanketing than α CMa. A comparison will be made soon.

Carruthers: Would you care to comment on the relative velocity shifts in the P Cygni line profiles in γ Vel as compared to ζ Pup?

Stecher: They appear to be about half that of ζ Pup. I believe you and Morton can measure this better than I can.

Underhill: The fact that γ^2 Vel is classified as WC8 and ζ Pup as O5 f indicates that the shell of γ^2 Vel is considerably more dense than that of ζ Pup.

Stecher: Yes, I agree. This makes it most useful to compare the two and then attempt to explain the difference.

Houziaux: We have seen many lines in your spectra, despite the quoted resolving power of 10 Å. Could you state what is the shape of the instrumental profile?

Stecher: The geometric slit width.

Houziaux and Butler (eds.), Ultraviolet Stellar Spectra and Ground-Based Observations, 163.

ROCKET SPECTROSCOPY OF ζ PUPPIS BELOW 1100 Å

ANDREW M. SMITH

NASA, Goddard Space Flight Center, Greenbelt, Md., U.S.A.

Abstract. A spectrum of ζ Pup extending from 920 Å to 1360 Å with approximately 0.8 Å resolution has been recorded at rocket altitudes. Tentative identification of 38 multiplets below 1100 Å has been made from which it is concluded that all the lines appearing in a model atmosphere (Hickok and Morton, 1968) with $T_e = 37450$ K have been detected with the exception of those masked by telluric N_2 or strong P Cygni-like profiles. Additional absorption lines indicate a wide range of ionization and excitation entirely consistent with observations in the visible spectral region of similar type stars; they also appear to affect sensibly the energy distribution within the spectrum. From newly detected blue-shifted absorption features produced by the ions S VI (933.4 Å, 944.5 Å), N IV (955.3 Å), N III (991.0 Å), and O VI (1033.8 Å) mean radial velocities of 1200, 530, 1800 and 1900 km sec^{-1} respectively have been derived. It is pointed out that the transition in N IV (955.3 Å) does not originate in the ground state configuration as do the other P Cygni profile transitions, but from an excited level which can decay radiatively to the ground state. It seems likely, therefore, that the profile is generated close to the photosphere, and this, together with previously reported results, constitutes evidence for a positive velocity gradient in the detected portion of the circumstellar envelope. On the basis of available data, ions in the ground state of N V, N IV (by inference) and N III are all present in the circumstellar envelope, the abundance of the latter remaining large enough to produce a strong P Cygni profile.

1. Introduction

In recent years the star ζ Pup (O5 f) has come under close spectroscopic scrutiny in the rocket ultraviolet due in part to interest in its intrinsic physical characteristics and in part to the fact that it is one of the few nearby hot objects accessible to small aperture instruments. Some of the results are dramatic. For example, the observations of Carruthers (1968), Stecher (1968), and Morton *et al.* (1969) have revealed strong blue shifted absorption features accompanied by emission at the long wavelength edges similar to P Cygni profiles. The chief distinction of these combined absorption-emission features in the case of ζ Pup is that the radial velocities calculated from the profiles are considerably larger than those normally associated with P Cygni stars. Thus, the ions C III, C IV, N V, and Si IV have all been observed with mean radial velocities exceeding 1500 km sec^{-1}, and the extrapolated maximum radial velocities are approximately twice this value. Bearing in mind that for reasonable values of stellar mass and radius the escape velocity at the surface of the star is near 1100 km sec^{-1}, the data indicate a continual loss of mass from ζ Pup, and the rate of mass loss becomes an interesting evolutionary question.

Regarding the model atmospheres appropriate to ζ Pup (Hickok and Morton, 1968; Bradley and Morton, 1969) we may also inquire as to whether the selected absorption lines are sufficient to calculate the effect of line blanketing on the distribution of energy in the optical spectrum. The observations thus far have indicated that the existing models are adequate in this respect. However, for a star of this type with an effective temperature estimated to be 48000 K (Morton, 1969) serious line-blanketing effects would be expected at wavelengths below 1100 Å, and this region has not

Houziaux and Butler (eds.), Ultraviolet Stellar Spectra and Ground-Based Observations, 164–172.

as yet been studied. In any case the identification of weak unshifted lines can give a qualitative indication of the ionization and excitation in that part of the atmosphere with no detected residual radial motion.

A rocket observation of ζ Pup carried out in March, 1968 with a stigmatic mounting spectrograph has provided data extending the wavelength coverage of this star down to 920 Å. Several heretofore unobserved P Cygni-like profiles have been identified as have a number of weak, photospheric lines arising from excited levels in both abundant and relatively rare ions. In this report these data will be presented and discussed.

2. A Brief Experiment Description

A description of the instrument appears elsewhere (Smith, 1969); however, some of the salient features are presented here again. There is a single optical element namely, a concave grating ruled at 1200 lines mm^{-1} which defines an instrumental aperture of 14 cm^2. The focal length is about 25 cm, the linear dispersion is 33.4 Å mm^{-1}, and in flight the spectrograph provides approximately 0.8 Å resolution. The dimensions of the ζ Pup spectrogram were 13.5 mm in the dispersion plane by 0.13 mm normal to the dispersion plane. Kodak Pathé SC5 film was used, and was developed for 2 min in D19b according to the manufacturer's specifications. The rocket trajectory achieved an altitude of about 125 miles during which two exposures were made, one for 214 sec, the other for 68 sec.

The wavelength scale was determined by fitting the data to a quadratic function of position measured along the spectrogram with the bands of telluric N$_2$ (960 Å) and the interstellar lines of N I (1134.6 Å) and Si II (1304.4Å) serving as calibration points. The resulting scale is thought to be accurate to 0.3 Å.

3. Results

A reproduction of the two exposures with some of the more obvious lines indicated appears in Figure 1. The top spectrum, corresponding to the 68-sec exposure, suffered fogging from a light leak resulting from a launch–aggravated shutter malfunction. Nevertheless, it contains some useful data at wavelengths longer than 1200 Å. For the purposes of this report, however, attention is focussed on the spectrum below 1100 Å. Accordingly, Figure 2 contains a densitometer trace of the long exposure spectrogram from 910 Å to 1100 Å together with the tentative identifications of 28 multiplets. The Hickok and Morton theoretical spectrum for an atmosphere with $T_e = 37450$ K and $\log g = 4.00$ is exhibited over the recorded spectrum for comparison purposes.

In this particular wavelength region it is necessary to be aware of the effects of telluric N$_2$ absorption bands. The most important of these extend as indicated from approximately 950 Å to 992 Å strongly modifying the stellar spectrum at 960, 966, and 972 Å. It should also be noted that a strong resonance line due to interstellar C II at 1036 Å influences greatly a P Cygni-like profile in O VI near this same wavelength. A rather weak line at 1097 Å, attributed to interstellar Fe II, affects the stellar spectrum very

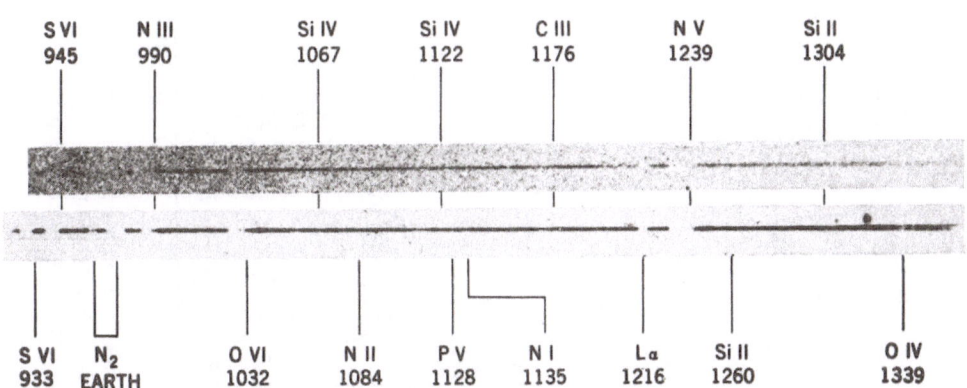

Fig. 1. Two spectrograms of ζ Pup. Exposure times of top and bottom spectrograms were 68 sec and 214 sec respectively. Some identifications of the stronger features are shown, the numbers indicating the approximate wavelengths in Å.

little. Water vapor evaporating from the rocket and payload probably accounts for the moderate sized feature at 1055 Å, and the feature at 1084.3 Å must be due partially to a resonant transition in N II (1083.98 Å) arising in the Strömgren sphere.

4. Discussion

Most of the bound-bound transitions used in the model can be identified with features appearing in the spectrogram; those for which no positive identification can be made occur in wavelength regions occupied either by N_2 absorption bands or strong P Cygni profiles. Some of the measured line strengths, however, may be vastly different from the model values. Such is the case for the S IV doublet at 1063 and 1074 Å. Also, the combined effects of H I and He II seem to be less than expected as is seen most clearly by the feature attributed at least in part to these absorbers at 950 Å.

There are weak lines in the observed spectrum, not included in the model, which reflect a wide range in ionization and excitation in the star's atmosphere. Included in this category are lines of N III, N IV, N V, Si III, Si IV, P III, P IV, P V, S III, S IV, Cl III, and Cl IV. Due to the ragged character of the spectrum the evidence for those transitions proceeding from excited states and revealed as weak lines is considered to be only mediocre. On the other hand, there is good evidence for ground state transitions in C III (977 Å), P IV (951 Å), S III (1012 Å), S IV (1063–74 Å), and Cl IV (973–86 Å). The identification of the resonance transitions in Cl III (1005–15 Å) is much less certain. The lines of C II (1036 Å) and Fe II (1097 Å) are probably interstellar in origin.

Table I presents the line identifications made in the spectrum for wavelengths between 923 Å and 1100 Å. Listed in column 1 is the ion identification, and directly under this the multiplet number found in either *An Ultraviolet Multiplet Table* or *Selected Tables of Atomic Spectra* both by Moore (1950, 1965). If a multiplet or line is not contained in this table two references of Kelly (n.d., 1968) signified by K1 and K2 respectively, are indicated. Columns 2, 3, 4 and 5 contain the laboratory measured

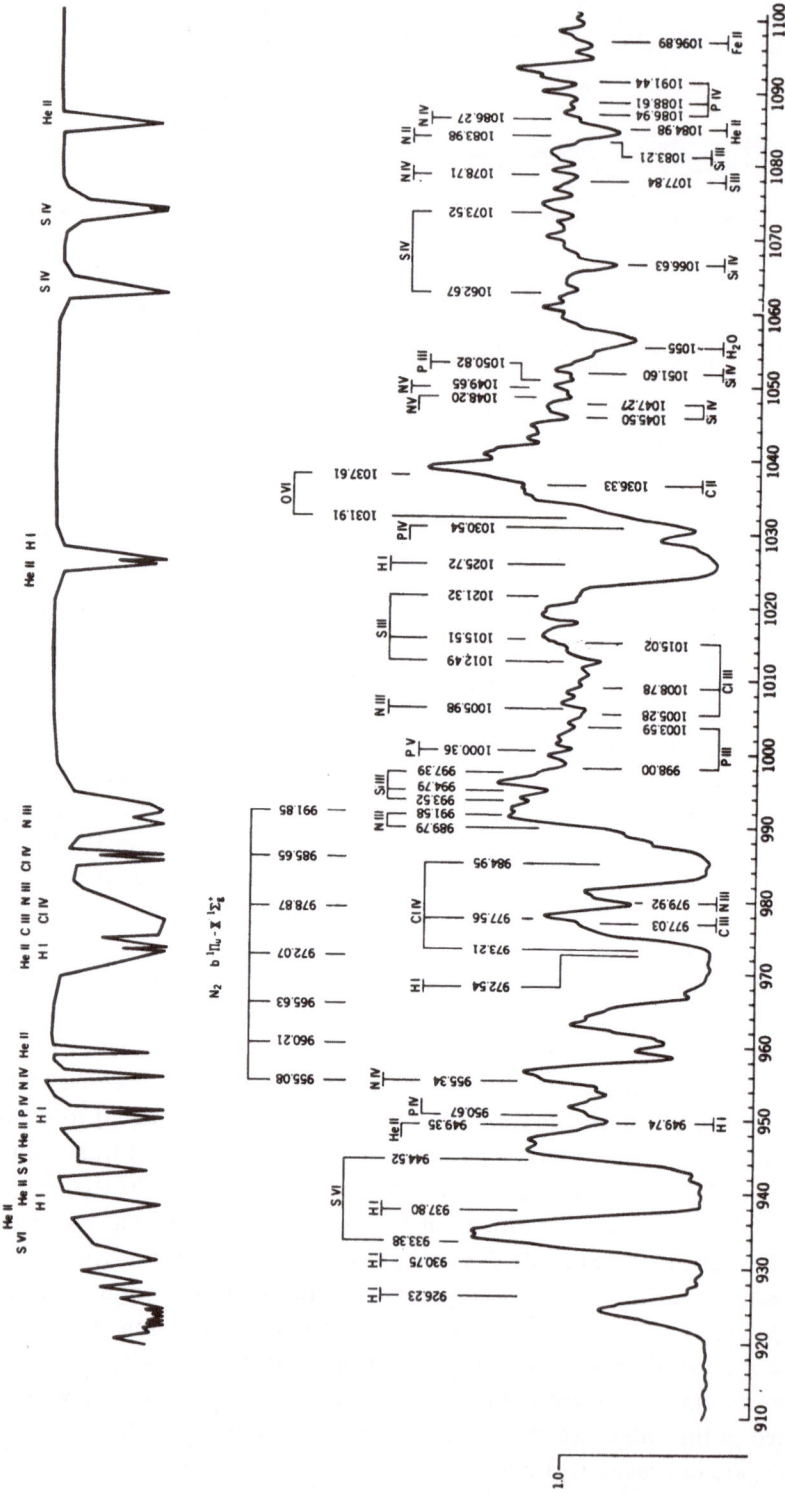

Fig. 2. Microdensitometer trace of the 214 sec exposure ζ Pup spectrogram. The abscissa is in Ångstrom units, the ordinate in density units. Drawn above the recorded spectrum is the model spectrum of Hickok and Morton (1968) in which the ordinate units are ergs cm^{-2} sec^{-1} (Hz)$^{-1}$. Vertical lines indicate transitions associated with various spectral features; horizontal lines connect members of the same multiplet.

TABLE I

Identification of lines in the UV spectrum of ζ Pup

Ion multiplet	Laboratory wavelength (Å)	Measured wavelength (Å)	χ(eV)	gf	Remarks
H I UV (4)	949.74	949.9	0.00	0.027 88	Moderate
He II UV (19)	949.30 949.35	949.9	40.64	0.030 808	Weak; probable
He II UV (14)	1084.91 1084.98	1084.3	40.64	0.357 36	Moderate
C II UV (2)	1036.33	1036.3	0.00	0.12	Weak; probable; interstellar
C III UV (1)	977.03	977	0.00	0.81	Weak; probable
N II UV (1)	1083.98 1084.57 1084.57 1085.54 1085.70	1084.3	0.00 0.01 0.01 0.02 0.02	0.17 0.13 0.39 0.12 0.70	Moderate
N III UV (12)	979.77 979.84 979.92 980.01	979.7	12.47 12.47 12.47 12.47	0.08 0.72 1.12 0.08	Weak; probable
N III UV (1)	989.79 991.51 991.58	985.0 990.3	0.00 0.02 0.02	0.36 0.07 0.65	Strong; P Cygni profile
N III UV (17)	1005.98 1006.03	1005.9	16.17	0.36 0.18	Weak; probable
N IV UV (8)	955.34	953.8 955.3	16.13	0.22	Moderate; P Cygni profile
N IV K2	1078.71	1078.4	52.98	–	Weak; possible
N IV K2	1086.08 1086.27 1086.69	1086.3	50.11 50.12 50.12	– – –	Very weak; possible
N V K2	1048.20	1048.7	76.28	–	Weak; possible
N V K2	1049.65	1048.7	76.29	–	Weak; possible
O VI UV (1)	1031.91 1037.61	1025.3 30.3 1034.2	0.00 0.00	0.26 0.13	Strong; P Cygni profile
Si III UV (6)	993.52 994.79 997.39	993.6 994.8 997.3	6.54 6.55 6.58	0.20 0.64 1.00	Weak; probable
Si III UV (23)	1083.21	1083.1	15.15	–	Very weak; possible

Ion multiplet	Laboratory wavelength (Å)	Measured wavelength (Å)	χ(eV)	gf	Remarks
Si IV UV (21)	1045.50	1045.8	27.06	–	Weak; possible
	1047.27	1048.2	27.08	–	
Si IV K1	1051.60	1051.1	–	–	Weak; possible
Si IV UV (11)	1066.63	1066.4	19.88	–	Moderate
P III UV (2)	998.00	998.7	0.00	–	Very weak; possible
	1003.59	1003.6	0.07	–	
P III K2	1049.82		9.29	–	Weak; probable
	1050.52	1051.0	9.29	–	
	1050.82		9.29	–	
P IV UV (1)	950.67	951.1	0.00	2.75	Weak; probable
P IV UV (2)	1025.58	–	8.41	–	Weak; possible
	1028.13	–	8.38	–	
	1030.54	1030.3	8.47	–	
	1033.14	–	8.41	–	
	1035.54	–	8.47	–	
P IV K2	1086.94	1086.3	23.47	–	Very weak; possible
	1088.61	1089.0	23.47	–	
	1091.44	1091.0	23.47	–	
P V K2	997.64	997.3	25.31	–	Very weak; possible
	1000.36	1000.6	25.31	–	
S III UV (2)	1012.49	1012.5	0.00	0.30	Weak; probable
	1015.51	1014.5	0.04	0.52	
	1015.76		0.04	0.38	
	1021.10	1021.3	0.10	0.38	
	1021.32		0.10	1.12	
S III UV (8)	1077.84	1078.4	1.40	–	Weak; possible
S IV UV (1)	1062.67	1063.0	0.00	0.94	Weak; probable
	1072.99	1073.0	0.12	0.19	
	1073.52		0.12	1.69	
S VI UV (1)	933.38	928.8	0.00	0.5	Strong; P Cygni profiles
		933.0			
	944.52	940.8	0.00	1.0	
		945.2			
Cl III UV (1)	1005.28	1005.0	0.00	0.56	Very weak; possible
	1008.78	1009.2	0.00	1.10	
	1015.02	1014.5	0.00	1.66	
Cl IV K2	973.21		0.00	0.55	Weak; probable
	977.56	977.0	0.06	1.24	
	977.90		0.06	0.42	
	984.95		0.17	2.32	
	985.75		0.17	0.44	
Fe II UV (18)	1096.89	1096.3	0.00	–	Weak; possible; interstellar

wavelength, the observed spectral feature wavelength, the excitation potential and the *gf* value respectively. In column 6 eye estimates of the magnitude of the observed spectral features are presented together with a certainty estimate of the identification. An interstellar line or P Cygni profile is also indicated in this column. In the case of the latter, two numbers are listed for each transition in column 3. The first is the wavelength at which the minimum residual intensity is found in the absorption feature; the second is the wavelength corresponding to the short wavelength edge of the emission feature. No blends were indicated as such because almost assuredly all observable features are blends of several transitions only some of which are known.

The recently detected P Cygni-like profiles occur for transitions in N III (991.0 Å), N IV (955.3 Å), O VI (1033.8 Å) and S VI (933.4, 944.5 Å). The mean radial velocities measured at the center of the blue shifted absorption feature and the extrapolated maximum radial velocities are listed in Table II.

TABLE II

Characteristic radial velocities of ions in the atmosphere of ζ Pup

Ion	N III	N IV	O VI	S VI
Central velocity (km sec⁻¹)	1800	530 (780)	1900	1200
Extrapolated maximum velocity (km sec⁻¹)	3300	1500	3500	2500

For the purposes of this table the appropriate velocities derived from each of the S VI line profiles have been averaged, and the contribution of the 985 Å line of Cl IV to the absorption component of the N III (991.0 Å) profile has been assumed negligible. The number appearing in parentheses is the mean velocity obtained by Morton *et al.* using a P Cygni profile in N IV found at 1719 Å. It is noteworthy that both the 955 Å and 1719 Å lines of N IV originate at the same excited level ($2p\ ^1P^0$, 16.13 eV) which can decay radiatively to the ground state ($2s^2\ ^1S$) with $gf = 0.64$. The most likely way that the $^1P^0$ level can be populated is by ordinary thermal processes. This implies that the radiation field can not be significantly diluted, and that the N IV ions must therefore be close to the photosphere. Morton (1969) has also reached this conclusion.

The P IV line at 1031 Å originates at an 8.47 eV level, and is expected to affect the spectrum only weakly. Thus, the moderate absorption feature at this wavelength can be interpreted as the absorption component of the P Cygni profile associated with the 1037.6 Å line in O VI, whereas the strong absorption feature at 1025.5 Å should be associated with the 1031.9 Å line. If this is the case the absorption components of the doublet are partially resolved, and may provide a basis for a mass loss estimate.

5. Summary and Conclusions

All lines used in the Hickok and Morton model have been detected except for those either masked by N_2 absorption bands or by strong P Cygni-like profiles. Lines identified in addition to the model lines are weak but closely packed together, and they would affect the energy distribution in the spectrum if they were included in the model. This is particularly noticeable near 1010 Å.

The weak lines also indicate the wide range of excitation and ionization which exists in the atmosphere of ζ Pup. As an example, Table I shows that lines of NIII (980 Å), NIV (1079 Å), and NV (1048–50 Å) originate at levels of 12.5, 53.0, and 76.3 volts respectively. These observations are completely consistent with those of similar type stars made in the visible spectral region.

None of the existing ultraviolet spectra of ζ Pup including the one presented here shows evidence for narrow, unshifted emission lines similar to the lines of HeII (4686 Å), NIII (4634–42 Å), and CIII (5696 Å) characteristic of Of stars. A probable explanation is that no selective excitation such as the familiar fluorescence mechanisms which require large fluxes of HeII (Lyman α) and HI (Lyman α) quanta are operating at wavelengths less than 1965 Å. As one might expect, however, it is very difficult to locate suitably a continuum flux level. It would be, therefore, impossible to detect any weak emission in the present data with reasonable certainty.

As has been pointed out it is likely that the P Cygni profile produced in NIV ions near 955 Å arises near the photosphere where dilution effects are weak. These ions exhibit a relatively small mean radial velocity (530 km sec^{-1}). On the other hand, the data of this experiment and that of Morton *et al.* reveal a P Cygni profile associated with a transition from a metastable triplet state in CIII at 1175.7 Å. The mean radial velocity of the CIII ions in this case is about 1700 km sec^{-1}. Further, it is known that such a situation can prevail only when the radiation is dilute, that is, when the CIII ions are at distances of the order of several stellar radii from the center of the star. They would be further from the photosphere than the NIV ions which produced the 955 Å profile. Another point is that for any plausible atmospheric temperatures and pressures there should be no appreciable abundance of OVI, and yet the data indicate a large abundance of this ion in the circumstellar envelope. It is reasonable to assume that OVI ions with a mean radial velocity of 1900 km sec^{-1} are created well beyond the photosphere where suitable conditions exist. These data are therefore interpreted as evidence for a positive velocity gradient in the detectable part of the circumstellar envelope.

The degree of ionization appears to increase with increasing distance from the photosphere, but the lower levels of ionization as evidenced by the strong NIII P Cygni profile at 991 Å are not completely depleted. The existence of a strong P Cygni profile at 1240 Å produced by NV ions is well established, and it is reasonable to expect that NIV exists in the ground state at roughly the same distances and velocities as NIII and NV. We do not detect it because the resonance transitions lie at wavelengths less than 912 Å which are hidden from view by interstellar atomic hydrogen.

References

Bradley, P. T. and Morton D. C.: 1969, *Astrophys. J.* **156**, 687.
Carruthers, G. R.: 1968, *Astrophys. J.* **151**, 269.
Hickok, F. R. and Morton, D. C.: 1968, *Astrophys. J.* **152**, 203.
Kelly, R. L.: n.d., *Table of Emission lines in the Vacuum Ultraviolet for a l.Elements*, University of Calif. Radiation Laboratory, No. 5612.
Kelly, R. L.; 1968, *Atomic Emission Lines Below 2000 Å*, U.S. Naval Research Laboratory Report No. 6648.
Moore, C. E.: 1950, N.B.S. Circ., No. 488, Sec. 1.
Moore, C. E.: 1965, N.S.R.D.S.-N.B.S. **3**, Sec. 1.
Morton, D. C.: 1969a, preprint, Princeton University.
Morton, D. C.: 1969b, *Astrophys. Space Sci.* **3**, 117.
Morton, D. C., Jenkins, E. B., and Brooks, N. H.: 1969, *Astrophys. J.* **155**, 875.
Smith, A. M.: 1969, *Astrophys. J.* **156**, 93.
Stecher, T. P.: 1968, in *Wolf-Rayet Stars* (ed. by K. B. Gebbie and R. N. Thomas), National Bureau of Standards Special Publication 307, Washington, D.C.

Discussion

Burton: Do you observe the ^1S–^3P intersystem line of Ov at 1218 Å? This line is similar to the 1909 Å Cⅲ line which has been observed with high intensity in the spectrum of γ Vel by Stecher. Since Oⅳ and Oⅵ lines are seen in your spectrum of ζ Pup, it is possible that [Ov] 1218 Å will also be observed.

Smith: No, there seems to be no indication of this transition. Lyman α absorption due to interstellar atomic hydrogen predominates at this wavelength, and perhaps there is also a contribution to the observed absorption due to water vapor evaporating from the rocket and payload surfaces.

Morton: What is the basis for your wavelength scale and how accurately do you know your wavelengths?

Smith: I used the interstellar lines of Siⅱ(1304.4 Å) and Nⅰ(1134.6) in addition to the telluric bands of N_2(960.2–5.6 Å) to establish the scale. I believe the scale to be accurate to 0.3 Å.

Burton: (1) What type of photographic film was used for your observations? (2) What was the width of your spectrum (perpendicular to the direction of dispersion)?

Smith: In the case of the first question the answer is Kodak Pathé SC5. The width of the recorded spectrum is between 0.10 and 0.13 mm.

OBSERVATIONS OF STRONG STELLAR LINES WITH THE OAO

A. D. CODE and R. C. BLESS

Space Astronomy Laboratory, Washburn Observatory,
University of Wisconsin, Madison, Wis., U.S.A.

Abstract. This paper reports on preliminary analysis of spectral scans of early-type stars obtained with the Orbiting Astronomical Observatory. The discussion is confined to the spectra of 50 stars observed with a resolution of approximately 10 Å over the spectral interval from 1050 Å to 2000 Å. Following a qualitative description of the spectra, observed equivalent widths of the Si iv (1400 Å) and C iv (1550 Å) lines are compared with model atmosphere calculations. The results suggest an upward revision of the stellar temperature scale for stars earlier than B3 V.

The Wisconsin instrumentation on the OAO-A2 spacecraft includes an objective grating scanning spectrometer sensitive to the spectral interval from 1050 Å to 2000 Å. Spectral scans are accomplished by rotating the grating in discrete 10-Å steps. An Ascop 541F photomultiplier, mounted behind a 10-Å slot in the focal plane, is operated in both a pulse-counting and direct-current mode. This paper describes some preliminary results from the data for approximately 50 early-type stars. Repeated scans of many of the objects over intervals of time as great as several months show internal agreement of the order of 2% for those stars not known to be variable. This is probably a realistic value of the photometric accuracy of the data.

Figure 1 shows the variation of the stronger spectral line features with spectral type for main-sequence stars where the digital counts are plotted as a function of wavelength. The C iv resonance line at 1550 Å is the strongest stellar feature in this spectral region for stars earlier than B0. The Si iv line at 1400 Å and the C iv lines at 1550 Å are equal at type B0.5 V and by B1 the Si iv resonance line dominates. At B3 V Si iv is just detectable and Si iii reaches a maximum. The stellar Lyman-α line and Si iii blend give a large absorption at Lyman-α in stars later than B3. The contribution to Lyman-α in earlier types is primarily due to interstellar hydrogen and stellar N v.

The spectra show a strong luminosity effect for supergiants, which is illustrated at type O9.5 in Figure 2. The Si iv line, the C iv line, and the Fe ii + C iii blend at 1610 Å all increase markedly in equivalent width among the supergiants, while maintaining about the same ratio among themselves. These lines display emission components in the red wings of the absorption lines and even at this low resolution the violet displacements of the absorption lines relative to Lyman-α observed by Morton (1967) are apparent. The variations in the strength of Lyman-α shown in the figure are due to variations in the interstellar absorption.

The measured equivalent widths of the C iv and Si iv resonance lines as a function of spectral type are shown in Figure 3. These curves are based on data from approximately 40 stars which show a mean scatter of about 10%. The main sequence and giants have about the same equivalent widths. The supergiants in general have considerably larger equivalent widths as was indicated in Figure 2. For main-sequence

Houziaux and Butler (eds.), Ultraviolet Stellar Spectra and Ground-Based Observations, 173–177.

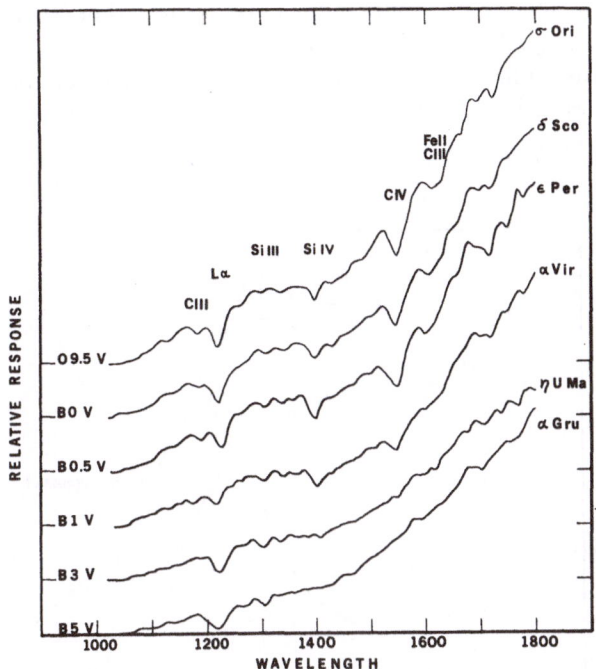

Fig. 1.   Spectral scans of early-type stars from 1000–1800 Å. The relative response is plotted against
wavelength for main-sequence stars of different spectral types.

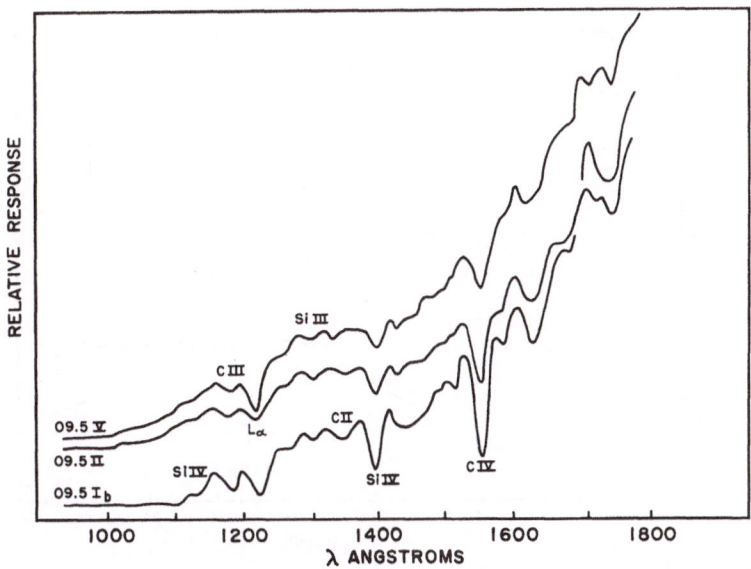

Fig. 2.   Spectral scans illustrating the luminosity effect exhibited
by the strong stellar lines at spectral type O9.5.

stars, the C IV line reaches a maximum in the O-type stars, while S IV peaks at about B1. The two lines are of equal strength near B0.5 V.

Some model atmosphere calculations have been carried out for these resonance line profiles. The model atmospheres program was developed by Klinglesmith (1969) and includes hydrogen line blanketing for Lyman and Balmer lines to $n = 40$. The line

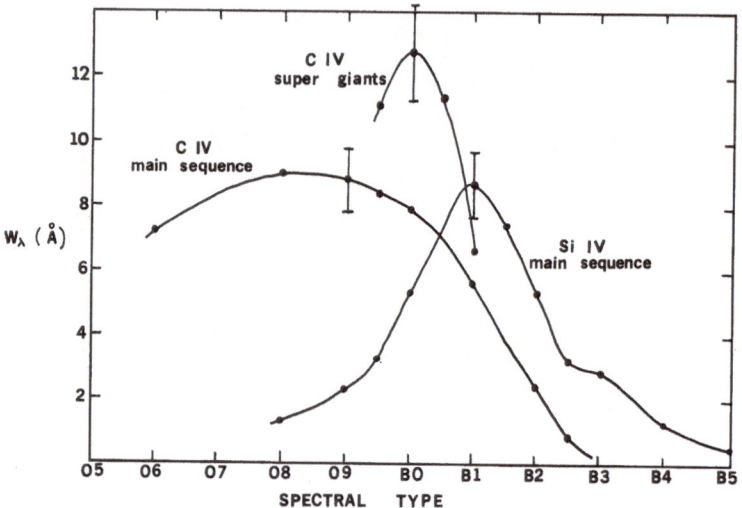

Fig. 3. Mean equivalent widths of the C IV (1550 Å) and Si IV (1400 Å) lines as a function of spectral type, determined from OAO scans.

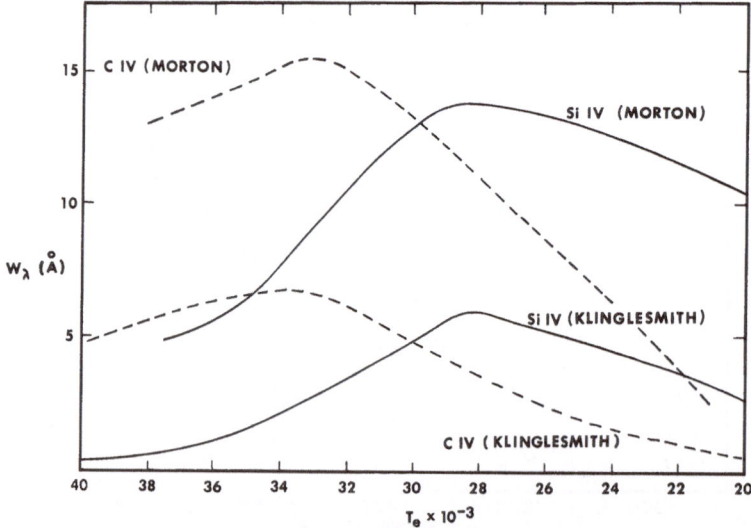

Fig. 4. Theoretical equivalent widths as a function of effective temperature for model atmospheres in local thermodynamic equilibrium for the C IV and Si IV resonance lines. The upper curves are from Bradley and Morton (1969), the lower curves from Klinglesmith (1969).

broadening mechanisms for the Si IV and C IV lines include electron impact, ion broadening, self-broadening, and radiation damping. The electron collision widths were computed by Oertel on the basis of a revised GKBO theory (Cooper and Oertel, 1967) which gives good agreement for those ions with measured widths. Figure 4 shows the equivalent widths of C IV and Si IV as a function of effective temperature for $\log g = 4.0$ models and abundances of 4×10^{-3} and 5×10^{-4} by weight relative to hydrogen. For comparison, also shown are the equivalent widths determined from the family of models by Morton and associates (cf. Bradley and Morton, 1969) in which a damping constant ten times classical damping was employed, which yields widths about twice those observed. The more detailed calculations yield equivalent widths that are too small by a factor of about 1.5 and would require a factor of 2 increase in abundance. Both sets of curves, however, show the same variations with temperature. The Si IV peaks at about 28000 K and the C IV at about 34000 K. These calculations assume LTE, which may not be valid for these lines.

In Figure 5 the observed equivalent widths are plotted against effective temperature using Morton's (1969) most recent temperature scale. The curve is distorted relative to the model atmosphere results, suggesting too large a temperature difference between B0 and B1. The Si IV peaks at 22500 K in contradiction to the model atmosphere calculations even for unrealistically low values of $\log g$.

A comparison of the continuum fluxes from the OAO spectral scans provides a preliminary temperature scale. If one assumes a temperature of 17800 K for a B3 V star, then the effective temperatures of a B0.5 V and an O9.5 V star are about 28500 K and 33000 K, respectively. These temperatures are in agreement with a temperature scale indicated from the line spectra. Further analysis is currently being carried out for both the continuum fluxes and line strengths measured with the OAO.

Morton's temperature scale for the hotter stars is essentially a Zanstra temperature based on radio emission measures and Hα intensities in diffuse nebulae. This is a mini-

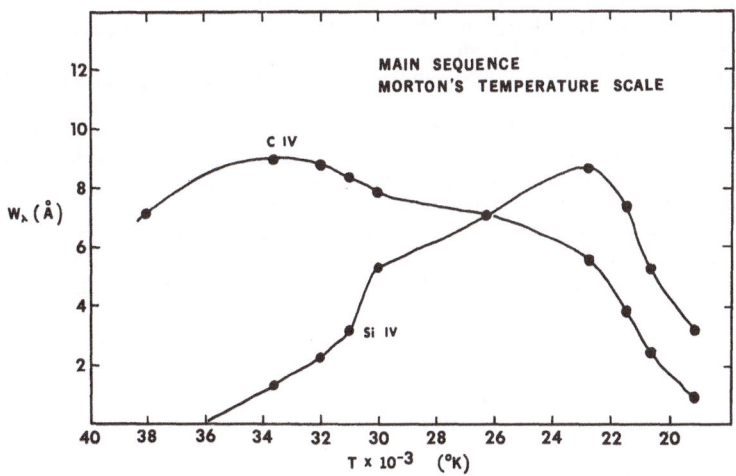

Fig. 5. Observed mean equivalent widths vs. Morton's temperature scale (Morton, 1969).

mum temperature from the standpoint that some radiation in the Lyman continuum may escape and some ionizing photons will be absorbed by dust grains. In addition, if recent calculations of non-LTE effects by Auer and Mihalas (1969) are valid for the earliest-type stars, higher temperature models must be chosen to obtain sufficient Lyman radiation.

The discussion above is based upon preliminary data reduction of only a part of the OAO spectrophotometry obtained to date. The conclusions presented here are subject to revision according to the results of more extensive analysis.

Acknowledgements

We wish to thank G. Oertel and D. Klinglesmith for providing, in advance of publication, information on electron collision widths and model atmosphere calculations, respectively.

References

Auer, L. H. and Mihalas, D.: 1969, *Astrophys. J.* **156**, 681.
Bradley, P. T. and Morton, D. C.: 1969, *Astrophys. J.* **156**, 687.
Cooper, J. and Oertel, G. K.: 1967, *Phys. Rev. Letters* **18**, 985.
Klinglesmith, D.: 1969, private communication.
Morton, D. C.: 1967, *Astrophys. J.* **147**, 1017.
Morton, D. C.: 1969, *Astrophys. J.* **158**, 629.

Discussion

Carruthers: Are theoretical calculations available for the equivalent width of the Nv line, which is very strong in the O stars?

Bless: I know of only those by Morton, which are approximate.

Morton: Your report demonstrates how important it is to have reliable damping constants for the ultraviolet resonance lines.

Bless: Indeed it does. We hope more will be forthcoming for the strong UV lines.

Houziaux: I noticed an interesting luminosity effect for the Civ line dependence vs. spectral type. Did you work any interpretation of this effect?

Bless: No. The problem, of course, is what kind of model atmosphere is appropriate for O and B-type supergiants.

Müller: Referring to the slide that Bless showed us in which predicted equivalent widths are plotted versus temperature I should like to ask Morton what particular reason he had, which made him increase the damping by a factor of ten in calculating equivalent widths.

Bless: The factor 10 came from an analysis of the solar spectrum by Minnaert and Maldera (*Z. Astrophys.* **2** (1931), 165). In the OB stars the source of broadening must be electrons and protons rather than the van der Waals forces of the hydrogen atoms but one had no better estimates for the broadening of the lines of ionized carbon, nitrogen, silicon etc. when the calculations were started some years ago.

THE FAR-ULTRAVIOLET SPECTRUM OF γ CASSIOPEIAE

DONALD C. MORTON, EDWARD B. JENKINS and RALPH C. BOHLIN

Princeton University Observatory, Princeton, N.J., U.S.A.

Abstract. The ultraviolet spectrum of γ Cas from 1060 to 2120 Å is described.

This is a brief report of the ultraviolet spectrum of γ Cas obtained on November 15, 1968 with the Princeton all-reflective rocket spectrograph. This star is famous for its sudden brightening from 2.2 to 1.6 magnitudes in 1936–37 and the accompanying appearance of pronounced shell lines. Since then the star has been relatively quiet until 1966, when Shelus (1967) observed that a veiling of the spectrum had occurred, filling in most of the visual absorption lines. Hα, β, and γ currently show emission profiles with double peaks.

The rocket spectrum has about 1 Å resolution from 1060 to 2120 Å. The principal absorption features are the interstellar Lyman α line and stellar lines of N II (1085 Å), C III (1175 Å), C II (1335 Å), Si IV (1394 Å, 1403 Å), and C IV (1550 Å). The C IV line has an emission component on its long wavelength edge, and C II may have weak emission at both edges of the absorption feature.

The interstellar line is 8 Å wide corresponding to 1.2×10^{20} hydrogen atoms cm^{-2} in the line of sight for broadening by pure radiation damping. If we adopt a distance of 200 pc to γ Cas, the average volume density is only 0.2 atoms cm^{-3}, similar to the low densities reported in the direction of Orion (Jenkins and Morton, 1967) and ζ Pup (Morton *et al.*, 1969).

A preliminary analysis of the ultraviolet stellar spectrum has yielded two immediate results. First, there are no outstanding emission lines in this wavelength region. Even the C II and C IV lines are not as strong relative to the continuum as the Hβ emission. Secondly, there are no significant velocity shifts except for the C IV absorption line, which may be displaced about 400 km sec^{-1} towards the observer. The spectrum is rather similar to the normal main-sequence stars δ Sco (B0V) and π Sco (B1V) reported by Morton and Spitzer (1966), though some of the weaker lines found in the Scorpius stars seem to be absent in γ Cas, probably as a result of the latter exposure having more background fog and poorer resolution.

Acknowledgements

This research was supported by contract NSr-31-001-901 with the United States National Aeronautics and Space Administration. A detailed report will be submitted to the *Astrophysical Journal* by R. Bohlin.

References

Jenkins, E. B. and Morton, D. C.: 1967, *Nature* **215**, 1257.

Morton, D. C. and Spitzer, L.: 1966, *Astrophys. J.* **144**, 1.
Morton, D. C., Jenkins, E. B., and Brooks, N. H.: 1969, *Astrophys. J.* **155**, 875.
Shelus, P. J.: 1967, *Sky and Telescope* **33**, 220.

Discussion

Houziaux: As additional information, I may say that we recently observed the spectrum of γ Cas. at the Observatoire de Haute-Provence, at 20 Å/mm. Double broad emissions of Paschen lines are seen as well as a strong emission line of O I at 8446 Å.

UV SPECTROPHOTOMETRY OF CANOPUS FROM GEMINI XI

Y. KONDO, K. G. HENIZE*, and C. L. KOTILA

Astronomy Branch (TG4), NASA Manned Spacecraft Center, Houston, Texas, U.S.A.

Abstract. The UV energy curve of Canopus in the 2400–4000 Å wavelength range has been obtained from spectrograms having a dispersion of 183 Å mm^{-1}. This curve is in good agreement with previous observations and with a model atmosphere. The equivalent width of Mg II 2795 Å, 2802 Å is found to be 22 Å.

Objective-grating spectra of the F0 supergiant Canopus (α Car) were obtained by astronauts C. Conrad and R. Gordon during the 2-hour extravehicular activity on the Gemini XI manned space flight on September 14, 1966. The equipment and operating procedures have been described in detail by Henize *et al.* (1968). This paper reports the results of the photometric reduction of the UV spectrum of Canopus. A more detailed paper will be published elsewhere (Kondo *et al.*, 1970).

Figure 1 illustrates the best spectrum obtained and identifies the prominent absorption features. A discussion of these identifications has been given by Henize *et al.* (1967). The dispersion of the original spectrum is 183 Å mm and the resolution, estimated from the diameters of the zero-order images, is approximately 15 Å.

Figure 2 shows the composite energy curve derived from measures of the best-exposed regions of three spectra. The straight lines drawn in this figure represent our estimate of the position of the continuous spectrum unaffected by obvious selective absorption features. This estimated continuum is used for comparison with other data in Figure 3, in which our data are compared with a model atmosphere by Parsons and with observational data by Aller *et al.* (1966) and by Stecher (1969). Aller's data and the model atmosphere were fitted to our data at 4000 Å while Stecher's data were fitted to ours at 3000 Å. The model atmosphere data were kindly supplied by Parsons especially for this comparison using methods described in a previous paper (Parsons, 1969). The chosen parameters were $T_e = 6900$ K, $\log g = 2.0$ for a non-LTE, blanketed model.

In general, the Gemini XI data appear to be in good agreement with both the previous observations and the model atmosphere. Our measured value for the magnitude difference between 3640 Å and 4000 Å (1.32 mag) is in excellent agreement with the model atmosphere. The slight disagreement with Aller's data approximates the expected probable error of our photometry. The fit with Stecher's data is good if we note that Stecher has not interpolated the continuum across the large absorption complex from 2600 Å to 2900 Å as we have. This agreement tends to confirm Stecher's data and suggests that the anomalous behavior of his observed continuum in the 2000–2300 Å region deserves further investigation. The only significant dis-

* On leave of absence from Lindheimer Astronomical Research Center, Northwestern University, Ill., U.S.A.

Houziaux and Butler (eds.), Ultraviolet Stellar Spectra and Ground-Based Observations, 180–184.

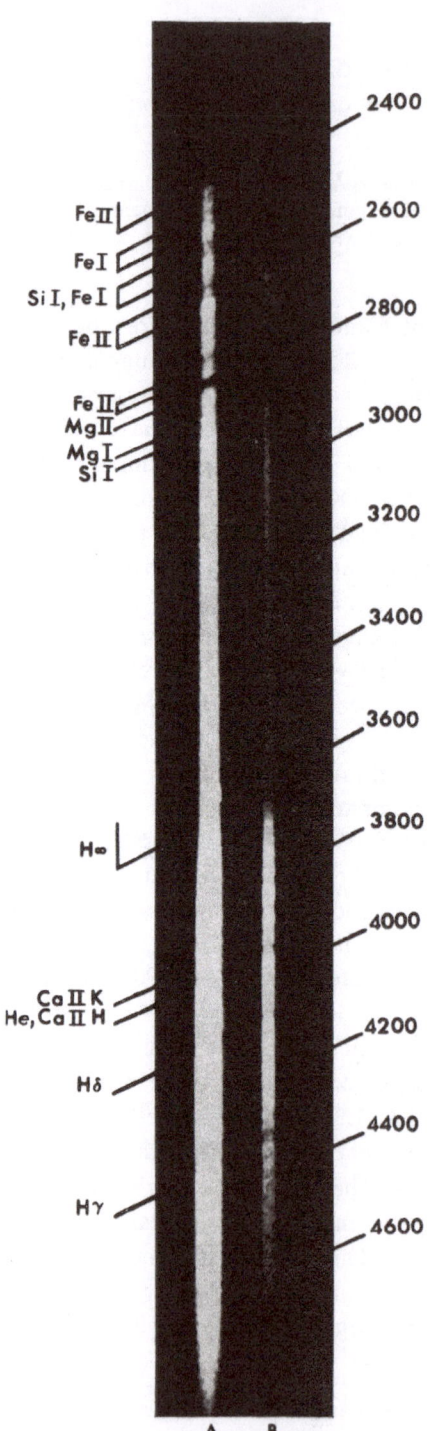

Fig. 1. The spectrum of Canopus on NASA
S-66-53102.

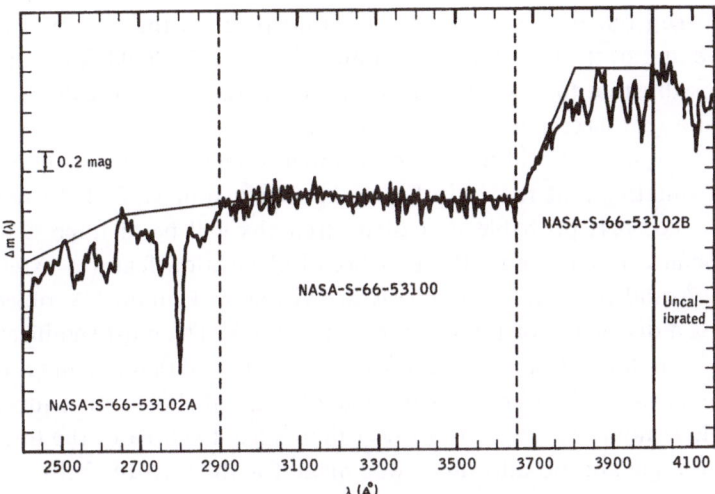

Fig. 2.   UV spectral energy curve of Canopus.

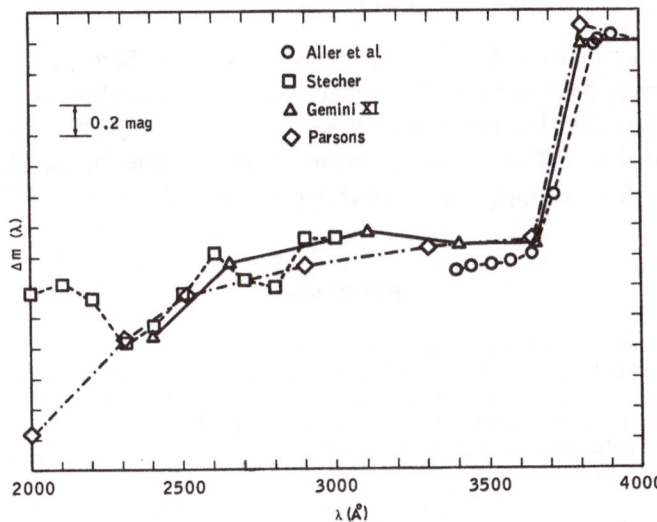

Fig. 3.   Comparison of UV observations of Canopus and a model atmosphere.

TABLE I

Equivalent widths in the UV spectrum of Canopus

Wavelength (Å)	Ident.	W (Å)	Remarks
2520–2560	blend of Fe I	15	Possibly includes Si II resonance lines
2580–2640	blend of Fe II	22	Possibly includes Mn II resonance lines
2735–2775	blend of Fe II	18	
2799	Mg II	22	Resonance doublet 2795 Å, 2802 Å
2852	Mg I	3.4	Resonance line
2882	Si I	2.0	Very weak, perhaps doubtful

crepancy between our data and the model atmosphere is the elevation of our continuum above that of the theoretical continuum in the 2550–3400 Å region. Although it is not impossible that this may be a result of inaccuracies in our calibration it seems probable that a real discrepancy with the model may exist.

Measures of equivalent widths, made on direct intensity plots using a continuum closely approximating that shown in Figure 2, are given in Table I. Blended features are included since it is probable that their strengths will be of interest in UV classification schemes. For example the two broad absorption features extending from 2520 to 2640 Å and from 2740 to 2900 Å are visible in Gemini UV objective-prism spectra having a dispersion of 1500 Å mm^{-1} at 2500 Å. The most significant measure is that of the MgII doublet, since this is the resonance line of that ion and is of sufficient strength to be measurable over a broad range of spectral classes. Its equivalent width given in Table I should be taken as a lower limit since the area of the line was taken as a simple triangle and no allowance was made for the possible existence of broad wings.

Acknowledgements

We wish to thank Roy C. Stokes, NASA Manned Spacecraft Center, and F. G. O'Callaghan, Northwestern University, for their contributions to this work. Stokes was responsible for the development of the spectrograph and participated in the data reduction. O'Callaghan was primarily responsible for the film calibration system. This work was supported by NASA contract NAS 9-4129 to Northwestern University.

References

Aller, L. H., Faulkner, D. J., and Norton, R. H.: 1966, *Astrophys. J.* **144**, 1073.
Henize, K. G., Wackerling, L. R., and O'Callaghan, F. G.: 1967, *Science* **155**, 1407.
Henize, K. G., Wray, J. D., and Wackerling, L. R.: 1968, *Bull. Astron. Inst. Czech.* **19**, 279.
Kondo, Y., Henize, K. G., and Kotila, C. L.: 1970, *Astrophys. J.* **159**.
Parsons, S. B.: 1969, *Astrophys. J. Suppl. Ser.* **18**, 127.
Stecher, T. P.: 1969, *Astron. J.* **74**, 98.

Discussion

Bonnet: It seems that your spectrum looks very much like the solar spectrum in the vicinity of the MgI discontinuity due to photo-ionization from the first excited level of Mg. More particularly there is a difference between the intensities of the two peaks appearing in your spectrum at 2500 Å and 2650 Å. Do you think that this difference might be due to MgI ionization as it is in the solar spectrum?

Henize: I have not considered the possible presence of the MgI discontinuity. I believe that the obvious absorption features in this region of the spectrum are adequately accounted for by blended iron lines but it is possible that the MgI discontinuity contributes to the feature extending shortward from 2500 Å.

Stecher: It is nice to be on the high side of theory for a change. The points are, of course, very weak at the short wavelengths and are dependent upon the background level.

Henize: Please pardon me for bringing undue attention to the disagreement between your data and the model. However, I was quite pleased by the agreement between your data and ours in our region of overlap and I feel that the discrepancy with theory suggested by your shorter wavelength data deserves further investigation.

Conti: (1) What was the exposure time for the Canopus spectrum? (2) What plans are underway to continue these spectrographic observations in future Apollo programmes?

Henize: (1) The longest exposure was 2 min. (2) At Northwestern University, we have designed and built a 6-inch aperture objective prism spectrograph which is scheduled to be flown on the first Apollo Applications Programmes flight. A calcium fluoride prism is the dispersing element. We expect to record spectra down to 1350 Å with a resolution of about 1 Å at 1500 Å.

B. GROUND-BASED OBSERVATIONS OF
SPECTRA RELEVANT TO THE ULTRAVIOLET

REVIEW OF GROUND-BASED OBSERVATIONS OF
SPECTRA RELEVANT TO THE ULTRAVIOLET

M. W. FEAST

Radcliffe Observatory, Pretoria, South Africa

1. Introduction

Up to the present the chief results of investigations of stellar line spectra in the extreme ultraviolet have been limited to two fairly distinct fields. In the case of the Sun the data have referred mainly to the chromosphere and also to the region of the temperature minimum at the top of the photosphere and to the corona. We may hope that observations will eventually be extended to the chromospheric ultraviolet spectra of other stars. In the case of the early type stars the most spectacular results so far obtained have concerned the evidence for mass loss by several of these stars. I have therefore chosen to spend most of the available time discussing some of the ways of studying from the ground, first the chromospheres of stars other than the Sun and secondly mass loss from stars.

2. Observations Relating to Stellar Chromospheres

A. Ca II H AND K EMISSION IN STARS

Probably the most comprehensive attack on the problem of stellar chromospheres has been O. C. Wilson's work on H and K emission in late type stars [1, 2, 3, 4, 5]. By analogy with the Sun it has been easy to suppose that the emission arises at the chromospheric level and that in many late type stars the degree of chromospheric activity is much greater than in the Sun. The main conclusions of Wilson's work are well known. There is a linear relation, independent of spectral type, extending over about 15 magnitudes between the K_2 emission line widths and the visual absolute magnitude (the Wilson-Bappu effect). For stars near the main sequence there is evidence that the intensity of Ca II emission decreases with age. This evidence comes both from intercomparing clusters of different ages and from stars in the general field whose relative ages can be fixed by the Stromgren-Perry photometry. Wilson also finds that chromospheric H and K generally become inconspicuous earlier than about F4 on the main sequence. This is just the point where rotational velocities of main sequence stars increase sharply and where theory predicts that main sequence stars should cease to have hydrogen convection zones. These results suggest therefore that the formation of a chromosphere in a normal star depends on the existence of a hydrogen convection zone. In addition stars with hydrogen convection zones are assumed by analogy with the Sun to have stellar winds. The interaction of these winds with stellar magnetic fields leads to rotational deceleration. This aspect of the problem has been carried a step further by Kraft [6] who finds evidence from a detailed investigation of the rota-

Houziaux and Butler (eds.), Ultraviolet Stellar Spectra and Ground-Based Observations, 187–198.
All Rights Reserved. Copyright © 1970 by the IAU.

tional velocities of solar type stars, that such stars are indeed rotationally decelerated after reaching the main sequence.

Further statistical work on H and K emission in late type stars should benefit considerably from the publication by Warner [7] of emission line strengths and widths in 200 southern G, K and M type stars.

Wilson [8] is engaged in a programme of accurate intensity measurements of H and K emission which is inspired by the desire to find stellar equivalents to the solar sunspot cycle. So far no significant variations have been found in the total flux determined photoelectrically. However changes (particularly changes in profile) of the K_{232} structure have been found by a number of workers including Griffin [9], Deutsch [10] and Liller [11]. Liller found evidence for variations in a number of G and K type stars. For Aldebaran (α Tau, K5 III) he found, in addition to general variability, a flare-like increase in the intensity of the violet component of K_2. The increase was about 19% and lasted half an hour. Deutsch, whose work extends over a long period, finds evidence for variable H and K profiles in a number of K and M giants as well as a good deal of fine structure in the lines. He has also looked for possible periodicities in the phenomena (this volume, pp. 199–208).

Not all stars showing Ca II emission obey the Wilson-Bappu effect. The transient Ca II emission lines in cepheids are considerably wider than would be predicted [12] (for example in X Cyg, 213 km sec^{-1} against an expected 140 km sec^{-1}). Perhaps associated with this is the fact that the K_3 reversal is much stronger in cepheids than in non-variable stars, probably indicating that the emission comes from an abnormally deep layer [13]. It will be recalled that there is good evidence that the complex emission spectrum found in Mira variables which can include H and K, arises at a rather deep level in the atmosphere. One might hesitate to call this spectrum chromospheric but it may well be basically a closely related phenomenon being associated with energy derived from the hydrogen convection zone. Because the basic mechanism of the H and K widths in stellar spectra remains somewhat obscure, progress may as well come from detailed studies of abnormal cases such as the cepheids or the Mira variables as from normal stars.

Whilst chromospheric H and K are, apparently, generally missing from early type stars a couple of interesting exceptions have been found by Warner [14, 15].* Canopus (F0 Ib) shows emission but this is much too narrow (50 km sec^{-1} against a predicted 150 km sec^{-1} for the Wilson-Bappu relationship. A particularly interesting case is the well known binary γ Vir with its two equal F0 V components. The following points are worth noticing. (1) Emission is found in γ Vir N with a width of 43 km sec^{-1} agreeing with the Wilson-Bappu relation.† (2) Since γ Vir S does not show emission there can

---

* See also the case of γ Boo (A7 III) mentioned by Mrs Praderie (this volume, p. 69).

† Warner notes that γ Vir N has a significant amount of rotation ($v \sin i \sim 16$ km sec^{-1}). This does not appear to be at all abnormal. Stars lower down the main sequence with $v \sin i \sim 35$ km sec^{-1} and strong H and K emission are known [5, 6]. In fact Kraft's principal conclusion was "for stars less massive than $M/M_\odot = 1.25$ the largest rotational velocities are associated with stars having active chromospheres".

apparently be a significant range of emission intensities in stars of the same mass and age. (3) γ Vir N has a magnetic field of -390 gauss [16] suggesting a connection between high magnetic fields and high chromospheric activity. (4) There is evidence for (abnormal?) surface nuclear activity in these stars. Zirin [17] finds evidence from HeI 10830 Å for both He^3 and He^4 in γ Vir N (though mass motions in the chromosphere *could* falsify this conclusion). Both stars have extreme overabundances of lithium (by a factor of about 100) [18]. The beryllium abundance on the other hand is low and the Li/Be ratio is consequently extremely high [19], about 640 in γ Vir N and half this in γ Vir S. This together with the low Li^6/Li^7 ratio in γ Vir N leads Conti to suggest that spallation has involved particles with energies of tens of MeV rather than the hundreds of MeV that appear necessary for spallation reactions in solar type stars. Work of this type gives hope that it will eventually be possible to draw together work on stellar chromospheres, stellar magnetic fields and nuclear reactions in stellar atmospheres; fields between which only vague relationships at present appear to exist.

B. OTHER CHROMOSPHERIC EMISSION LINES

There are a number of other ways of investigating the outer regions of stars from studies of emission lines, none of which have as yet been extensively exploited. Wilson [20] has shown that Hε emission occurs in many late type stars strengthening in general towards the later types. It seems most likely that this is chromospheric, Hε being most easily seen because it falls on the wing of the strong photospheric Ca II absorption.

Fe II emission lines are found in a considerable number of peculiar objects. However Herzberg [21] showed that permitted multiplets near 3200 Å occurred in emission in the normal M type supergiants α Her and α Sco, and more recently Bidelman [22] has noted them as a normal feature in M giants and supergiants. They also occur in dMe stars [23]. It seems most likely that these lines are chromospheric in origin and a quantitative study of them could be very valuable. For instance little is known about the line widths and shapes. Herzberg's reproductions suggest that the Fe II lines in α Her and α Sco are significantly narrower than the H and K emission lines which are about 150 km sec^{-1} wide in these stars. In α Ori, Weymann [24] made the significant discovery that the stronger of these Fe II emission lines is self reversed. Even so the widths appear to be significantly less than the H and K widths (roughly 149 km sec^{-1} as against 186 km sec^{-1}) whilst the weaker Fe II lines are not self reversed and are apparently somewhat narrower. There is evidently a strong indication here that (contrary to what has sometimes been supposed) the Wilson-Bappu effect cannot be interpreted as due to turbulent broadening in an optically thin model. Swings and Swings [25, 25a] have recently drawn attention to the importance of measurements of the relative intensities of these ~ 3200 Å Fe II lines for studies of the level of excitation in the emitting region. In quantitative work considerable care will probably be necessary in allowing for the effects of the superimposed photospheric spectrum and for any overlying absorption.

C. CHROMOSPHERIC ABSORPTION

Whilst chromospheric effects may be found in a considerable number of absorption lines there are one or two chromospheric absorption line problems which are of particular importance. Evidence from the Sun suggests that the cores of Hα absorption in late type stars are of chromospheric origin. In accord with this interpretation these cores show an increasing width with increasing luminosity, an absorption Wilson-Bappu effect [26]. An example of what may be achieved from detailed studies of Hα, at least in the case of a variable star, has been given recently by Rodgers and Bell [27]. In the long period (35.5 days) cepheid *l* Carinae they found evidence for multiple absorption cores. One of the components is observed throughout the cycle and has a velocity close to the γ-velocity of the star. They concluded that there is a chromospheric shell which does not take part in the cepheid pulsation.

In the Sun HeI 10830 Å absorption is known to be wholly chromospheric. Vaughan and Zirin [28] have studied this line in late type stars. Because of its high excitation potential the presence of the line gives evidence for a hot chromosphere ($T > 20000$ K). The line is found mainly in G and K type stars and its strength correlates roughly with K_2 emission intensity. There is also a luminosity effect, the absorption being weaker in the dwarfs. Evidence of temporal variations in the line intensity has been found and macroscopic motions in the absorbing region are indicated by velocity shifts which are generally negative (i.e. rising) and up to 50 km sec^{-1}. The line is generally fairly diffuse (~ 1.5 Å wide) but with no correlation of width and luminosity (no Wilson-Bappu effect).

The occurrence of HeI D_3 (5876 Å) in absorption in a few late type stars has been reported [29]. Further work on this line is very desirable though care has to be taken about some faint terrestrial lines in this region of the spectrum. Rather surprisingly Wilson and Aly find in the three stars they measured that the line is displaced longward (by about 0.3 Å) in contrast to the general shortward displacement of 10830 Å.

D. THE CHROMOSPHERES OF THE R CORONAE BOREALIS STARS

Amongst the various peculiar stars the R CrB variables seem worth mentioning here because they hold out prospects for studies of the outer atmospheres of stars which would probably not be otherwise possible. The two best studied stars of this class R CrB itself and RY Sgr are similar carbon rich, hydrogen poor, F-type supergiants which at irregular intervals undergo large and rapid decreases in light. At such times the stars show a very rich emission spectrum mostly of ionized metals (some 350 emission lines have been measured at medium dispersion during the recent minimum of RY Sgr). There is a sharp drop in continuum intensity, possibly due to a large increase in opacity and this allows the observation of emission from the chromosphere or upper photosphere. In RY Sgr an emission continuum has been observed which has been attributed to the electron attachment spectrum of CN [30]. Marked changes take place in the emission spectrum and we are apparently observing the decay of chromospheric emission when its source of excitation is cut off. Two further points about

R CrB stars are relevant to our general topic. Firstly, He I 5876 Å absorption has been identified in their spectra near maximum light. In view of the almost certain chromospheric origin of this line in a number of normal late type stars it is clearly dangerous to attempt a determination of helium abundance from the line. Secondly, simultaneous UBV photometry and spectroscopy shows that there are times when RY Sgr has a normal spectrum (i.e. as at maximum) but is faint and reddened. This is good evidence for the formation of an absorbing cloud of particles well clear of the stellar surface. Since this presents a unique opportunity to study variable circumstellar reddening in detail, attempts are expected to extend the photometry into the extreme ultra-violet. However caution should be exercised in interpreting any results that are obtained since chromospheric emission in the ultraviolet could well complicate the picture.

E. OTHER DATA

I am aware of having omitted a number of important aspects of the chromospheric problem. In particular I have no time to discuss the study of eclipsing systems which reveal the complex mass motions in the chromospheres of the ζ Aurigae type supergiants (cf. [31]).

3. Mass Loss from Stars

The other chief topic where there is much in common between ultraviolet and ground-based observations is that of mass loss. Here I shall be very restrictive, leaving out entirely the work of Deutsch and others on mass loss in late type stars. It has long been known that mass loss takes place from the atmospheres of some early type stars. There is an extensive literature on WR stars, P Cygni stars, Be stars etc. much of which is directly relevant to the present topic. However there are excellent recent general reviews of these fields available (for instance by Underhill [32, 33, 34] and I have chosen simply to discuss the most recent work in a little detail.

A. THE OB SUPERGIANTS

The work of the last few years which has seemed most closely related to the ultra-violet work on OB stars has been that of Hutchings [35, 36]. From studies of velocities of lines arising at different mean levels he has been able to deduce the existence of extended atmospheres for some early type supergiants in which matter is being accelerated outwards. Velocities up to 600 km sec^{-1} were found in the extreme outer parts of the stars. These are somewhat smaller than the velocities found in the ultra-violet observations but it has seemed that the various observations might fit together quite well.

A number of points arise which link this work with more general questions. It is reasonable to ask if the large amount of low and medium dispersion work on OB stars can tell us whether mass loss, observable from the ground, is a common occurrence. Hutchings studied the very luminous B1 Ia-O star ζ^1 Sco. For this star a normal radial velocity measurer would probably omit the lower Balmer lines which have P Cygni profiles but would almost certainly use many lines (He I, Mg II, O II, N II, etc.) for which substantial velocity shifts are found by Hutchings. Probably one would expect the measured velocity to be algebraically smaller than the true stellar velocity by 10 km sec^{-1} or so in this case. A direct comparison is not possible because the velocity of the star is variable, presumably due to the varying velocity fields in the atmosphere. Nevertheless if net outward velocities were present in the atmospheres of a substantial number of OB stars we should expect kinematical investigations to show significant negative K terms. Fairly extensive investigations of this type have in fact been carried out [37], dividing the stars according to type and luminosity class. No significant K terms were found except for a *positive* K term for O-type stars which is probably at least partly due to a gravitational red shift. If anything most K terms are slightly positive. This suggests that so far as ground-based observations are concerned, evidence for mass loss on the scale of that found for ζ^1 Sco will not be found for any large number of OB stars (including supergiants). This conclusion would however not hold if for some reason the stars showing mass loss effects were just the ones omitted in kinematical investigations because they showed variable radial velocities. Furthermore these results do not preclude the general occurrence of mass loss amongst supergiants of the very highest luminosity which probably form a small fraction of the total number of supergiants in the analysis. One might expect that radial velocities of early type stars in galactic clusters would throw some light on the mass loss problem. In fact it has been known for some years [38, 39] that early-type supergiants in clusters tend to show greater *positive* velocities than the giants. This effect is not yet understood but it may as well be due to some peculiar effect in the giants as in the supergiants.

The problem of Hα emission in early type supergiants is intimately connected with the problem of mass loss. Andrews [40] has shown that the absorption strength (measured photoelectrically) is a good luminosity indicator except amongst main sequence Be stars where the emission destroys the correlation. However Hα strength remains well correlated with luminosity amongst the supergiants even when the line is going over into emission. So the occurrence and strength of the emission is closely connected with luminosity. Similar conclusions were deduced by Abt [41, 42]. These results appear to fit naturally with the spectroscopic work on the early type supergiant members of the Magellanic Clouds. Here the emission line stars at a given spectral type are strongly concentrated amongst the most luminous stars [43]. Many of these stars show P Cygni effects and the Magellanic Clouds results show rather clearly that instability and consequent mass loss increases as one goes towards the supergiants of the highest luminosity. Apparently these effects are to be attributed to the increasing importance of radiation pressure in these atmospheres.

The mere presence of Balmer emission in a supergiant is apparently no guarantee of mass loss. Peterson [44] has recently found that one can predict from a non-LTE theory that bright early-type supergiants may show (double) Hα emission even under conditions of hydrostatic equilibrium. Of course both in the Galaxy and the Magellanic Clouds, the majority of the very luminous stars we are discussing show definite evidence of mass loss from P Cygni profiles. Nevertheless the effects predicted by Peterson may need taking into account in detailed interpretations of the Hα profiles of these stars.

Thus various ground-based observations give evidence for the general occurrence of mass loss amongst the most luminous early-type supergiants. There is as yet however little ground-based evidence for mass loss amongst somewhat fainter supergiants, some of which are shown by the ultraviolet observations to be ejecting matter (e.g. [45]).

The discussion of this section will require substantial modification if a recent suggestion by Underhill [46] receives confirmation. She has suggested that the early-type Ia supergiants have relatively low masses and large underabundances of hydrogen. If this is indeed the case these stars may no longer be regarded as simply an extension of the Ib supergiants to greater masses. They would apparently be in a very advanced evolutionary stage.

B. THE Of STARS

The Of stars are a little understood group of stars which are of particular interest at this symposium since high velocity mass ejection has been established for the Of star ζ Pup from ultraviolet work. Hutchings [35] has recently discussed the marked mass ejection effects in HD 152408. The high negative radial velocity of this star was interpreted long ago by Struve [47] as indicating a rapidly expanding shell. Spectroscopic workers have generally hesitated to regard the star as a normal Of star and Struve remarked that it might be classed as a hot P Cygni star. HD 151804 which was also studied by Hutchings appears a more normal Of star and the atmospheric velocity effects though present are less pronounced. In this case low dispersion work might indicate a velocity of 20 or 30 km sec^{-1} lower than the true stellar velocity.

It is of importance both for problems of mass loss and for problems of galactic kinematics to know whether the measured radial velocities of a substantial number of Of stars are falsified by atmospheric effects. This can be seen from Blaauw's discussion of the run-away stars [48]. Of the 5 stars of type O8 or earlier in his discussion, 3 are of type Of and one of these is HD 152408 which must be rejected in the light of the work of Struve and Hutchings. In Table I are listed the residual velocities (i.e. velocities corrected for solar motion and galactic rotation) for 20 galactic Of stars with published radial velocities and photometry. Distances were calculated using an absolute magnitude of −7.0 (see below) and the residuals were calculated using the $\omega(R)$ results for OB stars and interstellar gas [37]. Six of the 20 stars have residuals greater than 35 km sec^{-1}. That is, they are run-away stars if the measured velocities reflect the true stellar velocities (notice that HD 152408 is *not* included in the list). Four of the 6 high velocity stars have negative, and two positive, residuals. The existence of some positive residuals presumably indicates that some Of stars are real run-away objects. +60°2522 is only marginally in this category but HD 157857 is clearly in this class. Of the stars with large negative residuals one is the spectroscopic binary 29 CMa for which the emission lines give a much more positive γ-velocity (by 66 km sec^{-1}) than the absorption lines which were used for Table I [49]. The quoted

TABLE I
Of stars

HD	r (kpc)	Residual velocity (km sec^{-1})	
16691	4.6	+9	
14947	3.3	−13	
15570	2.5	+9	
108	4.0	−10	
57060	2.0	−51	29 CMa
66811	0.7	−69	ζ Pup
157857	2.9	+55	
166734	1.7	−2	
167971	1.8	+9	
171589	5.2	−14	
175754	4.8	−31	
188001	2.6	−1	9 Sge
192281	3.0	−50	
192639	2.8	+6	
193514	2.6	−11	
210839	1.2	−53	λ Cep
225160	4.8	+16	
+60°2522	5.0	+37	
190429br	3.5	−11	
148937	2.1	−29	

velocity is thus very suspect. However it does not follow that all the negative velocities must be rejected as probably badly affected by mass ejection.

A few radial velocities are available for stars in IC 1805 including an Of member. The velocities do not appear to be of high accuracy but they indicate that if anything the velocity of the Of star is somewhat positive with respect to the other stars. There is evidently need for detailed work on as many Of stars as possible to determine the influence of mass ejection effects on the measured velocities.

The 7th magnitude Of star HD 148937 is of considerable interest. As Henize [50] and Gum [51] noticed the star is centrally located between the two nebulosities NGC 6164-5. The object was at one time classed with the planetary nebulae and would in that case be very close (∼150 psc) [50]. However the star has the colour of a normal reddened O type star [52] and strong interstellar lines as well as a low proper motion, whilst the nebulosity is of low excitation. It appears now to be agreed [52, 53] that the star must be considered a normal, high luminosity, Of type. It is of considerable interest to find an Of star associated with nebulosity which gives the impression (see photographs in [52] and [54]) of having been ejected from the star. A full study of the kinematics of the nebulosity is clearly desirable.*

---

* In presenting this paper at Lunteren it was pointed out that the southern nebulosity had a radial velocity close to the published [55] high negative radial velocity of the star. This suggested that the system might have a high space motion. Subsequent work has confirmed this approximate agreement but has shown that the northern nebulosity has a much more positive velocity. Evidently the stellar velocity is affected by mass loss effects.

It might be supposed that, as in the case of the OB supergiants, the mass ejection from Of stars was the result of radiation pressure and this view has been taken by some workers (compare Hutchings and Solomon, this volume, pp. 209–214; 236–240). However at least two other effects may be suggested for the instability of these stars.

(1) It has been suggested that the Of stars are somewhat evolved massive stars and are related to main sequence O stars in the same way that the early-type Be stars are related to the main sequence OB stars [32, 56]. The early-type Be stars appear to be slightly evolved objects and to show the effects of rotational instability in evolving stars (e.g. [57, 58, 59]).

(2) The absolute magnitudes of the galactic Of stars have been discussed by a number of workers, Kopilov, Hack, Underhill and others. Whilst there remains some doubt as to the spread of luminosities amongst these objects there seems little doubt that some of them are highly luminous, M_V up to about -7. For instance the Of stars in the I Sco association have an absolute magnitude near -7.* Van den Bergh [56] quotes a value of $-6\overset{m}{.}6$ for the Of star in IC 1805 whilst the two known Of stars in the Magellanic Clouds [43] also have absolute magnitudes of about -7. If the Of stars are near main-sequence objects then these high luminosities indicate such high masses (of the order of 100 $M_\odot$) that the stars would be expected to show the effects of pulsational instability which are predicted to set the upper limit to masses on the main sequence [61, 62]. It may also be noted here that the agreement between the absolute magnitudes of the Of stars in the Galaxy and the Magellanic Clouds suggests that it is unnecessary to attempt to explain away the high luminosities of the Magellanic Cloud objects by regarding them as unresolved groups of stars [61].

Further work appears to be necessary to decide on the relative importances of rotational instability, pulsational instability and atmospheric radiation pressure in the case of the Of stars. In this type of theorizing however it is sobering to recall that Wilson [63] showed long ago that the nucleus of the planetary nebula NGC 2392 which is similar to an Of star but is presumed to be a sub-luminous object has an expanding atmosphere with outward acceleration closely similar to that found by Hutchings in the highly luminous Of's. It is not at all clear whether we have a number of different mechanisms producing similar observational effects or whether there is one mechanism working over a large range of stellar masses and luminosities.

C. RAPID CHANGES IN THE SPECTRA OF THE EARLY-TYPE STARS

Although mass loss is established for a number of early-type stars we know, as yet, little about the variations of the mass loss with time. Changes in radial velocity and line shape, especially at Hα, are known for several bright supergiants. More recently evidence has been obtained for quite rapid changes in the spectra of some early type stars (cf. [64, 65, 66]). Some of these effects such as variations within a few minutes of the emission structure of P Cygni seem directly connected with the mass loss problem.

---

* Absolute magnitudes about 1 magnitude fainter have recently been quoted for these stars [60] but this appears to be due to a numerical error.

A particularly interesting problem concerns the Of stars. Already some years ago Oke found evidence for variations in the intensities of the emission lines in Of stars in the course of a few hours [67]. It would be important to follow up this work in more detail since periods of about this length would be expected if the basic cause of the Of phenomenon is pulsational instability.

4. Ground-Based Predictions of Ultraviolet Spectra

To conclude I should mention briefly a quite different topic. There are a number of cases in which ground-based observations have been used to make definite predictions of ultraviolet spectra. One thinks particularly of Zanstra's work predicting ultraviolet continua of central stars of planetaries from the observed nebular line emission. Similarly Bowen's fluorescence mechanism to explain mutilated multiplets in nebulae requires strong HeII 304 Å and other emission lines in this region. Direct confirmation of these predictions may be difficult or impossible because of interstellar absorption. However a more tractable case is that of the Mira variables. The application of Bowen's fluorescence mechanism to the observed emission spectra of these stars by Thackeray [68] leads to the prediction of a number of strong ultraviolet emission lines in these stars particularly the MgII doublet at 2800 Å. It is even possible [69, 70] to predict from the observed emission lines that the MgII lines will be self reversed with the short wavelength emission wing strongest. One supposes that these ultraviolet lines must be abnormally intense in Mira variables. Direct observation of these lines is desirable not only to verify the predictions but to help understand the basic cause of line emission in Mira variables.

Acknowledgements

I am indebted to A. D. Thackeray and T. Lloyd Evans who kindly read and commented on an early draft of this paper.

References

[1] Wilson, O. C. and Bappu, M. K. V.: 1957, *Astrophys. J.* **125**, 661.
[2] Wilson, O. C.: 1959, *Astrophys. J.* **130**, 499.
[3] Wilson, O. C.: 1963, *Astrophys. J.* **138**, 832.
[4] Wilson, O. C. and Skumanich, A.: 1964, *Astrophys. J.* **140**, 1401.
[5] Wilson, O. C.: 1966, *Astrophys. J.* **144**, 695.
[6] Kraft, R. P.: 1967, *Astrophys. J.* **150**, 551.
[7] Warner, B.: 1969, *Monthly Notices Roy. Astron. Soc.* **144**, 333.
[8] Wilson, O. C.: 1968, *Astrophys. J.* **153**, 221.
[9] Griffin, R. F.: 1963, *Observatory* **83**, 255.
[10] Deutsch, A. J.: 1967, *Publ. Astron. Soc. Pacific* **79**, 431.
[11] Liller, W.: 1968, *Astrophys. J.* **151**, 589.
[12] Kraft, R. P.: 1960, in *Stars and Stellar Systems*, Vol. VI (ed. by J. L. Greenstein), University of Chicago Press, Chicago.
[13] Herbig, G. H.: 1952, *Astrophys. J.* **116**, 369.
[14] Warner, B.: 1966, *Observatory* **86**, 82.

[15] Warner, B.: 1968, *Observatory* **88**, 217.
[16] Babcock, H. W.: 1958, *Astrophys. J. Suppl. Ser.* **3**, 141.
[17] Zirin, H.: *Astrophys. J.* **152**, L177.
[18] Conti, P. S. and Danziger, I. J.: 1964, *Astrophys. J.* **146**, 383.
[19] Conti, P. S.: 1969, *Astrophys. J.* **155**, L167.
[20] Wilson, O. C.: 1957, *Astrophys. J.* **126**, 46.
[21] Herzberg, G.: 1948, *Astrophys. J.* **107**, 94.
[22] Bidelman, W. P.: 1961, *Astron. J.* **66**, 453; 1961 *Trans. I.A.U. XIA*, 305.
[23] Wildt, R.: 1951, *Astron. J.* **56**, 51.
[24] Weymann, R.: 1962, *Astrophys. J.* **136**, 844.
[25] Swings, J. P. and Swings, P.: 1967, *Astrophys. Letters* **1**, 54.
[25a] Swings, J. P.: 1969, *Astrophys. J.* **155**, 515.
[26] Kraft, R. P., Preston, G. W., and Wolff, S. C.: 1964, *Astrophys. J.* **140**, 235.
[27] Rodgers, A. W. and Bell, R. A.: 1968, *Monthly Notices Roy. Astron. Soc.* **138**, 23.
[28] Vaughan, A. H. and Zirin, H.: 1956, *Astrophys. J.* **152**, 123.
[29] Wilson, O. C. and Aly, M. K.: 1968, *Publ. Astron. Soc. Pacific* **68**, 149.
[30] Feast, M. W.: 1969, *Non-Periodic Phenomena in Variable Stars*, (ed. by L. Detre), Academic Press, Budapest p. 253.
[31] Wilson, O. C.: 1960, in *Stars and Stellar Systems*, Vol. VI: *Stellar Atmospheres* (ed. by J. L. Greenstein), University of Chicago Press, Chicago.
[32] Underhill, A. B.: 1966, *The Early-Type Stars*, Reidel, Dordrecht, The Netherlands.
[33] Underhill, A. B.: 1960, in *Stellar Atmospheres* (ed. by J. L. Greenstein), University of Chicago Press, Chicago.
[34] Underhill, A. B.: 1968, *Ann. Rev. Astron. Astrophys.* **6**, 39.
[35] Hutchings, J. B.: 1968, *Monthly Notices Roy. Astron. Soc.* **141**, 219.
[36] Hutchings, J. B.: 1968, *Monthly Notices Roy. Astron. Soc.* **141**, 329.
[37] Feast, M. W. and Shuttleworth, M.: 1965, *Monthly Notices Roy. Astron. Soc.* **130**, 245.
[38] Feast, M. W.: 1958, *Monthly Notices Roy. Astron. Soc.* **118**, 618.
[39] Feast, M. W.: 1963, *Monthly Notices Roy. Astron. Soc.* **126**, 11.
[40] Andrews, P. J.: 1968, *Mem. Roy. Astron. Soc.* **72**, 35; and Ph.D. Thesis Cambridge, 1966.
[41] Weymann, R. (quoting H. A. Abt): 1963, *Ann. Rev. Astron. Astrophys.* **1**, 97.
[42] Abt, H. A. and Golson, J. C.: 1966, *Astrophys. J.* **143**, 306.
[43] Feast, M. W., Thackeray, A. D., and Wesselink, A. J.: 1960, *Monthly Notices Roy. Astron. Soc.* **121**, 337.
[44] Peterson, D. M.: 1969, Smithsonian Astrophys. Obs. Special Report 293.
[45] Morton, D. C.: 1967, *Astrophys. J.* **150**, 535.
[46] Underhill, A. B.: 1969, *Astron. Astrophys.* **1**, 494.
[47] Struve, O.: 1944, *Astrophys. J.* **100**, 189.
[48] Blaauw, A.; 1961, *Bull. Astron. Inst. Neth.* **15**, 265.
[49] Struve, O. and Sherman, F.: 1941, *Astrophys. J.* **93**, 84 (see also *Astrophys. J.* (1958) **128**, 328).
[50] Henize, K. G.: 1959, *Astron. J.* **64**, 51.
[51] Gum, C. S.: 1955, *Mem. Roy. Astron. Soc.* **67**, 155.
[52] Westerlund, B.: 1961, *Arkiv Astron.* **2**, 467.
[53] Henize, K. G.: 1964, *Publ. Astron. Soc. Pacific* **76**, 385.
[54] Westerlund, B. E. and Henize, K. G.: 1967, *Astrophys. J. Suppl. Ser.* **14**, 154.
[55] Buscombe, W. and Morris, P. M.: 1960, *Monthly Notices Roy. Astron. Soc.* **121**, 263.
[56] Van den Bergh, S.: 1968, *Astrophys. J.* **151**, 1191.
[57] Crampin, J. and Hoyle, F.: 1960, *Monthly Notices Roy. Astron. Soc.* **120**, 33.
[58] Feast, M. W. and Lloyd Evans, T.: 1967, *Observatory*, **87**, 286.
[59] Slettebak, A.: 1968, *Astrophys. J.* **154**, 933.
[60] Feinstein, A. and Ferrer, O. E.: 1968, *Publ. Astron. Soc. Pacific* **80**, 410.
[61] Stothers, R. and Simon, N. R.: 1968, *Astrophys. J.* **152**, 233.
[62] Ledoux, P.: 1941, *Astrophys. J.* **94**, 537.
[63] Wilson, O. C.: 1948, *Astrophys. J.* **108**, 201; and Schwarzschild, M. and Härm, R.: 1959, *Astrophys. J.* **129**, 637.
[64] Van Helden, R.: 1966, *Bull. Astron. Inst. Netherl.* **18**, 367.
[65] Hutchings, J. B.: 1967, *Observatory* **87**, 289.

[66] Hutchings, J. B.: 1969, *Non-Periodic Phenomena in Variable Stars*, (ed. by L. Detre), Academic Press, Budapest p. 191.
[67] Oke, J. B.: 1954, *Astrophys. J.* **120**, 22.
[68] Thackeray, A. D.: 1937, *Astrophys. J.* **86**, 499.
[69] Merrill, P. W.: 1947, *Astrophys. J.* **106**, 274.
[70] Herbig, G. H.: 1969 (in press).

Discussion

Underhill: The M_v of a star is determined by the radius of the photosphere and the brightness temperature in the V band. To find a spectroscopic criterion for absolute magnitude one must find lines where strengths are sensitive to the size of the photosphere and the density of gas there. This type of line has not been isolated yet in O type spectra, thus one cannot separate bright Of stars from less bright Of stars. Certainly the 'f' characteristics do not permit such a separation.

Viotti: In the P Cygni-like star AG Car, there is also some evidence for an outward acceleration of matter. This is a very interesting object in the same celestial region of η Car, that would require a wide study in the visible and ultraviolet.

CHROMOSPHERIC ACTIVITY IN RED GIANTS,
AND RELATED PHENOMENA

ARMIN J. DEUTSCH[†]

Mount Wilson and Palomar Observatories,
Carnegie Institution of Washington,
California Institute of Technology

Abstract. Normal red giants of a given spectral type are shown to be heterogeneous with respect to the following chromospheric features: the Balmer absorption lines, the emission line at Hε, and the double-reversed emission lines at Ca II H and K. These chromospheric lines are also shown to be strongly time variable, in at least some red giants, on a time scale of a few months or years. Other chromospheric features that require study lie in the infrared (He I 10830), the near ultraviolet (Fe II emission lines), and the vacuum ultraviolet.

When one intercompares spectrograms of normal stars having spectral types and luminosities that are accurately the same, one often finds the intensities of certain lines to differ appreciably from one object to another. Similar differences also occur among stars having the same color and absolute magnitude. As is well known, abundance anomalies are responsible for a large part of this heterogeneity of spectra among stars having the same values of effective temperature T_e and surface gravity g. Similarly, in late-type stars with the same classification, chromospheric effects are also responsible for appreciable spectroscopic differences.

In Figure 1, the Balmer lines are shown on enlargements from 10 Å mm^{-1} spectrograms of two M2 III stars, λ Aqr and φ Aqr. In both objects, the lower Balmer lines are far stronger than those to be expected from the reversing layer, where the metallic lines originate. The magnitude of the anomaly can be easily estimated if we note that in the spectrum of λ Aqr Hγ ($\chi = 10.15$ eV) has an equivalent width slightly larger than does Fe I 4376 Å ($\chi = 0$). If we may take the lines to be of roughly equal strength, and if both lines lie on the same curve of growth, they will then have approximately the same abscissa, $\log \eta_0$. In the notation of Aller (1963), we have

$$\log \eta_0 = -1.824 + \log N + \log gf - \theta\chi - \log K_\lambda Vu(T).$$

Since $T_e = 3050°$ at M2 III, we may neglect ionization in both species. We may also neglect the relatively small differences in $\log \kappa_\lambda Vu(T)$. Then

$$\log \frac{N(\mathrm{H})}{N(\mathrm{Fe})} \simeq \theta\left[\chi(\lambda 4340) - \chi(\lambda 4376)\right] - \log \frac{(gf)_{\lambda 4340}}{(gf)_{\lambda 4376}} \simeq 14.6.$$

But in the Sun, $\log N(\mathrm{H})/N(\mathrm{Fe}) \simeq 5.4$ (Goldberg *et al.*, 1960). The calculation shows that if Hγ were formed in the reversing layer, its overabundance relative to iron would be 10^9 times! Alternatively, if the abundances are normal, the strength of 4340 Å would indicate $T_{\mathrm{exc}} = 6600°$, or more than double T_{eff}.

[†] Deceased.

Houziaux and Butler (eds.), Ultraviolet Stellar Spectra and Ground-Based Observations, 199–208.

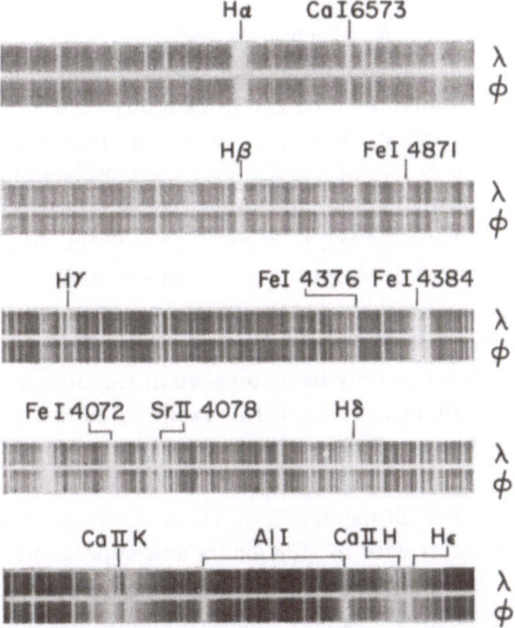

Fig. 1.   The Balmer lines in λ Aqr (M2 III) and φ Aqr (M2.5 III). In the top strip, the plates are Pc 10883 (λ, 1968, Dec. 11) and Pc 10884 (φ, 1968, Dec. 12). In the next three strips the plates are Pc 10873 (λ, 1968, Dec. 10) and Pc 10874 (φ, 1968, Dec. 10). In the bottom strip the plates are Pc 10960 (λ, 1969, Jan. 8) and Pc 10809 (φ, 1968, Oct. 14).

The Balmer lines in M-type giants therefore represent an extreme case of super-excitation, as was first noted by Adams and Russell (1928). In discussing the spectrum of α Ori (M2 Iab), Spitzer (1939) found excitation temperatures of 2100° for the strong iron lines and 17000° for the Balmer lines. Many subsequent studies of red giants and supergiants have also made it clear that the Balmer lines are formed in a high-temperature region which lies above the reversing layer, and which is the stellar analogue of the solar chromosphere.

The spectrograms reproduced in Figure 1 show that the Balmer decrement is far steeper in φ Aqr than in λ Aqr. In the former star, Hδ has nearly disappeared, and Hε has gone into conspicuous emission. This regularity suggests the possibility that emission from some parts of the stellar chromosphere 'fills in' the absorption lines from other parts, and that the emission actually 'overfills' the weak absorption line expected at Hε. As yet, no quantitative formulation of this hypothesis has been attempted. Wilson (1957) has illustrated the Hε emission line in the spectra of a number of late-type giants.

Figure 1 also shows that these two M giants differ systematically in the metallic lines that are on the damping part of the curve of growth. These are stronger in φ Aqr, the star with the weaker Balmer lines. The difference cannot be attributed to a simple error in spectral classification, for the TiO bands are very nearly equal in the two stars. Also, there is close agreement in the intensity ratio of the relatively weak lines at 4020.4 Å and 4121.8 Å, which are sensitive criteria of temperature (Deutsch *et al.*, 1969). One can nearly match the strong, damped lines of φ Aqr with those of the M4 III star 51 Gem (actually, the lines are still slightly stronger in the former object); but then the TiO bands are much weaker in φ Aqr. According to the catalogue of Johnson *et al.* (1966), the colors of λ and φ Aqr are closely similar, with λ slightly the cooler. The weakness of the metallic lines in λ Aqr was noted earlier by Keenan, in Deutsch *et al.* (1969); in this paper, he slightly revised the classification to M2.5 III.

From intercomparison of 10 Å mm^{-1} spectrograms of about 15 other bright giants near M2 III, it has now been established that differences commonly are found which are similar to those illustrated in Figure 1. Few regularities have yet emerged from these observations; indeed, one can only be astonished at the ubiquity, the variety, and the amplitude of the effects that can be seen. For example, in the spectrum of α Cet (M1.5 III) Hγ and the damped metallic lines are nearly the same as in λ Aqr; but Hε is an emission line in α Cet, nearly equal to the one in φ Aqr.

These anomalies are not altogether new. Thus, Kraft *et al.* (1964) found that the half-width of Hα correlates with M_v for giants and supergiants earlier than M0, but not for the M stars. Again, Wilson (1962) has found that in K-type dwarfs "...the hydrogen lines are frequently erratic in behavior and have intensities which would indicate a spectral type somewhat different from that derived from the metallic lines". He has also noted discrepancies in the bands of CN and SiH. Metallic lines that are *weak* for the type are well known, of course, in many late-type stars of the halo population and the disk; these metal-deficient stars occur among giants and dwarfs, alike. Recent studies, e.g. by Spinrad and Taylor (1969) and by Taylor (1969), have recently established the existence of numerous late-type stars, both giants and dwarfs, which have metallic lines *stronger than normal* and appear to be 'super-metal-rich'. The relations are still not clear between these spectroscopic anomalies and the ones that have now emerged among the M giants.

A degree of caution is necessary before attributing all these line-intensity anomalies to abundance irregularities. For the Balmer lines, recent work has established that

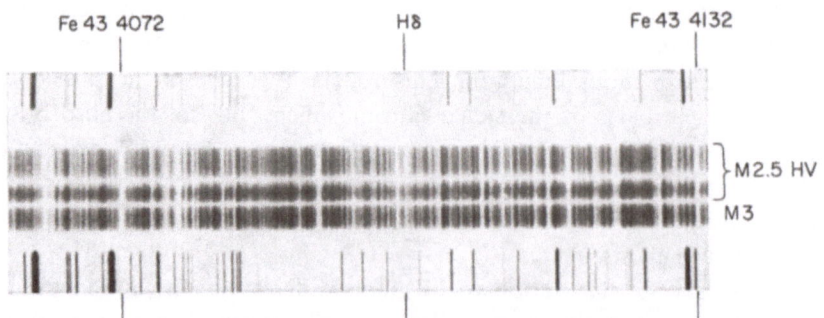

Fig. 2. Variation of Hδ in HR 6128 (M2.5 III). From top to bottom, the plates are Ce 13488 (1960, May 14), Pc 2477 (1956, Mar. 2), and Ce 11850 (1958, Apr. 9). The M3 III star is μ Gem. (Courtesy from *Astrophys. J.*, University of Chicago Press.)

the intensities are strongly variable with time in at least some of the M giants (Deutsch *et al.*, 1969). The spectra of Figure 2 show this effect. Similar changes of the Balmer lines have now been found in nearly half of the early M-type giants for which two or more suitable 10 Å mm^{-1} plates have been obtained during the last year. The time-scale of the changes is of the order of a few months. In a few stars – e.g. μ Gem (M3 III) and π Leo (M2 III) – the plates appear to show intensity variations of the damped metallic lines, as well. However, the changes in the metallic lines cannot yet be considered securely established.

Figure 1 shows the chromospheric components of Ca II H and K in λ and φ Aqr. These lines have also been illustrated in spectra of numerous late-type giants by Wilson and Bappu (1957). Deutsch (1960) has noted the occurrence of transitory chromospheric absorption components in some K-type giants, and of deep, non-variable circumstellar absorption components in most M-type giants. In Figure 1, the circumstellar lines are well shown in λ Aqr, but they cannot be seen in φ Aqr. The plates at hand suggest, but do not establish, that the profiles at H and K are generally uncorrelated with the strengths of the Balmer lines in K and M giants.

Following an unsuccessful search by Wilson (1954) for time-variations of the Ca II emission lines in a number of red giants, Griffin (1963) discovered that large changes are visible on higher-dispersion spectrograms of the K4p giant α Boo. Similar changes were subsequently found by Deutsch (1967a, b), Liller (1968), and Vaughan (1966) in other red giants; and by Wilson (1969) in a few red dwarfs.

Figure 3 shows the region of the K-line on four spectrograms of α Tau (K5 III).

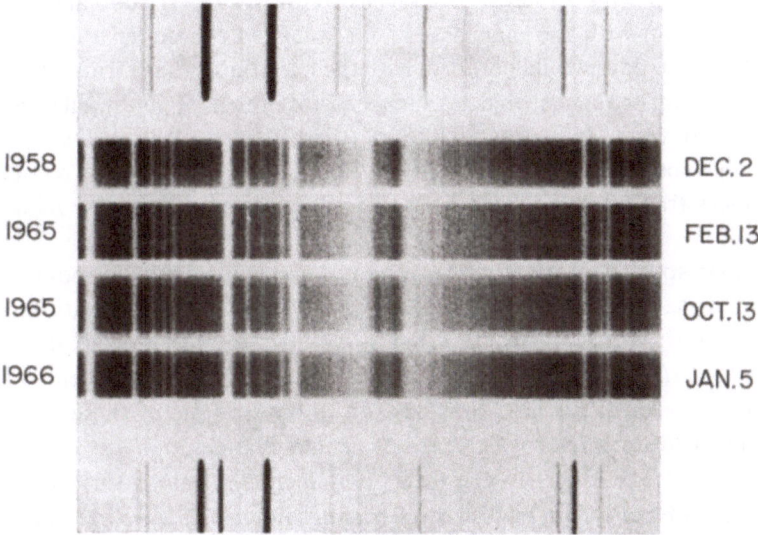

α Tauri K5 III

Fig. 3.   The K line in α Tau (K5 III). The two strongest comparison lines are 2.4 Å apart; Wilson and Bappu (1957) measured the width of K_2 as 1.13 Å. The spectrograms are Pb 4214, Pb 8522 Pb 9012 and Pb 9121.

Adopting the notation of solar spectroscopy, we may designate the wide absorption wings as $K_1 V$ (for violet) and $K_1 R$ (for red); the emission peaks as $K_2 V$ and $K_2 R$; and the central dip as K_3. These features change with time approximately in the way described by Liller for α Tau, and by Griffin for α Boo. As may be seen in Figure 3, the largest variations occur in the intensity of $K_2 V$ and in the depth of $K_1 V$.

The spectrograms enlarged in Figure 3 have a dispersion of 4.5 Å mm^{-1} and a resolution of about 0.12 Å. Similar plates have been obtained during the last 10 years for about 30 of the brightest red giants. The number of spectrograms suitably exposed at H and K now ranges downwards from about 40 for α Tau, to only two or three for the least-frequently observed stars. For about 10 of the stars observed, existing plates show conspicuous variations more or less like those shown in Figure 3. For about half the remaining stars observed, the plates show similar variations of smaller amplitude; the other half of these stars have shown no changes large enough for reliable detection. Probably large-amplitude changes will eventually be recognized in many or most of the stars where they have not yet been established. In this connection we may note that, on the 15 Palomar spectrograms obtained in the last 4 years, the star α Boo has exhibited only relatively small profile variations, although this is the star in which Griffin first demonstrated large changes in K_2.

The time scale of the changes seen at H and K in the Palomar spectrograms is of the order of a few weeks or months. The α Tau observations appear to show a cyclical phenomenon with a quasi-period of ~ 350 days. Since K_2 is modulated in the integrated solar spectrum by solar rotation (Bumba and Růžičková-Topolová, 1967; Sheeley, 1967), the possibility exists that 350 days is the rotation period of α Tau. The equatorial velocity of rotation would then be 3.6 km sec^{-1} for a K5 III star, and this velocity would be consistent with the sharpness of the weaker Fraunhofer lines. A rotational period of about 350 days would conform with the hypothesis that α Tau is a metamorph of an A-type main-sequence object which has lost most of its angular momentum through a stellar wind, in a process like that described by Wilson (1966) and Kraft (1967) for late-type dwarfs. Equally well, the 350-day period could be reconciled with an evolutionary track starting from the position of a dwarf star on the main sequence near F5 or G0.

The Palomar spectrograms of α Tau required exposure times of about 30 min, on the average. These plates therefore could not show the intensity fluctuations of $\sim 20\%$ in the peak intensity at $K_2 V$, which Liller sometimes found to occur in α Tau on a time scale of ~ 15 min. Some of the Palomar observations were repeated after time intervals of the order of 0.1 day, or 1 day, or 10 days. None of these closely spaced spectrograms yield unequivocal evidence of profile changes.

Microphotometer tracings (in the transmission mode) have been obtained for 23 of the spectrograms of α Tau, and on these tracings intensities have been determined at the intensity minima in $K_1 V$, K_3, and $K_1 R$; and at the maxima in $K_2 V$ and $K_2 R$. Similar measurements have also been made on tracings of 14 plates of γ Aql (K3 II), another star (Figure 4) that has exhibited high activity at H and K. On both sets of tracings, the intensity was taken as unity in a conventional 'continuum' which is well-

ARMIN J. DEUTSCH

the intensities are strongly variable with time in at least some of the M giants (Deutsch *et al.*, 1969). The spectra of Figure 2 show this effect. Similar changes of the Balmer lines have now been found in nearly half of the early M-type giants for which two or more suitable 10 Å mm^{-1} plates have been obtained during the last year. The time-scale of the changes is of the order of a few months. In a few stars – e.g. μ Gem (M3 III) and π Leo (M2 III) – the plates appear to show intensity variations of the damped metallic lines, as well. However, the changes in the metallic lines cannot yet be considered securely established.

Figure 1 shows the chromospheric components of Ca II H and K in λ and φ Aqr. These lines have also been illustrated in spectra of numerous late-type giants by Wilson and Bappu (1957). Deutsch (1960) has noted the occurrence of transitory chromospheric absorption components in some K-type giants, and of deep, non-variable circumstellar absorption components in most M-type giants. In Figure 1, the circumstellar lines are well shown in λ Aqr, but they cannot be seen in φ Aqr. The plates at hand suggest, but do not establish, that the profiles at H and K are generally uncorrelated with the strengths of the Balmer lines in K and M giants.

Following an unsuccessful search by Wilson (1954) for time-variations of the Ca II emission lines in a number of red giants, Griffin (1963) discovered that large changes are visible on higher-dispersion spectrograms of the K4p giant α Boo. Similar changes were subsequently found by Deutsch (1967a, b), Liller (1968), and Vaughan (1966) in other red giants; and by Wilson (1969) in a few red dwarfs.

Figure 3 shows the region of the K-line on four spectrograms of α Tau (K5 III).

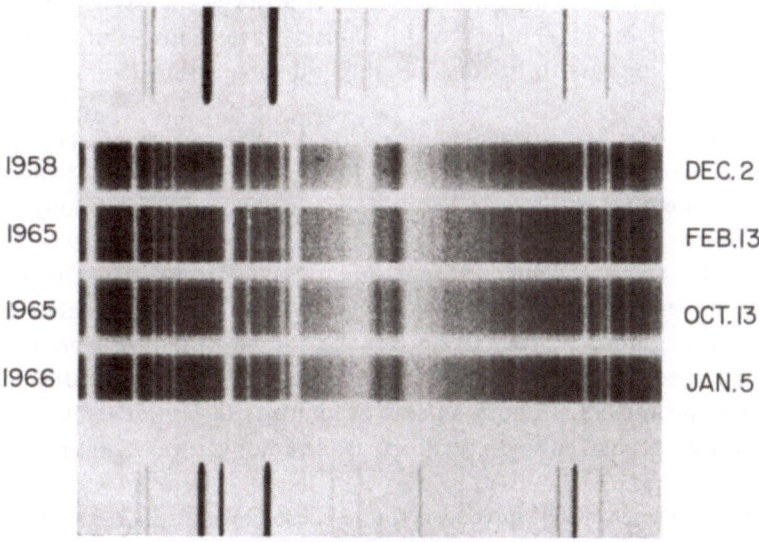

α Tauri K5 III

Fig. 3. The K line in α Tau (K5 III). The two strongest comparison lines are 2.4 Å apart; Wilson and Bappu (1957) measured the width of K_2 as 1.13 Å. The spectrograms are Pb 4214, Pb 8522 Pb 9012 and Pb 9121.

(1968) has pointed out. Together, these two effects can easily account for the prominence of K_2 in late-type giants relative to the weak emission in G2 dwarfs like the Sun. With respect to α Tau, e.g., one can show that if K_2 originates chiefly in chromospheric masses that have the same monochromatic surface brightness at K as do plages in the solar chromosphere, then these stellar plages need cover only a far smaller fraction of the photosphere than solar plages cover near sunspot maximum – about 20%, according to Sheeley (1967).

In many of the red giants observed at H and K, the Palomar spectrograms at 4.5 Å mm^{-1} show fine structure in the profile of K_2 and K_3, down to the resolution limit of the plates. This is about 0.12 Å, or 12% of width of K_2 in luminosity class III (Wilson and Bappu, 1957). Of course, it is still unknown what processes and structures in red-giant chromospheres are responsible for this fine structure. The intensity measures in the H and K profiles of α Tau and γ Aql confirm the visual impression that even the principal features of the chromospheric profiles are subject to complex variations, which are not easily subsumed by any simple description. In particular, one finds little correlation between intensities measured at different points in the profile.

The absorption line He I 10830 provides additional evidence for chromospheric activity in stars. Vaughan and Zirin (1968) have reported finding this feature in a 'substantial number' of G and K giants, and preferentially in those that have intense emission lines at H and K. Wilson and Aly (1956) have found the absorption line 5876 Å of He I in at least two of nine late-type stars which they examined. Both these He I lines are very weak or absent in integrated sunlight, but both are often conspicuous absorption lines over plages or along the borders of the chromospheric network. The infrared helium line is usually wider than other lines, as it is in the solar spectrum. In most G and K giants, it is Doppler shifted towards shorter wavelengths, and in some of these stars the intensity appears to be time variable, on some spectrograms the line going over into emission.

In the ground-accessible ultraviolet, near 3200 Å in the spectra of α Her (M5 II) and α Sco (M1 Ib), Herzberg (1948) found permitted emission lines arising from Fe II levels at excitation potentials of about 5 eV. In reporting these lines, he conjectured that "...if one were to investigate the solar spectrum in the far ultraviolet, where the intensity of the continuous spectrum corresponding to the temperature of the photosphere is negligible, it seems certain that one would find it to consist of coronal emission lines of rather high excitation. In stars like α Her the corona would be expected to have a much lower temperature than that of the Sun, so that Fe II lines (rather than Fe X) might become prominent. In addition, on account of the lower temperature of the photosphere, the coronal emission becomes visible at longer wavelengths than it would for the Sun."

Bidelman and Pyper (1963) have also found these chromospheric emission lines in various cool stars, including β Peg (M2 + II − III) and other objects of similar luminosity. In β Peg and in α Ori (M2 Iab), high-dispersion plates have shown that the Fe emission lines present wide and complex profiles that are reminiscent of the structure seen in H and K (Weymann, 1962). It comes as no surprise, therefore, to find that some of the Fe II emission lines are strongly time-variable. Figure 5 shows some effects of chromospheric activity in the near-ultraviolet spectrum of α Ori.

Few generalizations can yet be made about the behaviour of chromospheric features in stars of various types. From the data at hand, it appears that K giants show large profile variations at K_2 more often than M's. As compared with the G and K giants, the M's also have less intense lines at 10830 Å, if any. However circumstellar absorption lines appear in most M-giant spectra, and not in the K's. At a given spectral type, rough correlations appear to exist between the intensities of chromospheric lines arising from different elements. However, the time-variations seem not to be syn-

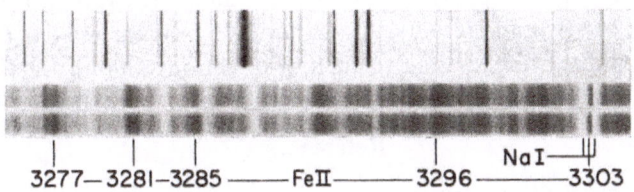

Fig. 5.   Part of the near-ultraviolet spectrum of α Ori (M2 Iab). The spectrograms are Ce 14132[1]
(1960, Dec. 31) and Pb 8283 (1964, Oct. 22).

chronous in Ca II and H, e.g., as we would expect them to be if the lines arise together in long-lived plages which rigid stellar rotation carries across the visible hemisphere.

The possibility should not be disregarded that related time-variations occur in the reversing layers and photospheres of some red giants. We have already cited some evidence for such effects in the profiles of strong damped metallic lines. Very recently, Eggen (1969) has shown that, among red giants of a given spectral type, there is a considerable temperature range, as indicated by the red-infrared colors. Wilson (1962) has demonstrated similar effects in the red dwarfs. In Eggen's report, HR6128 (Figure 2) appears to be the coolest of the four M2 III stars he has observed – as cool, to judge from the color index, as some M4 III stars. In addition, Eggen confirms an early result by Stebbins and Huffer (1930) that light variation occurs, with amplitude greater than 0.^{m}05, in virtually all giants having black body color temperatures lower than 3400° (spectral type about M2 III). Since colors and spectra are not usually observed simultaneously, time-variations in both quantities may account for a significant part of the dispersion found in the correlations between them.

In all probability the chromospheres of red giants are no more nearly homogeneous than is that of the Sun. In a given star, temperatures, densities, velocities, and magnetic fields are likely to vary appreciably with location and with time. Arguing by analogy with the Sun wherever this is possible, we may nevertheless hope to discover the mean structure of these atmospheric layers; the scales of their irregularities; and the nature of their coupling with the reversing layer below and the circumstellar wind (corona?) above.

For this program, our observational data are still very fragmentary. It is clear that much remains to be done in ground-based spectroscopy and photometry, to elucidate the apparent heterogeneity of these cool stars. In addition, we may now look ahead to observations in the vacuum ultraviolet, of a kind that will facilitate the comparison of stellar chromospheres with the solar chromosphere. The strongest chromospheric emission line in the Sun is Lyman α, with a mean flux at 1 AU of ~ 5 erg cm^{-2} sec^{-1} (Tousey, 1967). At a distance of 10 pc, this flux would amount to ~ 0.1 photons cm^{-2} sec^{-1}. This is the same order of magnitude as the faintest ultraviolet fluxes that have been measured in OB stars by the Wisconsin experiment in the Orbiting Astronomical Observatory (Bless, 1969). In the brightest red giants, the Lyman α flux ought to be of the order of ~ 10 photons cm^{-2} sec^{-1}; and there is the realistic prospect of observing in them a number of the other strong chromospheric lines that characterize active regions of the solar chromosphere. These would include the MgII D-lines near 2800 Å, CII 1335 Å, HeI 584 Å, and HeII 304 Å. Scattering by the interstellar medium should represent no significant impediment to such observations among stars that lie within 100 pc.

References

Adams, W. S. and Russell, H. N.: 1928, *Astrophys. J.* **68**, 9.

Aller, L. H.: 1963, *Astrophysics: The Atmospheres of the Sun and Stars*, second edition, The Ronald Press Company, New York, p. 377.

Bidelman, W. P. and Pyper, D. M.: 1963, *Pub. Astron. Soc. Pacific* **75**, 389.

Bless, R. C.: 1969, private communication.

Bumba, V. and Růžičková-Topolová, B.: 1967, *Solar Phys.* **1**, 216.

Deutsch, A. J.: 1960, in *Stars and Stellar Systems*, Vol VI: *Stellar Atmospheres* (ed. by J. L. Greenstein), University of Chicago Press, Chicago, p. 543.

Deutsch, A. J.: 1967a, Reported by H. W. Babcock in *Carnegie Inst. Yearbook* **65**, 139.

Deutsch, A. J.: 1967b, *Publ. Astron. Soc. Pacific* **79**, 431.

Deutsch, A. J., Wilson, O. C., and Keenan, P. C.: 1969, *Astrophys. J.* **156**, 107.

Eggen, O.: 1969, *Konkoly Bull.* No. 355.

Goldberg, L., Müller, E. A., and Aller, L. H.: 1960, *Astrophys. J. Suppl. Ser.* **5**, 1.

Griffin, R. F.: 1963, *Observatory* **83**, 255.

Herzberg, G.: 1948, *Astrophys. J.* **107**, 94.

Johnson, H. L., Mitchell, R. I., Iriarte, B., and Wisniewski, W. Z.: 1966, *Commun. Lunar and Planetary Lab.* **4**, 99.

Kraft, R. P.: 1967, *Astrophys. J.* **150**, 551.

Kraft, R. P., Preston, G. W., and Wolff, S. C.: 1964, *Astrophys. J.* **140**, 235.

Liller, W.: 1967, private communication.

Liller, W.: 1968, *Astrophys. J.* **151**, 589.

Linsky, J. L.: 1968, Smithsonian Astrophys. Obs., *Special Report* 274.

Sheeley, N. R., Jr.: 1967, *Astrophys. J.* **147**, 1106.

Spinrad, H. and Taylor, B. J.: 1969, in preparation. (abstract) 1967, *Astr. J.* **72**, 320.

Spitzer, L. Jr.: 1939, *Astrophys. J.* **90**, 494.

Stebbins, J. and Huffer, C. M.: 1930, *Publ. Washburn Obs.* **15**, 139.

Taylor, B. J.: 1969, in preparation.

Tousey, R.: 1967, *Astrophys. J.* **149**, 239.

Vaughan, A. H., Jr.: 1966, private communication.

Vaughan, A. H., Jr. and Zirin, H.: 1968, *Astrophys. J.* **152**, 123.

Weymann, R. J.: 1962, *Astrophys. J.* **136**, 844.

Wilson, O. C.: 1954, 'The Atmospheres of Giant and Supergiant Stars', in *Proc. Nat. Science Foundation Conference on Stellar Atmospheres*, Indiana University, p. 147.

Wilson, O. C.: 1957, *Astrophys. J.* **126**, 46.

Wilson, O. C.: 1962, *Astrophys. J.* **136**, 793.

Wilson, O. C.: 1966, *Astrophys. J.* **144**, 695.

Wilson, O. C.: 1969, in *Low-Luminosity Stars* (ed. by S. S. Kumar), Gordon and Breach, London,
    p. 103.
Wilson, O. C. and Aly, M. K.: 1956, *Publ. Astron. Soc. Pacific* **68**, 149.
Wilson, O. C. and Bappu, M. K. V.: 1957, *Astrophys. J.* **125**, 661.

Discussion

Severny: Do you not think that the effect of asymmetry of emission in the K-line that you found in
late type-stars could be ascribed to the St. John effect in flocculae which is connected with the pre-
dominant downward motions?

Deutsch: The velocities associated with the flow in the chromospheric network are only 1 or 2 km/sec
in the Sun. But the width of K_2 is ~ 35 km sec^{-1}. It would therefore suppose that this explanation is
not possible, even in the Sun. On Aldebaran, the width of K_2 is nearly 70 km/sec, and velocities as
large as this would be quite unexpected.

Feast: Do you know anything about possible light variations in the two M giants that you con-
trasted?

Deutsch: No. (But any possible light variations would have to be small – less than 0.3 magnitudes,
in all probability.)

Glushneva: Some years ago Essipov and myself (Sternberg Institute, Moscow) observed the emission
line He I 10830 in the spectrum of Algol. It appeared during the primary minimum of Algol. But it
was absent during our observations of Algol in the primary minimum the next year. It is probable that
this line could have a chromospheric origin.

Henize: When we are studying mass ejection from high luminosity stars, it is of interest to consider
those extreme examples in which the ejected gas has formed a visible nebulosity. I would like to call
attention to three stars in the southern hemisphere which probably fall in this category. These are (1)
the Of central star of NGC 6164–65 which Dr Feast has already mentioned, (2) AG Car, a P Cygni-like
star which is the central star of a ring planetary nebula, and (3) HD 88 643 which shows a nova-like
spectrum, even though there has been little change in brightness in 50 years, and which lies in a
nebulosity the morphology of which suggests that it was ejected from the star. It would be very
interesting to have UV spectra of each of these stars.

Gingerich: On Griffins' atlas of the Arcturus spectrum, Hε is very nicely in emission even though the
other Balmer lines are in absorption. This arises because the combined opacity of the Ca$^+$ and Hε
lines move the depth of formation out into the chromosphere. Does the Hε emission show a correlation with the Wilson-Bappu effect?

Deutsch: I think not. But Wilson has found that the strength of K_2 correlates loosely with the
strength of Hε. My places show that K_2 can vary with time, while Hε shows no appreciable change.

MASS LOSS FROM EARLY-TYPE STARS

J. B. HUTCHINGS

Dominion Astrophysical Observatory, Victoria, B.C., Canada

Abstract. Following the detailed study of four very high luminosity OB stars, a survey has been made for spectroscopic evidence of mass loss in a number of early-type supergiants. A list of spectroscopic criteria is given and the mass loss estimates for 24 stars plotted on the HR diagram. The dependence of the phenomenon on spectral type and luminosity is discussed as well as its significance in terms of stellar evolution.

Since Feast has mentioned the results of the analysis of three very luminous early-type stars showing evidence for extended accelerating envelopes and since the details have been published (Hutchings, 1968), I shall not go over this work again, but rather report on a less detailed survey of high luminosity early-type stars in the northern sky, to try to find how general the phenomenon of mass loss is from such stars. As a background to this discussion I shall just say very briefly that the earlier work was based on a detailed analysis of individual line velocities and profiles which was used in conjunction with theoretical work (Wellman, 1951; Menzel, 1937) on the behaviour of HeI and H lines in moderate radiation dilution. It was possible to relate certain lines (and thus expansion velocities) with radiation dilution ranges and hence build up a crude picture of an expanding envelope. This is used as the starting-point for an iterative method of computing strong line profiles, which converges to a structure of

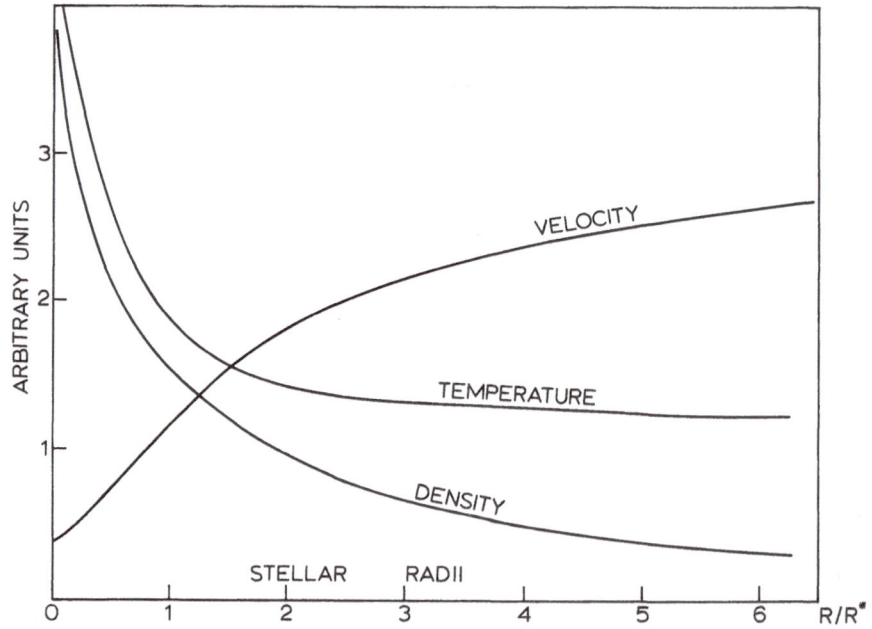

Fig. 1.   Typical structure of OB supergiant outer atmosphere.

Houziaux and Butler (eds.), Ultraviolet Stellar Spectra and Ground-Based Observations 209–212.
All Rights Reserved. Copyright © 1970 by the IAU.

which a typical example is shown in Figure 1. All the stars show this structure with wide variation in the maximum velocity, density and extent of the envelope. Mass loss rates for stars of spectral type about B0 and $M_v \sim -7.5$ are about $10^{-5}\ M_\odot\ \mathrm{yr}^{-1}$. In addition to the normal spectrographic evidence used in these investigations we must bear in mind the now well-known observations of the ultraviolet resonance lines which indicate velocities of expansion of some thousands of km sec^{-1} in stars whose mass loss rates are much lower than those referred to above. We will discuss the ultraviolet lines in more detail in the next session.

TABLE I

Expanding atmosphere criteria

(1) Hα emission present. (UV resonance line shift and emission?)
(2) Hβ emission present, Heɪ 5875 Å emission present.
(3) Balmer velocity progression.
(4) Velocity-excitation relation present.
(5) Heɪ 4471 Å, Mgɪɪ 4481 Å, Heɪ 4026 Å, Cɪɪ 4267 Å velocities separate out.
(6) Heɪ 3888 Å separates from H8.
(7) Hγ emission, other Heɪ emission, further emission lines.

TABLE II

Mass loss strength for stars studied

Star H.D.	Type	M_v	Mass loss
7583	A0Ia	−7.1	(3)
14134	B3Ia	−6.8	3
14143	B2Ia	−7.4	(3)
14818	B2Ia	−6.4	(3)
24398	B1Ib	−5.8	0
33579	A3Ia0	−9.1	(5)
36486	O9.5II	−5.8	½
37128	B0Ia	−6.6	1
37742	O9.5Ib	−6.2	1
41117	B2Ia	−7.3	4
45314	O9pe	−6.8	(4)
91316	B1Ib	−5.9	0
151804	O9f	−7.2	6
152236	B1Ia0	−8.0	7
152408	O8f	−7.1	7
164353	B5Ib	−5.2	0
169454	B1Ia+	−7.7	5
183143	B7Ia	−7.5	2
188001	O8f	−6.5	4
190603	B1.5Ia	−7.5	6
193237	B1p	−8.4	7+
198478	B3Ia	−7.2	2
210839	O6p	−6.8	(2)
223385	A3Ia	−8.6	(3)

From the earlier investigations and the present one it became evident that there is an increase in types of spectroscopic evidence for mass loss of this nature as the phenomenon becomes more marked, i.e. as the density and expansion velocity of the envelope increase. Thus it is possible to compile a list of observable effects in approximate order, as shown in Table I.

The phenomena 3–5 do not necessarily appear in the order given as they depend on a number of factors in the atmospheric structure. However, a good guide to the strength of mass loss is simply the number of these criteria which show up in any given star after a careful analysis has been made of several plates.

Before mentioning specific results it should also be noted that the stars showing these effects all appear to be time variable so that criteria vary in strength, and velocities change irregularly. It is thus desirable to have several well spaced observations of each star.

A survey was made on plates of dispersion 15 Å mm^{-1} or better on stars selected for high luminosity and early type, and detailed measurements revealed mass loss evidence in many of them.

A careful assessment was made of the M_v of each star, using Hα calibration (Andrews, 1968), Hγ (Petrie, 1965; Walker and Hodge, 1968), Hγ (Hutchings, 1966; Kopylov, 1958) and Si III (Hutchings, 1967), and Table II shows the results for all the stars.

Figure 2 shows a plot on the H-R diagram of these stars. It is evident that the mass

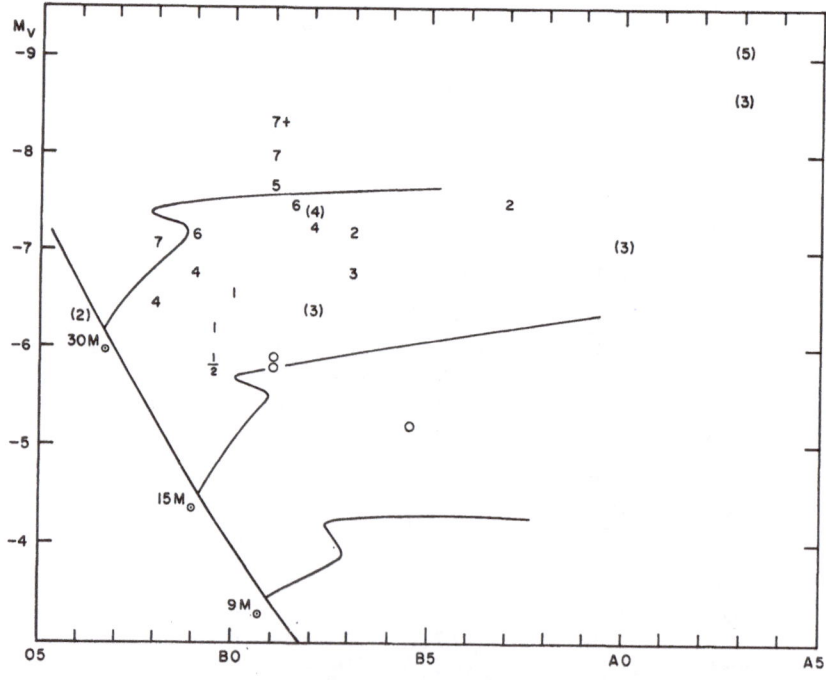

Fig. 2.   H-R diagram showing mass loss criteria.

loss depends on both luminosity and spectral type among the OB stars. The bracketed figures are those derived from less reliable plates, or the A stars, where extrapolation of the interpretative arguments for OB stars may be questionable.

The detailed models made so far indicate a mass loss rate of $10^{-5}\,M_\odot\,\mathrm{yr}^{-1}$ or greater for stars of some 30 $M_\odot$, and 10^{-7}–$10^{-6}\,M_\odot\,\mathrm{yr}^{-1}$ for 15 $M_\odot$ stars. No detailed work has yet been done on A stars and this is now being undertaken. It is evident that the mass lost may be a large portion of the total for high mass stars and therefore the problem has important stellar evolutionary consequences. In particular it is important to know whether mass loss begins before hydrogen core burning ends, as this may affect the stars' evolutionary track markedly (Hartwick, 1967). It is also important to see whether the ultraviolet data increase in strength as strikingly as the visual, as this may provide evidence on the mass loss processes. Such scans have been made with the present Orbiting Astronomical Observatory.

References

Andrews, P. J.: 1968, *Mem. Roy. Astron. Soc.* **72**, 35.
Hartwick, F. D. A.: 1967, *Astrophys. J.* **150**, 953.
Hutchings, J. B.: 1966, *Monthly Notices Roy. Astron. Soc.* **132**, 433.
Hutchings, J. B.: 1967, Thesis, Cambridge University.
Hutchings, J. B.: 1968, *Monthly Notices Roy. Astron. Soc.* **141**, 219 and 329.
Kopylov, I. M.: 1958, *Izv. Krymsk. Astrofiz. Observ.* **20**, 156.
Menzel, D. H.: 1937, *Astrophys. J.* **85**, 330.
Petrie, R. M.: 1965, *Publ. Dominion Astrophys. Obs.* **12**, 317.
Walker, G. A. H. and Hodge, S. M.: 1968, *Publ. Astron. Soc. Pacific*, **80**, 290.
Wellman, P.: 1951, *Z. Astrophys.* **30**, 71 and 88.

Discussion

Conti: Only one star in your diagram of 'Mass loss' is near the main sequence; which star is this?

Hutchings: It is an O6 star, HD 210839, for which I have slightly inferior data at present.

Peytremann: Which is the mechanism responsible for the mass loss? Is it radiation pressure or rotation of the star?

Hutchings: As discussions in the next session should show, it looks like radiation pressure. The detailed analysis I have done on a few stars indicate a spherically symmetrical envelope and a rotational velocity of 80 km/sec or less. There is no indication of the type of profile found in the equatorially extended atmospheres around fast rotating main sequence Be stars.

C. THEORY RELEVANT TO UV SPECTRA

A DISCUSSION OF THE THEORY FOR INTERPRETING
ULTRAVIOLET STELLAR SPECTRA

ANNE B. UNDERHILL

The Astronomical Institute, Utrecht, The Netherlands

Abstract. The observed spectral distribution from a star will, in principle, give information about the parts of the atmosphere from which the radiation comes. Most detailed theoretical studies assume that the atmosphere may be represented by plane parallel layers, but in the case of very strong absorption lines and weak continuous absorption, a case which is encountered in the ultraviolet spectral region of B stars, the radiation received from the centers of the lines may come from an entirely different part of the star than does that from the far wings of the lines or from the continuous spectrum. In the case of strong lines, because stars are three dimensional objects and because a field of motion may occur in the outermost part of the atmosphere, the appearance of the ultraviolet line spectrum may be poorly predicted using the hypothesis of stationary plane parallel layers.

Some numerical examples are presented demonstrating that with the UV resonance lines the opacity in the centre of a line may exceed the continuous opacity by a factor 10^6–10^8. A summary is given of the chief factors which should be taken into account in any theory of line formation when the hypothesis of LTE is not valid. A detailed description of the interactions between radiation and matter which can occur is necessary. Some examples of the distribution in energy of the lower energy levels of the ion are presented for typical ions of interest. This distribution and the presence of metastable levels are important factors in determining the significance of non-LTE physics. The paper concludes with remarks indicating that ultraviolet spectra of stars should be obtained with a spectral purity of at east 0.1 Å if progress is to be made in understanding the physical state of stars.

1. Introduction

Information about the physical conditions in stellar atmospheres is found by comparing observed stellar spectra with spectra predicted using a model of the stellar atmosphere and a theory of the process of line formation. The model is usually specified by its composition, effective temperature and $\log g$. The shape and strengths of the predicted line profiles depend upon the assumed abundance of the carrier atom (ion) and the type of expression used to represent the interactions between radiation and matter which lead to the formation of the stellar absorption or emission lines. When satisfactory agreement has been obtained between predicted and observed profiles and line strengths, the star is said to be like the model. There is no guarantee that a model which explains some spectroscopic details successfully will explain the full spectrum of the star.

It is convenient to divide the UV spectral region of B stars into two parts:

(i) The far ultraviolet which lies between 911.6 Å (the Lyman limit) and 1900 Å.

(ii) The near ultraviolet which lies between 1900 Å and 3100 Å, the effective ozone cut-off.

Many resonance lines and strong lines from low-lying levels in the first, second, third and fourth spectra of the light elements occur in the far ultraviolet region and here the continuous opacity due to H, He I and He II is significantly smaller than that

Houziaux and Butler (eds.), Ultraviolet Stellar Spectra and Ground-Based Observations, 215–225.

in the blue-violet spectral region. The near ultraviolet region contains the resonance lines of Mg II and Fe II and it is full of lines, mostly subordinate, from the second and third spectra of the metals and light elements. Here the continuous opacity is about the same as in the normally observed blue-violet spectral region. This means that a model which gives a satisfactory representation of the continuous spectrum in the blue-violet region will also represent well the continuous spectrum in the near ultraviolet region. The only spectra from relatively abundant atoms which do not have resonance lines at wavelengths longer than 911.6 Å are He I, He II, O II, O III, Ne I and Ne II. It is expected that interstellar absorption in the Lyman continuum will prevent observation of the spectra of B stars at wavelengths shorter than 911. 6 Å.

2. Geometric Considerations

The simplest configuration is plane parallel layers in hydrostatic equilibrium which is the configuration for which most detailed studies have been made. The examples quoted here are for this case.

The part of the atmosphere most important for determining the value of the emergent flux F_v at any frequency is that lying outside the geometric level z_1, where

$$t_v = - \int_{z_1}^{\infty} (\kappa_v + l_v + \sigma)\, \varrho dz \approx 0.4 . \tag{1}$$

This is an empirical relationship which can be derived from the quadrature formulas used to calculate the emergent flux from model atmospheres when spectrum formation takes place as though LTE exists. In this expression the geometric coordinate is measured outward from the center of the star, the total continuous absorption coefficient at frequency v is represented by κ_v, the total absorption coefficient due to line sources of opacity by l_v, and the electron scattering coefficient by σ. These coefficients are per unit mass; the density is represented by ϱ. At far ultraviolet wavelengths the continuous opacity κ_v is small and z_1 will lie deep in the atmosphere for fluxes in the continuum. On the other hand, near the centre of strong lines l_v is very large and z_1 will lie at a high level in the atmosphere.

These considerations give one a feeling for which layers of a model atmosphere one observes when one studies the emergent spectrum in different wavelengths. From the characteristics of the model one can estimate the particle density and electron temperature in the significant layers for each spectral feature and from these quantities one can check whether the exploratory hypothesis of LTE is likely to be valid for understanding the spectral details in question. To demonstrate what is involved, Table I gives the monochromatic optical depth at the centers of the C II, C III and C IV lines expected to be strong, at the centre of Hγ and in the continuous spectrum at 1458 Å and at 4340.5 Å for layers in three model atmospheres at which the characteristic optical depth, τ, is 0.001, 0.01, 0.10 and 0.50. The selected models represent fairly well main-sequence stars of types B6, B1.5 and O9 in so far as the normally observed con-

TABLE I

Monochromatic optical depths in the centres of some lines and in the continuum

τ	Wavelength (Å)	B6 V[a]		B1.5 V[b]		O9 V[c]	
0.001	C$_{II}$ 1335.3	.2255	$+5$	–		.4040	$+1$
	C$_{III}$ 1175.8	.3505	-3	–		.3378	$+2$
	C$_{IV}$ 1550.8	.5581	-3	–		.1280	$+5$
	cont. 1458.0	.4996	-3	–		.1016	-2
	Hγ 4340.5	.1776	$+2$	–		.3061	-0
	cont. 4340.5	.2696	-2	–		.1022	-2
0.01	C$_{II}$ 1335.3	.1643	$+6$	.1814	$+6$	.2040	$+2$
	C$_{III}$ 1175.8	.3249	-2	.1046	$+3$	.6355	$+3$
	C$_{IV}$ 1550.8	.5487	-2	.2714	-1	.1170	$+6$
	cont. 1458.0	.4848	-2	.6508	-2	.8764	-2
	Hγ 4340.5	.1214	$+3$	.3768	$+2$	.3668	$+1$
	cont. 4340.5	.2899	-1	.1066	-1	.9739	-2
0.10	C$_{II}$ 1335.3	.8476	$+6$	.4913	$+6$	.2804	$+3$
	C$_{III}$ 1175.8	.2861	-1	.2719	$+4$	.5784	$+4$
	C$_{IV}$ 1550.8	.5153	-1	.5284	$+2$	.7578	$+6$
	cont. 1458.0	.4471	-1	.7207	-1	.7161	-1
	Hγ 4340.5	.5194	$+3$	.2170	$+3$	.4623	$+2$
	cont. 4340.5	.4590	$+0$	.1586	-0	.1206	-0
0.50	C$_{II}$ 1335.3	.3054	$+7$	.6833	$+6$	.5045	$+3$
	C$_{III}$ 1175.8	.4428	$+1$	.1234	$+5$	.1111	$+5$
	C$_{IV}$ 1550.8	.2801	-0	.1782	$+5$	.2853	$+7$
	cont. 1458.0	.2452	-0	.3799	-0	.3289	-0
	Hγ 4340.5	.1426	$+4$	.6216	$+3$	.1898	$+3$
	cont. 4340.5	.4377	$+1$	.1097	$+1$	.8028	$+0$

[a] Represented by model P13, see Underhill (1968b).
[b] Represented by model B13, see Guillaume (1966).
[c] Represented by model PPB40, see Underhill (1968a).

tinuous spectrum is concerned. Line blanketing has been taken into account for the B1.5 and O9 models. The monochromatic optical depths have been calculated from an assumed fractional abundance of carbon by weight of 3.25×10^{-3}; the fractional abundance of hydrogen is 0.68, that of helium is 0.32. The degree of ionization and excitation has been calculated by the Saha and Boltzmann laws.

In all three models and at all levels considered, the monochromatic optical depth at the centre of Hγ is at least 100 times that in the continuum at the same wavelength. This ratio is typical for strong lines in the normally observed spectral region. Since models are constructed to represent the continuous spectrum well in the normally observed spectrum region, they usually are made in insufficient detail to permit one to calculate the centre of Hγ accurately. This is certainly true for late B type models. Furthermore in the outermost layers of models the electron density is sufficiently low, usually less than 10^{12}, that the assumption of the Saha and Boltzmann laws is of doubtful validity.

In the ultraviolet region the continuous absorption coefficient is small with the result that layers near or deeper than $\tau = 0.5$ are most significant for forming the continuous spectrum. On the other hand, in the centres of resonance lines the monochromatic opacity may become very large, some 10^6–10^8 times that in the neighbouring continuous spectrum; in many cases the model may be opaque in the line before one has traversed a significant opacity in any continuum frequency. In that case the model defined by properties of the continuous spectrum and so identified with a star is probably not at all relevant for predicting the profile of the absorption line. One is observing an entirely different source in the resonance line from what one is observing in the normally observed parts of the spectrum.

The numerical results shown in Table I, taken at their face value, suggest that the Cii lines at 1335 Å will be very strong in B6 stars, but the Ciii and Civ lines will be weak. At type B1.5 one expects all three groups of lines to be strong, but the Cii lines will be stronger than Ciii or Civ. At type O9 the pattern will be reversed, the Civ lines being strongest. Such predictions are of little real meaning because (a) the models are not defined in sufficient detail to represent accurately the outermost layers of main-sequence B type stars, and (b) the adopted LTE theory is certainly not valid in the outermost layers of these stars, i.e. in the region where the resonance lines are formed.

When the ratio $l_\nu/(\kappa_\nu + l_\nu + \sigma)$ is very large, the hypothesis of plane parallel layers becomes of doubtful validity and one should take into account the fact that a star is a three-dimensional object. The observed radiation in the centre of a strong line will give information about physical conditions in an outer spherical shell which may have a radius significantly larger than that of the photosphere which produces the continuous radiation. If there is a differential field of motion in the extended spherical atmosphere, for instance a uniform expansion, the spectral lines will have a P Cygni shape consisting of an emission component flanked by a shortward displaced absorption component. Simple examples of typical P Cygni line profiles were first predicted by Beals (1929, 1934); recent applications of this type of geometric theory to interpreting the spectra of supergiants have been made by Hutchings (1968). In a purely geometric theory usually a very simple assumption is made about the source function at each part in the atmosphere; often this is that the source function is the Planck function at an assigned local temperature.

3. The Physics of Line Formation

For the following discussion the geometry of the model is plane parallel layers. It is assumed that the distribution of the atoms over their various possible stages of ionization and over the individual energy levels of each stage of ionization may be found from the requirement that statistical equilibrium exist. This constraint is represented by a set of simultaneous equations, one for each state of the atom, expressing the condition that the number of particles entering each energy level per unit time is equal to the number leaving. The processes which cause transfer of particles from one level to another are collisions (chiefly with electrons), spontaneous emission of radiation, impressed emission, photoexcitation, photoionization and recombination. To obtain a quantitative measure of the effect of these processes, the collision and radiative rates must be known. These data are provided by atomic physics. The collision rates depend upon the density of particles; some of the radiative rates depend upon the photon density.

In principle each atom and ion possesses an infinite number of levels; in practice the number of levels is limited by Stark broadening and the consequent disappearance of high levels owing to perturbations by the surrounding electrons and ions. In addition, the number of levels which must be considered separately can be restricted by realising that levels close to the continuum will have Boltzmann populations relative to the continuum population owing to the dominance of collisional processes between these high-lying levels. The art of handling the radiative transfer process in stellar atmospheres is to find out in the relevent situation to how low a level one may use the Boltzmann law without seriously misrepresenting the physics of spectral line formation. Since the Boltzmann law is an explicit formula which depends only on the local electron temperature, its use simplifies the calculation of level populations.

The transfer of radiation of frequency v in a stellar atmosphere consisting of plane parallel layers may be represented by the equation

$$\frac{\mu dI(t, \vartheta, v)}{dt} = \frac{(\kappa_v + l_0 \varphi_v + \sigma)}{\kappa} I(t, \vartheta, v) - S(t).$$ (2)

Here the monochromatic line absorption coefficient has been written as $l_0 \varphi_v$ where l_0 is an appropriate combination of physical constants and the population, n_1, of the energy level (denoted by the subscript 1) from which the absorption line arises and φ_v is a normalised shape function.

Thus

$$\int_0^\infty \varphi_v dv \equiv 1$$ (3)

at each point in the atmosphere. In general φ_v is a function of the local electron temperature and density. A characteristic opacity coefficient κ has been selected to define a characteristic optical depth, t, by which to measure geometric depth in the atmosphere.

By definition, the source function $S(t)$ is the ratio of the emissivity at frequency v to the characteristic opacity coefficient used to define the depth scale. The line source function is found by considering in detail what can happen to a photon which is absorbed and excites an atom from the lower level of the transition being considered, to level 2, the upper level of the line.

It is not appropriate here to go deeply into the details of the theory for the resonance and other intrinsically strong lines which occur in the ultraviolet spectral region. These problems of physics and of numerical analysis have engaged the attention of a number of workers in recent years. The notation of Hummer (1962, 1964, 1968) and of Avrett and Hummer (1965) will be followed here, with the change that the quantity they call ε

will be called λ in order to reduce confusion with a related quantity called ε by Thomas (1957, 1965) and by Jefferies and Thomas (1958). According to Avrett and Hummer it is useful to define a quantity λ ($\equiv \varepsilon$) which gives the probability that a photon in line frequencies is absorbed and then disappears. The disappearance may be due to collisional de-excitation of the atom or to radiative de-excitation via another line. For resonance lines or for strong subordinate lines from low-lying levels when the line-formation process takes place in layers of the stellar atmosphere which are not particularly dense, the particle density being of the order of 10^{11}–10^{13}, λ may take values of the order of 10^{-4}–10^{-8}.

Let us ignore, for simplicity, the continuous opacity κ_v and the opacity due to electron scattering, σ. Then one can write the source function in line frequencies as

$$S_L(t) = \{1 - \lambda(t)\} \int_0^\infty \varphi_v(t) \, J_v(t) \, \mathrm{d}v + \lambda^*(t) \, B_v(t). \tag{4}$$

Here

$$\mathrm{d}t = -l_0\varrho(z) \, \mathrm{d}z, \tag{5}$$

where $\varrho(z)$ is the density and $\lambda^*(t)B_v(t)$ is the contribution to the line source function arising from the excitation of atoms to level 2 by all other processes than photoexcitation from level 1. The second term on the right side of Equation (5) is written as a coefficient, λ^*, multiplied by the Planck function $B_v(t)$ in order to stress the formal resemblance between Equation (4) and the forms for the source function which are obtained when the line-formation problem is simplified by assuming either that the line-formation process is one of coherent, isotropic scattering or that it takes place as if local thermodynamic equilibrium, LTE, existed at each point in the atmosphere. In each case the Saha and Boltzmann equations are used to estimate the level populations.

The first term on the right side of Equation (4) represents non-coherent scattering with complete redistribution, while the second term is equivalent to using Kirchhoff's law to estimate the emissivity. Avrett and Hummer (1965) and Hummer (1968) have studied in detail how to solve a transfer equation in which the source function has a form like Equation (4) and they have given some results for isothermal finite atmospheres and for isothermal semi-infinite atmospheres. A result of importance for our discussion which may be extracted from their work is that in semi-infinite atmospheres strong absorption lines from low-lying or ground levels will have significantly deeper cores than would be predicted using the hypothesis of LTE, the number of atoms in level 1 being the same in both cases. Thus it may be expected that when an improved theory of line formation is used, stronger lines will be found than are found using the hypothesis of LTE.

In the case of line formation in a fairly dense atmosphere, the particle density being of the order of 10^{15}–10^{16}, the theory of line formation sketched above goes over to the well known, simplified theory of line formation obtained when the hypothesis of LTE is made. Then $\lambda \to 1$ and $\lambda^* \to 1$. Whether one is justified in adopting a simple LTE theory or not depends upon the characteristics of the atom, that is upon the collision cross-sections for transfer to and from the levels of interest for the line under study and upon the radiative cross-sections as well as upon the particle density and the photon density in the atmosphere.

Very many of the strong lines in the ultraviolet spectral region come from low-lying, rather isolated levels. Because the lines are intrinsically strong these lines are formed high in the atmosphere where the particle density is low. Consequently one may expect that if one is to obtain a meaningful interpretation of the spectrum it will be necessary to use a rather general theory of line formation in which the restrictive hypothesis of LTE is not made.

The possibility of interpreting stellar line profiles using the hypothesis that line formation occurs as if isotropic, coherent scattering occurred while the level populations are given by Saha's and Boltzmann's laws will not be considered, for these hypotheses are incompatible. Such a formulation of the problem of line formation leads to deeper lines, for the same number of atoms, than does the hypothesis of LTE.

The only predictions of ultraviolet stellar absorption lines available at present are those by Gaustad and Spitzer (1961), Morton (1965), Guillaume, Van Rensbergen and Underhill (1965), Guillaume (1966), Elst (1966), and Underhill (1968a). The work of Gaustad and Spitzer gives a schematic survey which is essentially replaced by the work of Morton who like Elst, Guillaume, Van Rensbergen and Underhill has used the hypothesis of LTE in the stellar atmosphere. This hypothesis can certainly not be justified for most of the lines studied, but the results give some idea of what sort of absorption lines may be expected in the ultraviolet spectrum of a main-sequence star of type B. No similar studies have yet been made of supergiant atmospheres and none are likely to be made because it is thoroughly understood that the restrictive hypothesis of line formation in LTE in plane parallel static atmospheres is unsuited to supergiants. The LTE computations have shown that the blanketing by strong lines is significant at wavelengths shorter than 1900 Å (Morton, 1965) while at wavelengths between 1900 and 3000 Å it is of the order of 0.1 mag per 100 Å (Elst, 1966).

4. The Spectroscopic Description of Some Typical Ultraviolet Multiplets

Partial, scaled energy-level diagrams are shown in Figure 1 for some of the ions that have strong lines in the far ultraviolet. Lines have been drawn indicating the most conspicuous multiplets. The spectroscopic notation and wavelengths of these multiplets are given in Table II. Profiles of many of these lines have been predicted by Morton (1965) using a model atmosphere representing stars of about type B1.5V. Morton's calculations were done assuming that the level populations may be found using the Saha and Boltzmann laws.

Two important conclusions can be drawn from Figure 1:

(1) The energy levels between which most of the strong ultraviolet lines occur are relatively isolated. That is their separation in energy from each other is several times greater than the average energy of electrons at temperatures near 15000° which is a typical value for the outer layers of an early B type star.

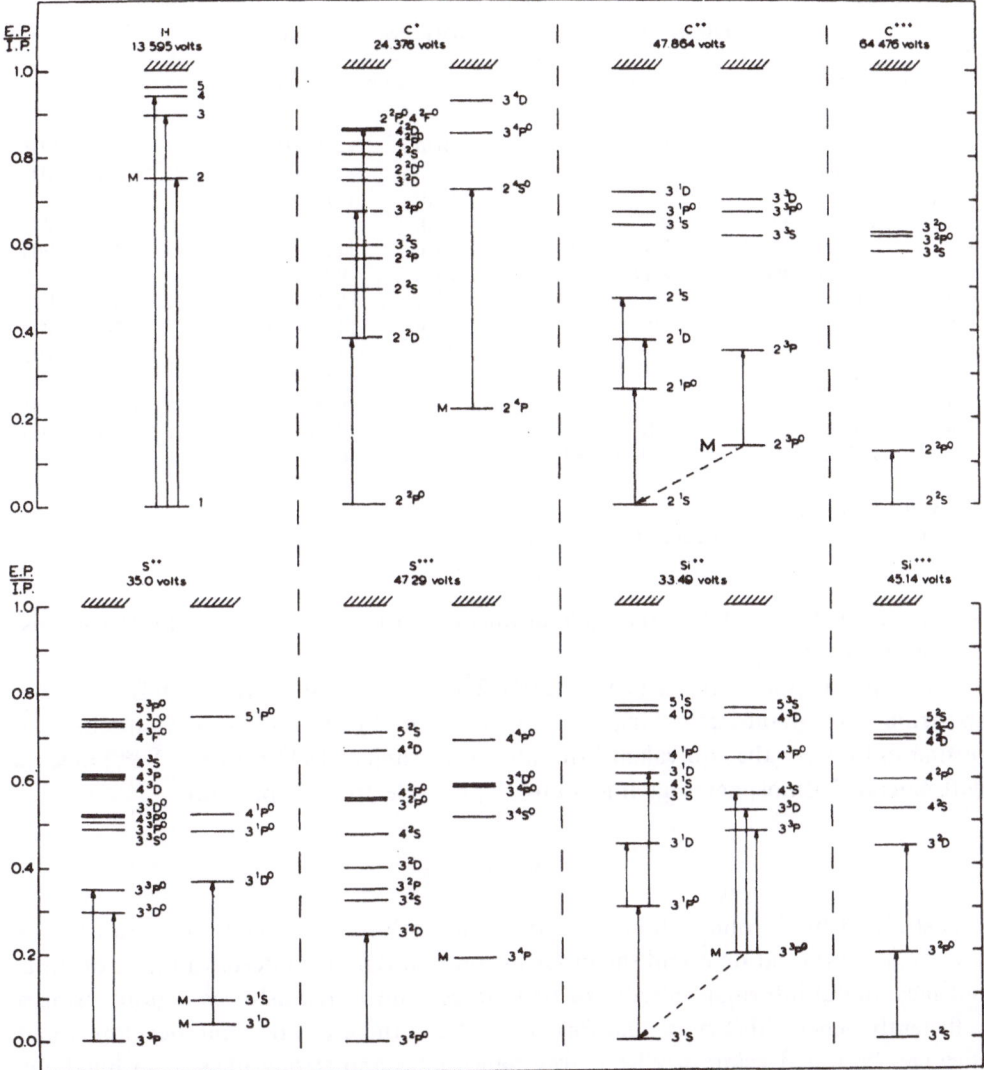

Fig. 1.   Partial, scaled energy-level diagrams. More levels exist above the highest level drawn but no known levels have been left out at lower energies. The primary ionization energy is indicated below the symbol for each ion. Strong ultraviolet multiplets are represented by vertical arrows. Intersystem lines are represented by broken lines.

(2) Metastable levels occur in many of the energy-level schemes and intrinsically strong lines are observed from these levels.

These facts indicate that the populations of the levels from which the strong ultra-violet lines come should be found by solving the equations of statistical equilibrium. Boltzmann populations relative to the ground level will only be obtained if collisional excitation predominates or if there is detailed radiative balance in the line. It is not

TABLE II

Ultraviolet lines and multiplets shown in Figure 2

Spectrum	λ_{vac} (Å)	Transition	Spectrum	λ_{vac} (Å)	Transition
H	1215.7	1 − 2	SiIII	1206.51[b]	$3^1S - 3^1P^o$
H	1025.7	1 − 3	SiIII	2541.8	$3^1P^o - 3^1D$
H	972.5	1 − 4	SiIII	1417.2	$3^1P^o - 3^1S$
CII	1334.5, 1335.7	$2^2P^o - 2^2D$	SiIII	1300	$3^3P^o - 3^3P$
CII	1760.4, 1760.8	$2^2D - 3^2P^o$	SiIII	1110	$3^3P^o - 3^3D$
CII	1065.9, 1066.1	$2^2D - 2^2P^o$	SiIII	996	$3^3P^o - 4^3S$
CII	1010	$2^4P - 2^4S^o$	[SiIII]	1892.0	$3^1S - 3^3P^o$
CIII	977.0	$2^1S - 2^1P^o$	SiIV	1393.8, 1402.8	$3^2S - 3^2P^o$
CIII	2296.9[a]	$2^1P^o - 2^1D$	SiIV	1128	$3^2P^o - 3^2D$
CIII	1247.4	$2^1P^o - 2^1S$	SIII	1194	$3^3P - 3^3D^o$
CIII	1176	$2^3P^o - 2^3P$	SIII	1015	$3^3P - 3^3P^o$
[CIII]	1909	$2^1S - 2^3P^o$	SIII	1077.8	$3^1D - 3^1D^o$
CIV	1548.2, 1550.8	$2^2S - 2^2P^o$	SIV	1072	$3^2P^o - 3^2D$

[a] Wavelength in air.
[b] Will be observed blended with 1206.53 3^1P^o–3^1D.

very probable that either of these conditions is satisfied in the extreme outer atmospheres of B stars.

The CIII intersystem line at 1909 Å (2^1S_0–$2^3P_1^o$) has been observed in emission in the spectrum of the Wolf-Rayet binary star γ_2 Velorum (Stecher, 1968). It would be interesting to look for the equivalent SiIII intersystem line at 1892 Å (3^1S_0–$3^3P_1^o$) in stars with extended atmospheres having not too high an electron temperature.

5. Discussion

The study of line profiles obtained from high-resolution ultraviolet spectra of stars should give information about the physical conditions in the outermost layers of stars, but meaningful information will not be obtained unless the analysis is done using a sufficiently general theory of line formation. Such theories are being developed and they can be tested using profiles of resonance lines and strong lines from low-lying levels which occur in the spectral region accessible from the surface of the earth. It is to be hoped that adequate theories will be available by the time adequate profiles of the strong ultraviolet lines have been obtained.

Because the shapes and depths of the cores of resonance lines and of strong lines from low-lying levels are sensitive to the density and geometric extent of the atmosphere, the strong ultraviolet lines may have rather different strengths in B stars having similar spectra in the normally observed region. Although the lines used for spectral classification are strong, and consequently somewhat sensitive to the density and extent of the atmosphere, they are not expected to be as sensitive as the far ultraviolet lines. It is rather probable that stars having the same spectrum in the 3900–5000 Å region will not have identical ultraviolet spectra. If this speculation is proved true by observation, then it may be rather difficult to obtain a reliable estimate of the shape of the interstellar extinction law in the far ultraviolet.

Many B type stars rotate rapidly, the rotation producing broad shallow absorption lines. Since the broadening or displacement of spectral lines by motion varies as $v\lambda/c$, where v is the velocity causing the broadening or displacement, λ is the wavelength, and c is the velocity of light, the broadening or shift of the ultraviolet lines will be reduced by a factor between 2 and 3 from that of lines in the normally observed spectral region. Consequently, unless observations are obtained with a resolution better than about 0.2 Å, only the most rapidly rotating stars will show the well known dish-shaped lines. Similarly it will only be possible to separate the spectra of double-lined spectroscopic binaries in the ultraviolet when the velocity difference is large. Rather high effective resolution (projected slit width in the spectrum) will be required to obtain true profiles of ultraviolet lines, particularly of the core of the line which is the part most sensitive to the physical conditions in the atmosphere.

References

Avrett, E. H. and Hummer, D. G.: 1965, *Monthly Notices Roy. Astron. Soc.* **130**, 295.
Beals, C. S.: 1929, *Monthly Notices Roy. Astron. Soc.* **90**, 202.
Beals, C. S.: 1934, *Publ. Dom. Astrophys. Obs. Victoria* **6**, 93.
Elst, E. W.: 1966, *Bull. Astron. Inst. Netherl.* **19**, 90.
Gaustad, J. E. and Spitzer, L.: 1961, *Astrophys. J.* **134**, 771.
Guillaume, C.: 1966, *Bull. Astron. Inst. Netherl.* **18**, 175.
Guillaume, C., Van Rensbergen, W. and Underhill, A. B.: 1965, *Bull. Astron. Inst. Netherl.* **18**, 106.
Hummer, D. G.: 1962, *Monthly Notices Roy. Astron. Soc.* **125**, 21.
Hummer, D. G.: 1964, *Astrophys. J.* **140**, 276.
Hummer, D. G.: 1968, *Monthly Notices Roy. Astron. Soc.* **138**, 73.
Hutchings, J. B.: 1968, *Monthly Notices Roy. Astron. Soc.* **141**, 329.
Jefferies, J. T. and Thomas, R. N.: 1958, *Astrophys. J.* **127**, 667.
Morton, D. C.: 1965, *Astrophys. J.* **141**, 73.
Stecher, T. P.: 1968, in *Wolf-Rayet Stars* (ed. by K. B. Gebbie and R. N. Thomas), N.B.S. Special Pub. 307, p. 65.
Thomas, R. N.: 1957, *Astrophys. J.* **125**, 260.
Thomas, R. N.: 1965, *Some Aspects of Non-Equilibrium Thermodynamics in the Presence of a Radiation Field*, University of Colorado Press, Boulder, Colo. A full set of references to work in this field is found here.
Underhill, A. B.: 1968a, *Bull. Astron. Inst. Netherl.* **19**, 500.
Underhill, A. B.: 1968b, *Bull. Astron. Inst. Netherl.* **19**, 526.

Discussion

Gingerich: Underhill's remarks on non-LTE seem convincing for *cores* of lines, and they suggest that if we are only interested in UV resonance lines in terms of opacity and line blocking, we can calculate models with lines represented by rectangles of infinite opacity. On the other hand, the wings can be formed deep enough to be in LTE; since we need to know the width of the approximating rectangle, what we really need are damping parameters. Note that in using Hγ profiles to determine the gravity of A stars, we fit only the wings and ignore the core, so LTE still works satisfactorily.

Underhill: The critical question is how far out in the wings must you be before the LTE hypothesis becomes tenable. I suspect that in the ultraviolet you will by this time have run into another line which is very strong. Fortunately this does not occur at Hγ in A stars.

Hekela: There are some misunderstandings in the description of atmospheric structure and the following interpretation of lines. But the problem is not LTE or non-LTE, because both are methods dealing directly only with radiative processes. The question is to discover such a set of characteristic equations that include all radiative and dynamical processes. Such a set would be different for different astronomical objects. This is the only way, though, I am sorry, an extremely complicated one, by which it would be possible to overcome continuously accumulated discrepancies in astrophysics.

Further, from the theoretical standpoint, it is not desirable to observe mostly only equivalent widths, because of the small amount of physical information in them.

I hope recent developments of detection techniques will allow us to measure line profiles as a function of position on the disc for many objects. If it would be possible to observe some of them by methods of absolute spectrophotometry we would obtain excellent input for a set of characteristic equations.

Underhill: The question of LTE or non-LTE is not only one of radiation processes for one must take into account collisional excitations and deexcitations as well as radiative processes. Furthermore, the line shape and position of the line centre comes in. In the most general formulation the dependence of these on a velocity field can be worked into the set of equations for statistical equilibrium and for radiative transfer. Solution of such a coupled set of equations would be extremely difficult and it is certainly a problem that should be looked at in more detail than has yet been done. The dynamic flow (or lack of it) must be used to determine the pressure structure in an atmosphere. The radiation flow determines, in principle, the temperature structure. These are related through the perfect gas law.

POSSIBILITY OF FLUORESCENCE PHENOMENA IN THE ULTRAVIOLET SPECTRUM OF SYMBIOTIC STARS AND LONG PERIOD VARIABLES

J. P. SWINGS and P. SWINGS

Institut d'Astrophysique, Cointe-Ougrée, Belgium

Abstract. We discuss essentially the cases where molecular fluorescences may be excited by Lyman α and other strong discrete ultraviolet emissions, including lines beyond the Lyman limit. The stars involved are the symbiotic objects and the long period variables. The molecules are H_2, N_2, O_2, NO and CO which have their resonance systems in the ultraviolet.

1. Introduction

Atomic fluorescence phenomena are known to play a very important role in different celestial objects: in particular selected lines of OIII, NIII and HeII are present in nebulae and in high excitation bright-line stars. The excitation of these lines is due to resonance lines of HI and HeII. Such observations are of importance since they give information on certain emission lines which, lying beyond 912 Å, will never be directly detectable because of the interstellar absorption in the Lyman continuum (except in the case of quasars). Atomic fluorescence occurs also in other elements of hot stars, such as CIII, NIV, SiIII, SiII and possibly FeII* giving rise to 'selectivities' among the lines of these elements.

Long period variables also show abnormal relative intensities due to the fluorescent excitation by lines located in the region < 3000 Å, hence not observable from the ground. This is the case for FeI and MnI excited by MgII 2795 Å, and for others excited by MgI 2851 and SiI (2516–2519 Å).

We have examined the possibility of an excitation of molecular lines in certain types of stars by the discrete emissions of these stars. One could thus obtain information on molecules such as H_2, N_2, CO, ... for which the resonance series are located in the far ultraviolet. The exciting lines may lie shortward of 912 Å.

2. Emission of Molecular Lines in the UV Spectrum of Symbiotic Stars and Long Period Variables

Certain stars which are rich in molecules also possess strong emission lines. This is the case of long period variables near minimum; they show emission lines of hydrogen, iron, ... in the observable spectrum although the temperature is low, of the order of 2000K. The intensity of their continuous spectrum should be very weak in the UV.

---

* UV FeII: Fluorescence may possibly explain the quartet-sextet selectivity. Observation of P Cygni stars will thus be of particular interest, especially in the region 2000–3000 Å.

Houziaux and Butler (eds.), Ultraviolet Stellar Spectra and Ground-Based Observations, 226–231.

This would apply also to other cool stars, such as T Tauri stars. Symbiotic objects constitute another class of objects which combine emission lines and molecular absorption bands.

The characteristic difference between the spectra of long period variables and symbiotic objects concerns the excitation energy of the emission lines: this excitation is low in LPV's and may be very high in symbiotic stars.

The following considerations may be applied qualitatively to both groups. We shall examine them separately later.

In both groups the Lyman lines may be intense in emission since the Balmer series appears in emission. Evidently the Lyman series is not necessarily brighter than the Balmer series. On the other hand regions of the stellar atmospheres showing strong TiO or other molecular bands are certainly also rich in diatomic molecules (or sometimes triatomic, like H_2O) of cosmically abundant elements: H_2, N_2, O_2, NO, CO, ... All these molecules have their absorption electronic systems in the ultraviolet.

The absorption* of an emission line by a molecule having the appropriate values of the vibrational (v'') and rotational (K'') quantum numbers will give rise, through a fluorescence mechanism, to the emission of what is called a *resonance series* which will be composed of triplets, doublets or singlets depending on the type of the involved transition (see Swings, 1933).

If an emission line covers several absorption lines (this may be the case for a wide emission) the resonance series consists of multiplets resulting from the addition of triplets, doublets and singlets that characterize a unique absorption.

A. WHAT ARE THE MOLECULAR FLUORESCENCES WHICH WE MAY EXPECT?

A striking case results from the absorption of Lyman γ by the nitrogen molecule N_2. In order to observe Lyman γ in the solar UV spectrum one must launch rockets higher than approximately 140 km: otherwise Lyman γ is obscured by a telluric line of N_2. At the same altitude, solar Lyman α and Lyman β are already intense.

Let us examine a few possible cases of molecular fluorescence.

1. H_2 *molecule*

The Lyman system of H_2 may be excited by 8 emission lines as shown in Table I; the Werner system could give rise to resonance series of triplets excited by lines also given in Table I. The transition probabilities are known in these cases and one may thus predict the intensity distribution in these resonance series.

2. N_2 *molecule*

The absorption of Lyman γ by N_2 gives rise to a resonance series of triplets. The electronic transition of the Birge-Hopfield system is $b^1\Pi_u - X^1\Sigma_g^+$; the components of the triplets would be $P(6)$, $Q(5)$ and $R(4)$ of the series (3, $v''=0$, 1, 2, ...) (Table II).

---

* The relative radial velocities of the molecular zones and of the exciting line zones must be taken into account. The problem is similar to that existing in comets (Swings, 1965).

However, the intensity variation as a function of v'' is slow so that the energy available will be distributed through many triplets that individually will be weak.

Other resonance series of N_2 may be excited by the discrete emissions given in Table II.

TABLE I

Possible fluorescences of H_2

(a) Lyman system $B^1 \Sigma_u^+ - X^1 \Sigma_g^+$

λ 1085.1	He II	$P(6)-R(4)$ of the transitions $(2, v'')$
λ 1085.7	N II	$P(3)-R(1)$
λ 1037.61	O VI	$P(1)$ of $(5, v'')$ or $P(4)-R(2)$ of $(6, v'')$
λ 1031.9	O VI	$P(6)-R(4)$ of $(6, v'')$
λ 1025.7	Ly β	$P(1), P(3)-R(1)$ and $P(4)-R(2)$ of $(6, v'')$
λ 977.03	C III	$P(5)-R(3)$ of $(11, v'')$
λ 972.54	Ly γ	$P(3)-R(1)$ of $(11, v'')$
λ 949.74	Ly δ	$P(2)-R(0)$ of $(14, v'')$ and $P(7)-R(5)$ of $(15, v'')$

(b) Werner system $C^1 \Pi_u - X^1 \Sigma_g^+$

λ 989.8	N III	$P(4)-Q(3)-R(2)$ of $(1, v'')$
λ 977.03	C III	$P(5)-Q(4)-R(3)$ of $(2, v'')$ / $P(7)-Q(6)-R(5)$
λ 949.74	Ly δ	$P(6)-Q(5)-R(4)$ of $(3, v'')$ / $P(2)-Q(1)-R(0)$
λ 937.80	Ly ε	$P(4)-Q(3)-R(2)$ of $(4, v'')$
λ 930.75	Ly ζ	$P(2)-Q(1)-R(0)$ of $(4, v'')$

TABLE II

Possible fluorescences of N_2

(a) Resonance series excited by Lyman γ
$b^1 \Pi_u - X^1 \Sigma_g^+$ system (Birge-Hopfield)

Transition	$P(6)$	$Q(5)$	$R(4)$
(3,0)	972.54	972.32	972.13
(3,1)	995.09	994.85	994.65
(3,2)	1018.41	1018.16	1017.96
(3,3)	1042.55	1042.28	1042.07
(3,4)	1067.53	1067.25	1067.03
(3,5)	1093.39	1093.11	1092.87

(b) Resonance series excited by:

| λ 949.74 | Ly δ | $P(7)-Q(6)-R(5)$ of $(6, v'')$ in $b^1 \Pi_u - X^1 \Sigma_g^+$ |
| λ 937.80 | Ly ε | $P(2)-R(0)$ of $(4, v'')$ in $b'^1 \Sigma_u^+ - X^1 \Sigma_g^+$ |

One may not exclude an excitation of N_2 by the Lyman continuum; in such a case the fluorescent emissions would correspond to numerous rotation transitions and would look like laboratory emission bands.

3. O_2 molecule

O_2 will probably not be observable because of the predissociation effect in the Schumann-Runge system and the weakness of the transition probabilities.

4. NO molecule

The only possibility of a discrete excitation seems to be that by N_{III} (1098.11 Å) which falls in the δ system $c^2\Sigma-X^2\Pi$ (0,0) (band head at 1910 Å, degraded toward the violet).

5. CO molecule

(a) Fourth positive system $A^1\Pi-X^1\Sigma$. A strong C_{IV} line (1546 Å) is situated near the head of the (0,0) band. The corresponding rotational quantum number is $J=32$. One may expect an emission of triplets $(0, v'')$, especially (0,3) and (0,4).

Other possibilities of coincidences may be found on the basis of a new vibrational and rotational study of the fourth positive system of CO in the UV (Simmons et al., 1969): for example N_I (1200 Å) lies near 1199.67 Å (15,0) and C_{III} 1175 Å near 1173.15 Å (17,0).

(b) Fluorescence could also possibly be excited in the Birge-Hopfield systems: the only two possible coincidences seem to be:

N_{II} (1087.7 Å) (strong line) which falls in the band (0,0) of $c^1\Sigma^+-X^1\Sigma^+$; and Lyman δ (949.7 Å) which falls in $G^1\Pi-X^1\Sigma^+$.

However the rotational analysis of the CO bands is still too incomplete to enable a discussion of the fluorescence possibilities.

6. H_2O molecule

A possible excitation of polyatomic molecules may occur only in very cold stars, i.e. in the coldest part of symbiotic objects or of long period variables.

The far ultraviolet spectrum of H_2O has not yet been sufficiently analysed to enable a prediction of fluorescence phenomena. Furthermore the absorption spectrum of H_2O shows bands that seem diffuse or continuous: fluorescence excited in such regions would thus be weak or continuous. It should be noted that certain bands may simply have a continuous appearance, yet have a discrete structure.

The Lyman α line is absorbed by H_2O but we do not know whether or not this absorption is followed by a fluorescent emission of H_2O. It is possible that the absorption by H_2O leads to a dissociation or predissociation that would produce an emission of atomic lines or OH bands.

B. WHICH ARE THE LINES PRODUCING AN EXCITATION?

In the case of long period variables, only Lyman lines and, possibly, a few lines of rather low excitation such as O_I or N_I may be considered. On the contrary, in symbiotic objects, we may also consider an excitation by permitted resonance and recombination lines and forbidden lines of high excitation, such as HeI, HeII, OII, OIII, [OIII].... The possible exciting lines are summarized in Table III (from Moore, 1952). A list of UV lines of planetary nebulae has been published by Aller (1961).

TABLE III

Strongest exciting UV lines (resonance and recombination)

CI	CII	CIII	CIV
1657	1335	977	1548–50
1561	1037	386	312
1277	904	1175	420
1193	858	538	384
	687	459	
		2297	

NI	NII	NIII	NIV	NV
1200	1084	990	765	1238–42
1134	916	764	247	209
964	672	686	923	248
1493–95	645	452	322	
1743–45	776	374	283	
1412		772	1719	
1327–28			955	
1311			335	

OI	OII	OIII	OIV	OV	OVI
1355–58	833–34	833–35	788–90	630	1032–38
1302–06	539	702–04	609	172	150
1039–42	718	507–08	554	759–62	184
1026–28	538	374	280	192	173
990		306	239	220	
		603	625		
		600			
		526			
		396			
		328			
		321			
		598			
		435			
		345			

3. Conclusion

Long period variables, despite their low temperature, may reveal an ultraviolet spectrum rich in emission lines of molecular origin. Since the ultraviolet continuum of the LPV is most probably very weak, only excitation by discrete emission lines may be considered in these stars. Symbiotic objects may show a more complex UV spectrum because of the presence of a greater number of discrete exciting lines. An ultraviolet continuum, combined with discrete absorption and emission lines may also be present in symbiotic objects and give rise to more complex fluorescence emissions.

It is impossible to determine quantitatively the intensities of the molecular fluorescence lines: this is due to the lack of information on geometrical and physical models for the stars considered here and also on the relative radial velocities. One thing is however certain: the intensities of the molecular lines will be much weaker than those of the exciting lines. Even if the energy of an exciting line is entirely absorbed by a molecule the resonance series which is produced is composed of several lines, generally numerous lines. For instance the intensity of each line of the resonance series of N_2 excited by Lyman γ is only a small fraction of the intensity of Lyman γ.

In any case observations from rockets or satellites will be of great interest.

References

Aller, L. H.: 1961, *Mem. Soc. Roy. Sc. Liège* **4**, 535.

Moore, C. E.: 1952, A UV Multiplet Table, NBS Circ. 488.

Simmons, J. D., Bass, A. M., and Tilford, S. G.: 1969, *Astrophys. J.* **155**, 345.

Swings, P.: 1933, 'Spectres moléculaires', *Act. Sc. Ind.* **74**, Paris, Hermann.

Swings, P.: 1965, 'Astronautical Investigations of Comets', in *Advances in Space Science and Technology*, Vol. 7 (ed. by F. I. Ordway III), Academic Press, New York.

Discussion

Underhill: If the cool gas where the molecular bands originate was irradiated by an absorption-line UV spectrum from a hot object (perhaps in the case of symbiotic stars), would not the bands show dips of intensity at some places, the lines of relevant energy not being excited.

Swings: It seems that long period variables may be represented by a stratified model. Emissions of hydrogen, [Fe II] and AlO take place in different regions. This can be seen by measuring the relative intensities of the lines and comparing them to theoretical values. In symbiotic objects and in long period variables, it happens often that Hβ in emission is very weak because of its absorption by TiO. This means that hydrogen lines are emitted in lower layers. In this configuration it is easily noticed that atomic emission lines may excite molecules lying above them. A paper discusses this stratification problem of the emitting layers in long period variables (J. P. Swings: 1969, *Bull. Soc. Roy. Sc. Liège*, in press).

Malaise: I would like to stress a point that was mentioned in the paper. When discussing comparisons between observed emission spectra and model atmosphere results in the UV, it should be kept in mind that if molecules are present, they can absorb selectively one emission line and redistribute the absorbed photons over such a number of lines that they simply appear to be washed out. This process could change the observed intensity ratios of emission lines in the UV.

RADIATIVE ACCELERATION AND ULTRAVIOLET RESONANCE LINE PROFILES IN OB SUPERGIANTS

J. B. HUTCHINGS

Dominion Astrophysical Observatory, Victoria, B.C., Canada

1. The Observations

The rocket spectra taken by Morton *et al.* (1967) were used as a basis for the investigation. These are in the far ultraviolet region of the Orion supergiants δ, ε, ζ Orionis. Because of the similarity in spectral type and luminosity of these stars and the great strength of the shifted resonance lines it was possible to derive a mean P Cygni profile (standard error 5% of the continuum) representing a typical single ultraviolet resonance line for a star of spectral type $\sim$B0, $M_v \sim -6.2$. This is the 'observed' mean profile in Figures 2 and 3.

2. Line Profile Calculations

Profiles were calculated for models of extended envelopes similar to those derived for the very high luminosity stars with higher mass loss (Hutchings, 1968). Only Doppler-broadened radiative and scattering interactions are included, and the integration is performed out to several radii from the stellar surface. An accelerating velocity field is necessary to produce the extreme P Cygni profiles observed and the line shape is dominated by the form of the velocity field and the dependence of the resonance line ion density on height above the star. Proceeding empirically it became evident that the density-height relationship differed considerably from that expected from LTE considerations, and an optimum profile fit was obtained with the velocity field (1) and the shell density relation shown in Figure 1. It is seen that the resonance line ion population (primarily Si IV and C IV) increases outwards for a considerable distance from the stellar surface. The profile calculated is still not a good fit to the observations, but is improved greatly by the addition of a 'zero height line profile' at the base of the envelope, which accounts for the absorption line formed by the low-lying higher density layers of the atmosphere. This zero height profile was taken from calculations found in the literature. The initial and improved calculated profiles are compared with the observed in Figure 2.

3. Radiative Acceleration

Using the velocity field derived we may calculate the number of radiative excitations undergone by a typical resonance line ion at different heights as it passes out through the extended envelope, taking into account radiation dilution and the absorption line strength between the ion and the surface. From the limb darkening and the solid angle subtended by the photosphere at different heights, the net absorption of outward

Houziaux and Butler (eds.), Ultraviolet Stellar Spectra and Ground-Based Observations, 232–235.

momentum by the resonance line ions can be derived. The ratio of momentum absorbed to acceleration produced is seen to increase and level off in the outermost layers of the envelope. Assuming that the acceleration is produced by the radiation in this way, it is possible to calculate the rate of collisional de-excitation of the reso-

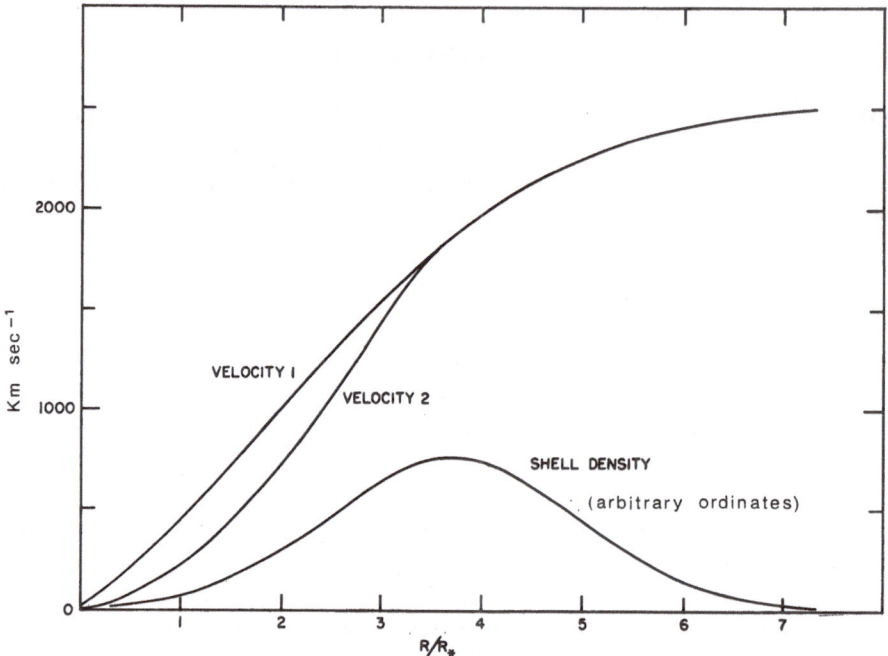

Fig. 1.   Velocity and resonance line ion density as a function of height. Velocity in km sec^{-1}; density in arbitrary units proportional to number of ions in a shell of constant thickness.

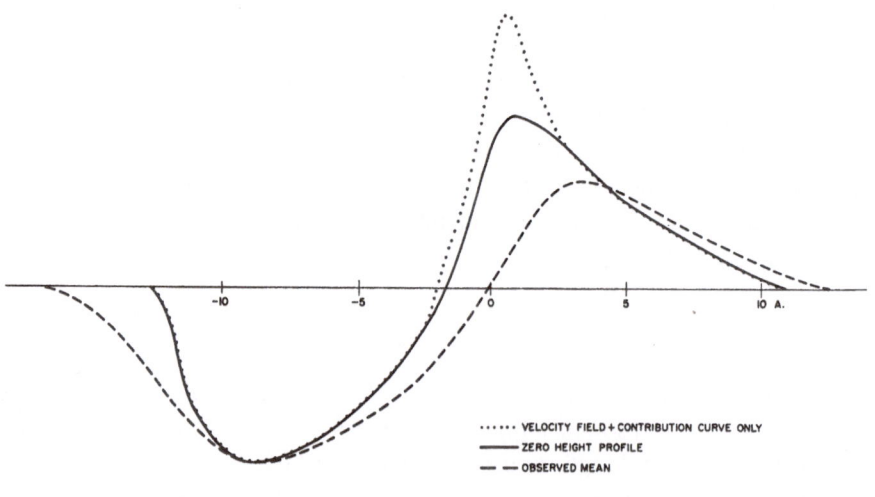

Fig. 2.   Observed and computed profiles (radiative de-excitation only).

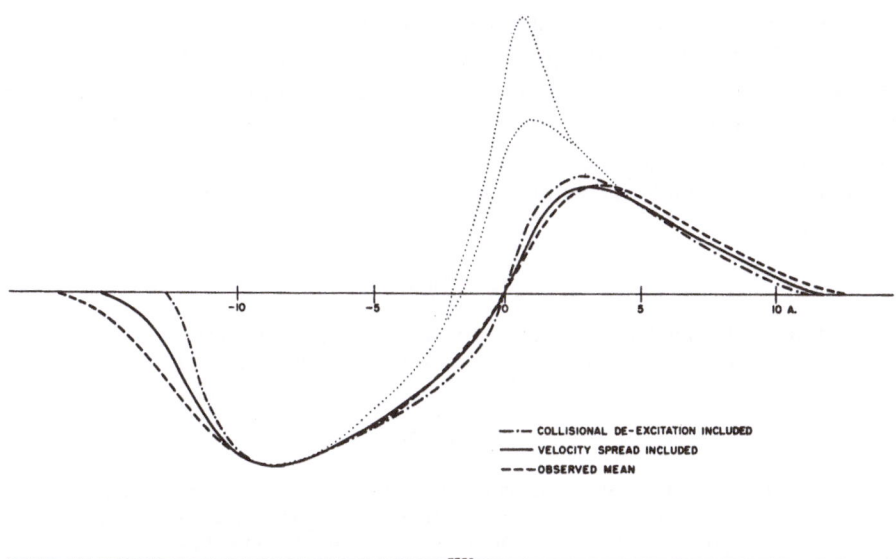

Fig. 3.   Observed and computed profiles taking into account collisions.

nance line ions as a function of height, and hence to derive the density of the total atmospheric mass as a function of height. From this the acceleration of the entire atmosphere by transfer of momentum from the radiation, via the resonance line ions, to the atmosphere as a whole, is calculated. The mean total density out to some 4 stellar radii is high enough that the whole mass must have the same velocity, so that the entire calculation described in this section has to be iterated until the resonance line ion and total velocity fields are the same out to this height. This results in the velocity field (2) in Figure 1, a mean total density similar to that derived for the high mass-loss stars, and an atmospheric structure whose mass-loss rate is the same at all heights.

A check on the self-consistency of the entire argument is obtained by calculating the resonance line profile with the new velocity field, and including the derived rate of collisional de-excitation. This is shown in Figure 3, and it is evident that a very satisfactory fit is achieved. It is therefore proposed that the atmospheric structure derived is a valid crude representation of the envelopes of the stars observed.

4. Mass Loss

At present it is not possible to fix a zero point to the density of the expanding atmosphere, from the resonance line calculations. However, high-quality ground-based observations are available of the Hα line in these stars and this line shows a weak P Cygni profile in the mean (it is variable) with a velocity displacement of some −150 km/sec. Previous calculations of Hα profiles (Hutchings, 1968) have indicated the density and radiation dilution at which this line starts to show emission, and fixing these figures to the derived velocity field provides a crude mass-loss rate of some 10^{-6} $M_\odot$/year. This figure is in agreement with that derived by Morton *et al.*, and is consistent with the higher values derived for the higher luminosity OB stars studied. It is therefore suggested once again that radiation pressure is responsible for the mass loss from massive early-type stars.

(A full account of this work will be published in *Monthly Notices Roy. Astron. Soc.* in due course.)

References

Hutchings, J. B.: 1968, *Monthly Notices Roy. Astron. Soc.* **141**, 219 and 329.
Morton, D. C., Jenkins, E. B., and Bohlin, R. C.: 1967, *Astrophys. J.* **154**, 661.

Discussion

Solomon: In any mass loss by radiative acceleration, the maximum rate of mass loss is determined by the maximum momentum available in the radiation which is doing the acceleration. In the case of resonance line radiation this is the momentum in a line-width when the flow is at the sonic point. For O and B supergiants this amounts to about 2×10^{-8} $M_\odot yr^{-1}$. Therefore the rates quoted by Hutchings of 10^{-6} $M_\odot$ yr^{-1} violate the conservation of momentum.

The models you presented have been fitted with Hα observations to determine absolute values of the density at a particular velocity which then gives the mass loss rate. However, if instead of Hα, the density is determined by some other ion such as N v (that is, the electron density is fixed by the recombination rate necessary to give the observed Nv) then the mass loss rate will be much less than 10^{-9} $M_\odot$ yr^{-1}. This makes it impossible to determine the mass loss rate in this manner since a large discrepancy of a factor of 10^3 is present in trying to reconcile two observed lines.

Hutchings: The observational results presented at this meeting have shown that there is not only one line which causes the atmospheric acceleration. In the case of the Orion supergiants there are six well established shifted lines and to judge on the results of Smith, who has found oxygen and sulphur lines as well, the assumption made in my paper of 10 lines seems a reasonable one. I am not clear how wide a linewidth is in Solomon's terminology but should point out that these lines have absorption widths of some 10 Å. Solomon's figure of 2×10^{-8} $M_\odot$ yr^{-1} should then be increased at least to 2×10^{-7} for the Orion stars, and in the case of more luminous stars or stars with P Cygni characteristics throughout the spectrum, to 10^{-5} or higher. While the zero point of my mass-loss figures leaves something to be desired I do not think they violate the conservation of momentum as suggested.

The fact that the mass loss rate is so much lower as calculated from the high-energy ion recombination rates serves to emphasize two points. Firstly that non-LTE equilibria apply in these extended envelopes, as is obvious from the ionisation-height relations, and secondly that most of the momentum is transferred from the high energy ions to the remainder of the atmosphere by collisional interactions.

Underhill: In the case of Wolf-Rayet stars a wide range of excitation is often seen among the emission lines and it is suspected that this is due to collisional excitation from high-energy particles emitted from the photosphere. Might not such a process be important in the stars you are discussing?

Hutchings: In the Orion supergiants the density of the outer envelope is low and collisional excitation is probably unimportant. It may be far more important in stars with greater mass loss such as the Of and W-R stars. As far as the relative population of high-energy ions is concerned there is no satisfactory theoretical approach at present. We must therefore simply accept that the populations of several high-energy ions reaches a peak in the outer layers (2–5 R_*), presumably, primarily by radiative processes.

Deutsch: If matter flows up through the base of the atmosphere, it will transport momentum which is additional to your radiation limit of L/c^2.

Hutchings: At present the sole mechanism of acceleration considered is momentum transfer from the resonance lines. The L/c^2 limit applies to each line which contributes to this process, so that we may increase the limit by the number of lines. In the extreme case of P Cygni there may be an acceleration mechanism action through some hundreds of spectral lines.

STELLAR-WIND THEORY FOR O AND B STARS

PHILIP M. SOLOMON

Columbia University, New York, N.Y., U.S.A.

Abstract. The rocket-ultraviolet observations of strong Doppler-shifted absorption lines of Si IV, C IV, N V and other ions in the spectrum of O and B supergiants clearly indicate a high velocity out-flow of matter from these stars. The presence of moderate ionisation stages in the stellar wind is conclusive evidence that the flow cannot be due to a high temperature corona as is the case for the solar wind. It is shown that the driving mechanism for the hot-star mass loss is radiation pressure exerted on the gas through absorption in resonance lines occurring at wavelengths near the maximum of the star's continuum flux. In the upper layers of these stars the outward force per gram of matter due to the radiation pressure can greatly exceed the gravitational acceleration making a static at-mosphere impossible.

The problem of a steady-state moving reversing layer is formulated and the solution leads to predictions of mass-loss rates as a function of effective temperature and gravity for all hot stars. These results are in substantial agreement with the observations.

Discussion

Morton: This mechanism certainly must be an important cause of mass ejection in the hot supergiants. However, there may be additional effects such as convective or turbulent motions in the photosphere which could contribute to the rate of mass loss.

Solomon: Any additional mechanism would require very high temperatures in the flow ($T > 10^7$ K) to account for the high velocities. These temperatures are clearly not present since no C III, C IV, Si III or Si IV could possibly exist at 10^7 K, where the ions would be completely stripped with only one electron remaining.

Herbig: I would think that there might be two possible, directly observable consequences of such high-velocity mass outflow:

(a) Either the region within a cluster such as NGC 6231 (which contains the three actively ejecting stars mentioned by Hutchings) would be scoured clean of interstellar matter, leaving a cavity. Is there any sign of this? Or,

(b) In the other extreme, this volume would contain a great deal of highly excited material that might be detected directly. Or to put this last in another way, is direct detection possible of this 1000–3000 km sec^{-1} material after it moves away from the immediate vicinity of these stars?

Solomon: The pressure in the flow would equal the interstellar pressure in an H II region at about 1 psc from the star. The most interesting region might be the interface between the normal H II region and the high velocity outflow. This might even yield detectable X-rays.

Hearn: How is the ionization balance determined in calculating the radiation pressure? Have the transient effects of the ionization balance been included? What is the distribution of the electron temperature and density in the outer layers?

Solomon: The ionization equilibrium is determined by assuming a steady state between radiative ionization and recombination. This assumption is valid in the subsonic branch of the flow, where the time-scale for recombination is much less than the flow time-scale (time for the matter to move to a region where the density is lower by a factor of e). The transient effects are certainly important in the supersonic flow and this has been accounted for by actually summing up all recombinations and ionizations in the flow for each point in the model. The net effect is, that after velocities > 100 km sec^{-1} are reached, the ionization remains almost constant since the time-scale for the flow becomes much less than the ionization time-scale.

Deutsch: Although you assert that your solution pertains only to the subsonic part of the flow, the line profile you have shown surely includes the absorption of the supersonic part. Can you say approximately how the different levels of supersonic flow are weighted in contributing to the line profile?

Houziaux and Butler (eds.), Ultraviolet Stellar Spectra and Ground-Based Observations, 236–237.

Solomon: The model gives the number of ions as a function of the velocity of outflow. The optical depth at a velocity V is inversely proportional to dV/dr, which is itself a function of V and is determined by the model.

I should emphasize that while we regard the subsonic flow and the mass loss rates as accurate, the supersonic flow shown here is only presented to demonstrate that line profiles not unlike those observed can be obtained form the mass-loss rates derived from the subsonic flow.

Underhill: I have a wicked postulate I should like you to comment on. Suppose that a star had lost most of its outer envelope so that it was nearly peeled down to its helium-rich core and thus had an enriched He/H content in its atmosphere, what do you think would happen to an expanding atmosphere generated in the manner you have suggested? Would the presumably hot He core contribute more momentum to the envelope by means of high-energy particles and strong UV radiation than reaches the outer atmosphere, when the core is enveloped in the normal deep radiative outer layers of a star?

Solomon: This postulate is too wicked for me to answer immediately. However, I expect that the star might adjust itself and that the He rich core would not be exposed. An enriched He atmosphere would probably not greatly effect the problem.

Praderie: (1) Do you include all resonance lines over the whole spectrum in the computation of the radiation pressure gravity term, and does this term vary with depth?

(2) A work by Bisnovetij-Kogan and Zeldovich considers the same type of hydrodynamical mass flow as Solomon, but the radiation pressure is that acting through the continuous spectrum; it is supposed that a sudden increase of the continuous absorption coefficient occurs near the surface of the star. Unhappily I cannot quote the values they find for the mass loss.

Solomon: We include all resonance lines at $\lambda > 912$ Å of C, N, and Si ions and it is the increase of this term outwards that leads to the flow. This increase is not postulated but results from the model atmosphere.

D. THE SUN – A TYPICAL G2V STAR

REVIEW OF ASTROPHYSICAL CONCLUSIONS FROM THE
UV SOLAR SPECTRA

S. R. POTTASCH

Kapteyn Astronomical Institute, Groningen

A good deal is known about the Sun from the measurement of its spectrum. The measurement of the ultraviolet spectrum has contributed to this knowledge and I shall try to summarize the results and problems of some of these investigations.

In a general way we can divide the discussion between the continuous radiation and the line radiation, and the line radiation can further be subdivided into the absorption spectrum and the emission spectrum, which can be discussed separately not only because they originate in different regions of the atmosphere, but because the physical conditions for line formation are quite different in these different regions.

Our goal in these investigations is to determine from the observed spectrum the structure of the atmosphere, i.e. the run of temperature and density as a function of height. Further we are interested in the chemical composition of the atmosphere, and to determine whether or not it is a function of height in the atmosphere.

It is interesting to know how far these investigations have proceeded not only from the point of view of studying the Sun, but also with a view to observing stars of spectral type not too different from the Sun. Because it is possible easily to study emission from various points on the disc of the Sun, and because of the great intensity of the observable emission from the Sun which enables a very detailed study of the spectrum, we can test theories and methods for general use on the Sun, and then apply them to neighbouring spectral types.

1. The Continuous Spectrum

The center-to-limb measurements in the visible part of the spectrum are used to determine the temperature as a function of optical depth at the frequency observed. In principle, if the limb darkening is precisely known, one can obtain the complete run of T vs. τ_λ throughout the atmosphere. In practice, the errors in the observed limb-darkening curve are such that the determination of T is only valid in the regions where most of the radiation comes from, i.e. the region $2 \geqslant \tau_\lambda \geqslant 0.2$. For this reason it is desirable to obtain limb-darkening measurements in other spectral regions. Since as we proceed into the ultraviolet part of the spectrum the absorption coefficient begins to increase strongly, this radiation comes from much further out in the atmosphere. In this way we can build a temperature model of the solar atmosphere which is probably reasonably reliable for $\tau_{5000} > 10^{-4}$. At observations made at 1500 Å or less, limb darkening is no longer seen: it is now limb brightening which is observed. This indicates that we are now looking at regions above the temperature minimum of the Sun, the region where the temperature is increasing outward.

Houziaux and Butler (eds.), Ultraviolet Stellar Spectra and Ground-Based Observations, 241–249.
All Rights Reserved. Copyright © 1970 by the IAU.

I may point out that detailed models which are given, e.g. the Bilderberg Continuum Atmosphere rely not only on ultraviolet measurements, but infrared and visual observations as well.

There is one difficulty in present studies of the solar continuum. When measuring limb darkening at two wavelengths we obtain two functions $T(\tau_\lambda)$, and we consider a given position in the atmosphere, let us say where the temperature $T = 5000$ K, then we have two values of τ_λ corresponding to the two wavelengths. Since the τ_λ's differ at that point because the absorption coefficient differs with wavelength, we can obtain the ratio of the two absorption coefficients. If we extend this argument to many wavelengths we can determine the variation of the absorption coefficient as a function of wavelength at a given temperature in the atmosphere. In the visual part of the spectrum, longward of 4000 Å, this empirically derived absorption coefficient agrees very well with that predicted theoretically on the basis of absorption by the H⁻ ion. Below 4000 Å there has always been an indication that an additional source of opacity is present. In the ultraviolet the situation becomes more difficult to interpret.

In Figure 1, taken from Bonnet (1968), the points show the empirically determined absorption coefficient in the wavelength region from 2000 to 3000 Å, while the solid line is a theoretical curve. The differences amount to an order of magnitude at some

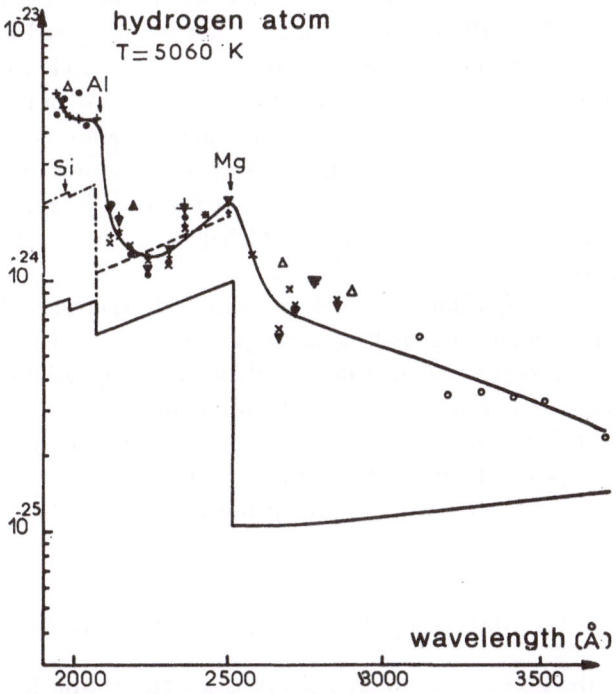

Fig. 1.   A comparison of the absorption coefficient in the solar atmosphere derived from limb-darkening measures (Bonnet, 1968) and a theoretical curve. The actual absorption is seen to be as much as an order of magnitude greater than the theoretical, indicating an important source of opacity has been omitted in the theoretical calculation. Iron may be this missing source.

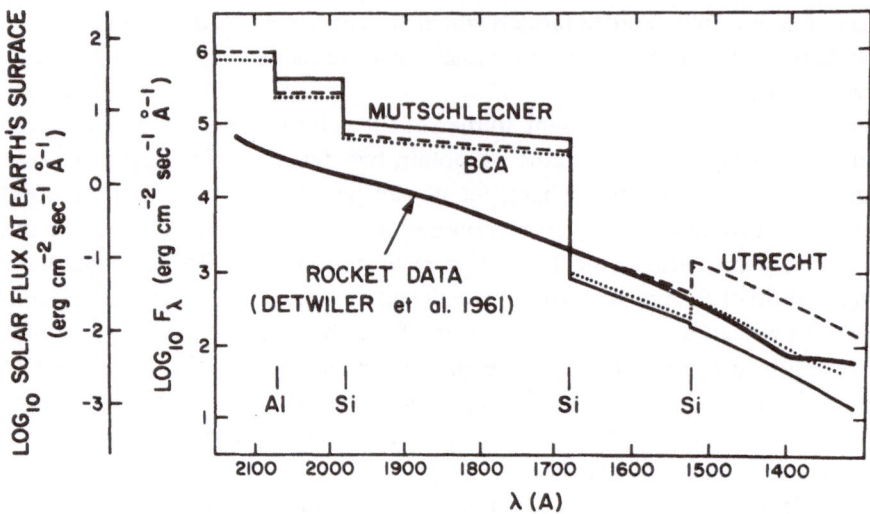

Fig. 2. A comparison of the observed flux with the theoretical prediction. The discrepancy is again
due to a missing opacity source.

points, and more striking, the predicted jump at 2510 Å due to MgI, does not appear
in nearly the predicted strength. This same result may be seen in a different form in
the spectral region 1400–2100 Å, in Figure 2, where the predicted flux is compared
to observed flux. The calculations are taken from Gingerich and Rich (1968). Again
differences of at least an order of magnitude appear, and a jump of intensity of almost
two orders of magnitude, due to SiI, is predicted and certainly not observed in this
strength. The reason for these discrepancies is that the source of the continuous
absorption is poorly understood below 4000 Å. Probably at least a part of the continuous absorption is due to neutral iron, which has not been taken into account
sufficiently, for two reasons:

(1) The absorption coefficient has not been measured experimentally and its theoretical calculation is based on a hydrogenic approximation which is known to give
errors greater than an order of magnitude in silicon and magnesium.

(2) The abundance of iron has previously been underestimated by a factor of
about 10–20. This brings the iron absorption in most of this spectral range to values
at least as high and probably higher than silicon and magnesium.

These are at present the most important problems in the ultraviolet continuous
solar spectrum.

2. The Absorption Line Spectrum and Line Profiles

The spectrum of the Sun above about 2000 Å is an absorption line spectrum very
similar to that in the visible region and the analysis also does not differ. This ultraviolet spectral region has some interest because special lines may be found here e.g.
the resonance lines of singly ionized iron, and lines of certain elements which are not

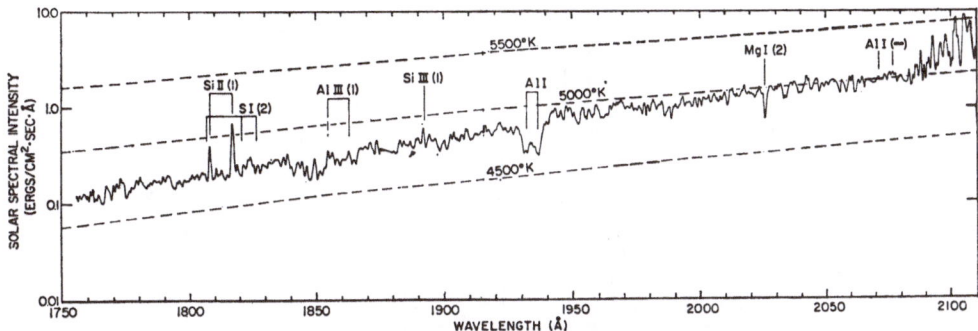

Fig. 3.   The observed solar spectrum (1750–2100 Å).

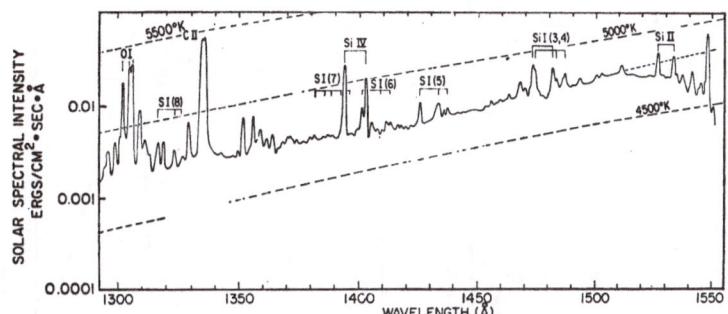

Fig. 4.   The observed solar spectrum (1300–1550 Å).

observed in the visual region are found in the ultraviolet. At present the spectrophoto-metry in this spectral region is of rather poor quality so that quantitative discussions are difficult.

A very special case is the MgII pair near 2800 Å, which have been well studied and for which high-resolution observations are available. These lines, which give similar information to the CaII lines and the Ly-α line, are being studied theoretically at present. These studies are not so far advanced for MgII as for CaII, but may eventually reinforce and supplement the latter work. The difficulty at present is that knowledge of the temperature structure in the upper chromosphere where these lines are formed, is almost completely lacking.

3. The Emission Line Spectrum of the Quiet Sun

Below 2000 Å the character of the solar spectrum changes rather suddenly from absorption line to emission line. This can easily be seen from Figures 3, 4 and 5, which illustrate the spectrum between 2100 Å down to 900 Å. The appearance of emission means that the source function is increasing in the lines relative to the continuum as one goes to higher layers in the atmosphere. This in turn occurs through a combination of factors:

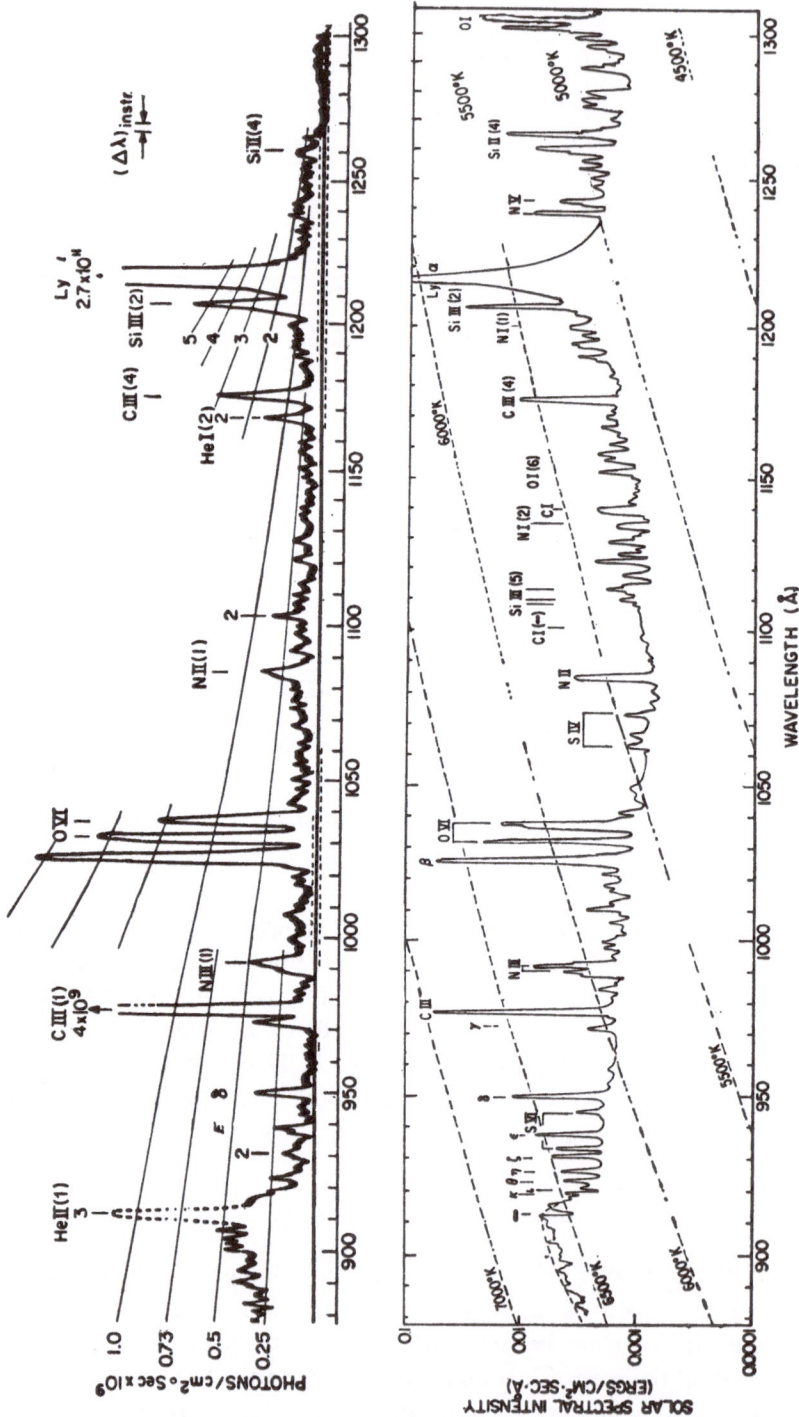

Fig. 5.   The observed solar spectrum (900–1300 Å).

(1) The electron temperature is increasing outward, so that collisions are rapid enough to populate the high excited levels;

(2) The continuum optical depth becomes so small that the higher layers contribute only a small fraction of the continuous radiation.

Lines of Si II are observed near 1800 Å and the state of ionization increases as we proceed further into the ultraviolet, as shown by Table I.

TABLE I

Wavelength interval (Å)	Average ionization potential
1000–2000	30 eV
600–1000	60 eV
300– 600	150 eV
40– 300	300 eV
15– 40	600 eV

The reason for this correlation is simply that most of the observed lines are resonance lines and the wavelength of the resonance line of a particular ion is roughly correlated with the ionization potential of that ion.

I shall discuss in a summary way the analysis of the emission line spectrum before I give the results, because this kind of analysis probably is directly applicable to many low temperature stars with coronae. Only the total intensity of the line need be observed, not the line profile. The key to this analysis is the following. Since in the chromosphere and corona, where these lines are formed, we cannot assume local thermodynamic equilibrium, we must know the process of formation of these lines. Of the possible contributing processes, direct collisional excitation of the emitting level is by far the most likely since temperatures up to several million degrees are present. In comparison, radiative recombinations, which have the same density dependence, are much slower. And absorption of radiation to higher levels followed by cascading is not an important process simply because there are very few photons in the far ultraviolet continuum. If this is true the energy emitted over the surface of the Sun is

$$I(\text{line}) \propto \int n_L n_e C_{LU}(T)\, dv,$$

where n_L is the ground state population of the emitting ion (almost all observed lines are resonance lines or at least have the ground state as lower level). The collisional excitation rate $n_e C_{LU}(T)$ is written so that the part dependent on the electron density, n_e, is separate from the temperature-dependent part. The integration is taken over all heights in the atmosphere, or if we consider an area greater than 1 cm^2, then the integral is taken over the entire volume.

Since the electrons are almost all due to the ionization of hydrogen, $n_e/n_H \approx 1$. Since the ionization is all collisional and the recombination all radiative, and both of these rates have the same density dependence, the state of ionization depends only on the temperature, thus

$$\frac{n_L}{n_{\text{element}}} \approx \frac{n_{\text{ion}}}{n_{\text{element}}} = g\,(T).$$

Thus

$$I\,(\text{line}) \propto \int \frac{n_{\text{ion}}}{n_{\text{element}}} \times \frac{n_{\text{element}}}{n_H} \frac{n_H}{n_e} n_e^2\, C_{LU}(T)\, dv$$

$$\propto A \int n_e^2\, f\,(T)\, dv,$$

where we have written the abundance n_{element}/n_H as A, and assumed that it remains constant in the atmosphere.

The function $f(T)$ can be computed for every ion considered. It is near zero for most of the temperature range and has an appreciable value only for a limited range of temperature. We make the approximation that it has a constant value for this limited temperature range (let us say, between T_1 and T_2), and is zero outside of this range. The physical reason for it being zero at certain temperatures is simply either that the stage of ionization is not present or that the collisional excitation rate is very small. The equation for the intensity then becomes

$$I\,(\text{line}) \propto A \int_{V(T_1)}^{V(T_2)} n_e^2\, dv,$$

where the integral is taken over all the material between the temperatures T_1 and T_2. One can thus see that for every ion whose intensity is observed, we can obtain a quantity which depends on the abundance of that element and on the structure of the atmosphere where the observed line is formed.

We can separate these quantities in the following manner. Suppose we make a plot of the above observed quantity

$$A \int_{V(T_1)}^{V(T_2)} n_e^2\, dv,$$

against the temperature T_1 and T_2 (or a mean value of T_1 and T_2) for each observed ion. An example of such a plot is shown in Figure 6 for many ions of sulphur and silicon. Notice on the diagram that curves drawn through either the silicon or sulphur ions are similar, but displaced. Since the integral which depends only on the structure of the atmosphere cannot be systematically different for the silicon and sulphur ions, the difference can only be explained by an abundance difference, i.e. silicon is about 3 times as abundant as sulphur.

This type of analysis can be extended to most of the elements observed in the ultraviolet spectrum. Only for those ions for which optical depth effects are important,

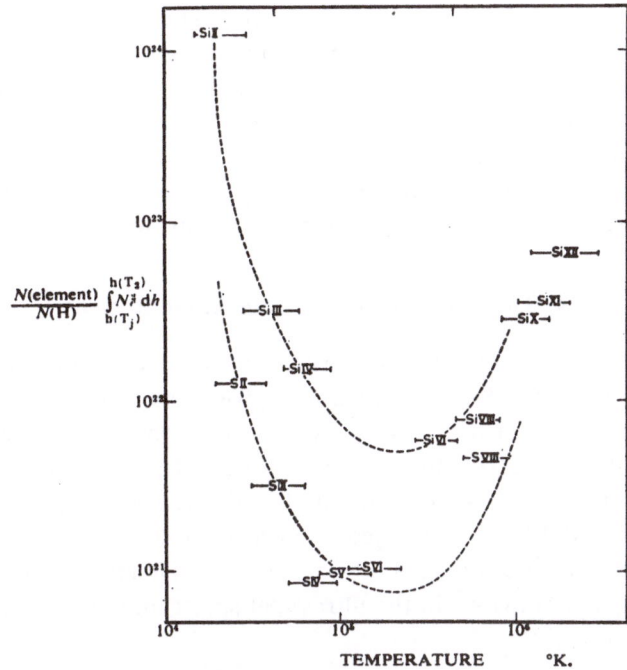

Fig. 6.   The value of $N(\text{element})/N(\text{H}) \int_R N_e^2 \, dh$ plotted against the temperature of the region of line formation for silicon and sulphur ions.

e.g. the hydrogen lines and HeI, is this method inapplicable. Thus the abundances relative to hydrogen cannot be determined solely from the ultraviolet data. Nevertheless interesting information about the abundances has been obtained in this way, e.g. that the abundances of silicon, iron and nitrogen are approximately equal in the outer solar atmosphere. More detailed abundance results are in the literature (e.g. Pottasch, 1967).

Once the abundances are known we are able to determine the value of the integral

$$\int_{V(T_1)}^{V(T_2)} n_e^2 \, dv,$$

which in principle describes the distribution of density and temperature in the atmosphere. There is not enough information, however, in the integral to be able to specify the density and temperature throughout the chromosphere-corona, where the ultraviolet lines are formed. This is true even if one assumes that the atmosphere is 'homogeneous', i.e. that the density decreases, and the temperature increases monotonically with height. In the case of a nearby homogeneous atmosphere one is able to conclude that the temperature gradient must be very steep between a temperature of 60 000 and 400 000 K, and these temperatures probably occur at densities several times 10^9 cm^{-3}.

If one makes an additional assumption about the outer atmosphere, e.g. that it is in hydrostatic equilibrium, one may then use the values of the integral to obtain a model of the temperature and density. Such solutions, while not unreasonable, are really not convincing, and what are really required to settle this important question are center-to-limb measurements in the ultraviolet, with sufficient angular resolution. These measurements are important not only for their direct application to the determination of the structure of the solar atmosphere, but as a guide to the interpretation of these measurements in stars, where center-to-limb observations are impossible.

4. Ultraviolet Emission from Active Regions

As one goes to shorter wavelengths, the emission from active regions begins to dominate the spectrum. Below 10–15 Å the active Sun gives many times the emission of the quiet Sun. It is probably mostly in the form of line emission. At 2 Å, lines of the helium-like Fexxv are observed. This indicates that for short periods of time temperatures of the order of 10–30×10^6 K are reached. The theoretical study of these regions is just beginning and will be greatly aided by ultraviolet and X-ray observations.

It may be that some stars have more important active regions than the Sun. This probably will be more obvious in the ultraviolet spectrum than in the visible.

References

Bonnet, R. M.: 1968, *Ann. Astrophys.* **31**, 597.
Gingerich, O. and Rich, J. C.: 1968, *Solar Phys.* **3**, 82.
Pottasch, S. R.: 1967, *Bull. Astron. Inst. Netherl.* **19**, 113.

Discussion

Swings: Is there no remaining problem on chromospheric iron abundance after your paper with Pecker where you used 'old' gf values?

Pottasch: The 'new' gf values of Garz and Koch for Fe I will change the neutral iron abundance, but most of the iron is still in the form of singly ionized iron, so the total iron abundance which Pecker and I have found, will not be affected.

Gingerich: Although the solar model I described yesterday was based explicitly on continuum intensities at the center of the disk, its construction was guided by a knowledge of the limb darkening, especially in the critical wavelength regions where the center-to-limb variations change from darkening to brightening. The model satisfies the continuum limb-darkening observations more successfully than any other model I have seen not only in infrared and ultraviolet, but also in the visual regions.

Pottasch: The importance of your method is that, unlike a model built up entirely of limb-darkening observations, it is applicable to stars other than the Sun. For spectral types similar to the sun, an initial model may be chosen by reference to the solar model. Although in the case of stars there is no check from the limb darkening observations, criteria may be determined from a complete study of the Sun as to how successful the model is.

RESONANCE LINES IN THE SOLAR CHROMOSPHERE

P. LEMAIRE

Laboratoire de Physique Stellaire et Planétaire, 91 Verrières-le-Buisson, France

Abstract. Stigmatic balloon spectra of the Sun in the vicinity of 2800 Å were obtained on September 22, 1968 and April 30, 1969. We compare the observed profiles of the H and K lines of ionized magnesium with the computed profiles of Athay and Skumanich and of Dumont. The discrepancy between observed and computed profiles of the Mg II lines is considerable. It is shown that this is a general fact for all chromospheric resonance lines of abundant elements. A brief review is given of different interpretations of this discrepancy.

1. Introduction

The observation of the H and K lines of ionized calcium has undoubtedly been the basis of most solar and stellar chromospheric investigations. In particular, the discovery made by Wilson and Bappu (1957) of a relation between the width of the central reversal of the lines in the spectra of several stars and the absolute magnitude of those stars increased the interest of such an investigation and initiated a considerable amount of work. The obtaining of high resolution profiles of these lines and other lines of chromospheric origin has been of great importance both for observers and for theoreticians.

Since the publication by Durand *et al.* (1949) of the first rocket solar spectra the investigations have been extended to the H and K lines of Mg II at 2800 Å and to the Lyman α and Lyman β lines of neutral hydrogen. All these lines show central self-reversals but the interpretation has, in general, only been partly successful.

Recent observations of solar Mg II lines by means of a balloon borne instrument are presented in this paper and their correlation with observations of other resonance lines of chromospheric origin is studied.

2. Observations of the H and K Lines of Mg II

The first profiles of the resonance lines of ionized magnesium at 2800 Å were obtained photographically during a rocket firing by Purcell *et al.* (1963) without spatial resolution on the disc. With a spectral resolution of some 3.10^{-2} Å both lines of the resonance doublet exhibit very strong central self-reversals similar to that observed at the centre of the H and K lines of ionized calcium (Figure 2a).

As a consequence of the rapid development of stratospheric balloon technology at the C.N.E.S. (National Centre for Space Research in France) a research programme was set up in France directed toward balloon spectrographic observations of the Mg II lines. A first stigmatic photographic spectrograph recorded centre-to-limb spectra of the lines with a resolving power of 5.10^{-2} Å and a resolution on the solar disc of the order of 2' of arc (Lemaire and Blamont, 1967).

Houziaux and Butler (eds.), Ultraviolet Stellar Spectra and Ground-Based Observations, 250–255.
All Rights Reserved. Copyright © 1970 by the IAU.

F Fig. 1.   15-sec exposure stigmatic balloon spectra in the neighbourhood of Mg II (2795, 2802 Å) lines
a across a solar diameter (flight of April 30, 1969). The black trace is caused by a guide mark used as a
re reference for the respective position of the Sun and the slit. The curvature of the line is an instrumental
effect. The shortest wavelengths of the spectrum are at the left of the photograph.

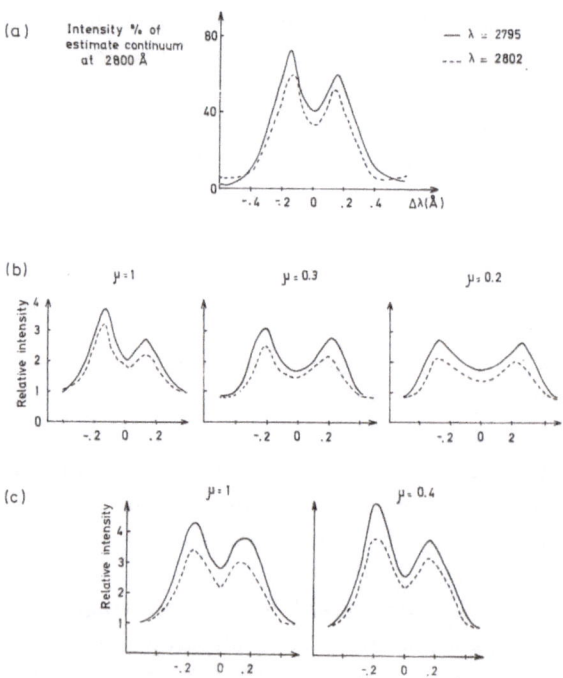

F Fig. 2.   Observed Mg II H and K profiles. (a) average over the central third of the solar disk on
A August 21, 1961 (Purcell *et al.*, 1963). – (b) and (c) Balloon spectra of April 1969 with a spatial
resolution of 10″ of arc. (b) over quiet regions and (c) over faculae.

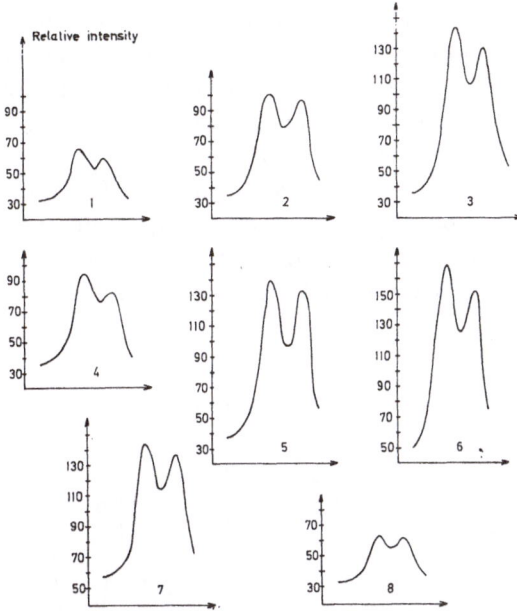

Fig. 3.   Observed Mg II K profiles over a sunspot during the flight of September 22, 1968. (1) and (8) are profiles of quiet Sun from both extremities of the spot. (2), (3), (5), (6), and (7) are profiles of the plage region around the spot. (4) is the profile from the umbra to the spot.

An improved instrument increased both spatial and spectral resolutions and two successful flights on September 22, 1968 and April 30, 1969, allowed us to obtain stigmatic profiles along a diameter of the solar disc (Lemaire, 1969).

Figure 1 shows a spectrum obtained on the latter flight with an angular resolution of $\pm 7''$ of arc and a spectral resolution of 35 mÅ.

The following characteristic features can be noticed on those spectra:

(i) The distance between the two emission maxima of a single line increases from the centre to the limb. This variation is also observed for the H and K lines of Ca II.

(ii) The intensity emitted in both lines varies considerably over the disc depending on whether the slit of the spectrograph has cut across active or quiet regions.

(iii) The intensities of the peaks of a single line are unequal at the centre of the disc but become gradually equal as one goes closer to the limb (Figure 2b).

(iv) In active regions this inequality can either completely vanish or be completely reversed.

It is to be noticed that the profiles given in Figure 3 are identical with those observed in the H and K lines of Ca II (Engvold, 1967). The Lyman-α profiles obtained by Tousey *et al.* (1964) are also in good agreement with these characteristics.

3. Comparison with Computations

Several successful computations of the H and K lines of Ca II and Mg II have been made, especially by Dumont (1967a, b) and Athay and Skumanich (1968a, b). These authors made the assumption that the chromosphere is homogeneous and computed the source function and optical depth at a given point in the lines by solving simultaneously the equations of statistical equilibrium and radiative transfer.

In Figure 4, we compare the results obtained by Athay and Skumanich and by Dumont with the observed profiles of the MgII lines. The profiles are of the centre of the disc. The distance between the two peaks appearing on the computed profiles is in good agreement with the observations. However, the lack of symmetry cannot be reproduced by the computation and the same is true for the width of the emission

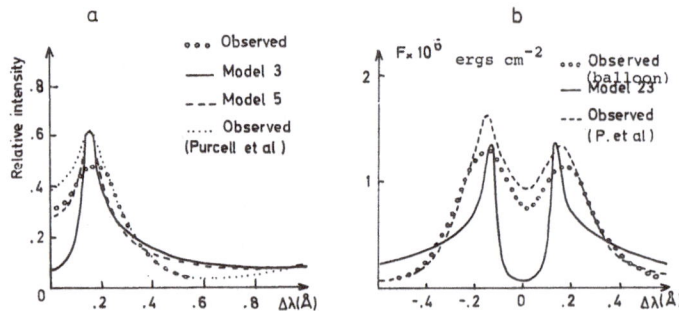

Fig. 4.   Comparison of computed and observed MgII K profiles. (a) Athay and Skumanich (1968). – (b) Dumont (1967).

peaks. Further, the computed centre-to-limb variation of the intensity ratio between the two peaks of a single line does not represent the progressive disappearance of the asymmetry toward the limb. Similar discrepancies in the theoretical and observed profiles appear for the CaII lines (Dumont, 1967a, b; Athay and Skumanich, 1968a, b) and for the Lyman-α line (Cuny, 1968).

The explanation for these discrepancies seems to be due to the use of a homogeneous chromospheric model which is obviously a very coarse approximation to the problem. If it is assumed that the emission and absorption occur in descending and ascending cells whose velocities are normal to the atmospheric layers of different temperatures, then it is possible to explain the appearance of non-symmetrical profiles.

This assumption is reinforced when one examines very high space-resolved profiles of the H and K lines of CaII such as those obtained by Zirin (1966) and Dodson-Prince and Mohler (1966). When the resolution is close to 1″ of arc, the emission occurs in totally inhomogeneous areas and the profiles for a single element are composed of a central emission line shifted preferentially toward the short wavelengths with reabsorption occurring in the wings. Each element therefore does show an unsymmetrical reversal and observing the disc with a resolution inadequate to separate these elements leads to a general blurring which gives the profile the appearance of a self-reversed line.

It is likely that the same thing occurs for both MgII and HI resonance lines but at

TABLE I

References related to observations and computations of H and K lines of Ca II

Observations	
Very high spatial resolution ($\simeq 1''$)	Other observations
Zirin (1966)	Goldberg *et al.* (1959)
Dodson-Prince and Mohler (1966)	Zirker (1968)
	Linsky (1968)
	White and Suemoto (1968)

Computations	
Unsymmetrical profiles	Symmetrical profiles
Miyamoto (1957)	*Thin chromosphere:*
Kulander (1967)	Zirker (1968)
	Linsky (1968)
	Thick chromosphere:
	Dumont (1967a, b, 1969)
	Athay and Skumanich (1968a, b)

a different altitude.* As a consequence, the interpretation of resonance line spectra emitted in the chromosphere in terms of only LTE departures seems highly questionable.

Table I reviews recent observations of the H and K lines of Ca II. We indicate also the different interpretations that have been advanced by different authors to explain the shapes of the lines. Among them two do account for unsymmetrical lines. Both introduce model atmospheres with macroscopic velocity fields:

(1) Miyamoto (1957) assumes that the upper part of the chromosphere is moving down with respect to the lower part.

(2) Kulander (1967) introduces a discontinuity in the variation of the macroscopic velocity with height, which corresponds to the level in the atmosphere where the lines originate.

The schematic models which do not take into account the chromospheric inhomogeneities could certainly be improved by introducing a network, spread all over the disc, of hot and cold cells, each pair of them possessing its own differential velocity or discontinuity in velocity. Such cells might well be the spicules and interspicular areas.

4. Conclusion

The observation of strong asymmetries in the shape of Ca II as well as Mg II and H I resonance lines cannot be explained with LTE departures only. It seems therefore necessary to introduce upward and downward moving cells, which is a reasonably

---

* According to Zirker (1968) and Thomas and Athay (1961) the Ca II, Mg II and H I resonance lines are emitted in layers located, on the average, at heights of 2000, 3000 and 4000 km respectively.

realistic hypothesis. It is more likely that a better agreement between observations and computations might be reached by solving simultaneously the equations of radiative transfer and statistical equilibrium for each cell of the chromospheric network. This is obviously a gigantic task.

We suggest that a better knowledge of both dynamical movements and physical properties of the chromospheric layers could be reached by

(i) High resolution ($\sim 1''$) simultaneous observations in CaII, MgII and HI resonance lines for a given point on the disc.

(ii) Computations of the shapes of the three lines, using inhomogeneous models, accounting simultaneously for the observations in the three kinds of lines.

This programme cannot be undertaken without the use of rocket or satellite observations.

Acknowledgments

We wish to acknowledge helpful discussions we have had with Dr. R. M. Bonnet and other scientists in the laboratory. The experiments have been supported by C.N.E.S. through grant no. 69 224.

References

Athay, R. G. and Skumanich, A.: 1968a, *Solar Phys.* **3**, 181.
Athay, R. G. and Skumanich, A.: 1968b, *Solar Phys.* **4**, 176.
Cuny, Y.: 1968, *Solar Phys.* **3**, 204.
Dodson-Prince, H. and Mohler, O. C.: 1966, *The Fine Structure of the Solar Atmosphere*, Franz Steiner Verlag, Wiesbaden, pp. 98–101.
Dumont, S.: 1967a, *Ann. Astrophys.* **30**, 861.
Dumont, S.: 1967b, *Ann. Astrophys.* **30**, 421.
Dumont, S.: 1969, *Astron. Astrophys.* **2**, 45.
Durand, E., Oberley, J. J., and Tousey, R.: 1949, *Astrophys. J.* **109**, 1.
Engvold, O.: 1967, *Solar Phys.* **2**, 234.
Goldberg, L., Mohler, O. C., and Müller, E. A.: 1959, *Astrophys. J.* **129**, 119.
Kulander, J. L.: 1967, *Astrophys. J.* **147**, 1063.
Lemaire, P. and Blamont, J. E.: 1967, *Astrophys. J.* **150**, L129.
Lemaire, P.: 1969, *Astrophys. Letters* **3**, 43.
Linsky, J. L.: 1968, Smithsonian Astrophysical Observatory, Special Report **274**.
Miyamoto, S.: 1957, *Publ. Astron. Soc. Japan* **9**, 146.
Purcell, J. D., Garrett, D. L., and Tousey, R.: 1963, in *Space Research*, vol. III, p. 781.
Thomas, R. N. and Athay, R. G.: 1961, *Physics of the Solar Chromosphere*, Interscience, New York.
Tousey, R., Purcell, J. D., Austin, W. E., Garrett, D. L., and Widing, K. G.: 1964, *Space Research*, vol. IV, p. 703.
White, O. R. and Suemoto, Z.: 1968, *Solar Phys.* **3**, 523.
Wilson, O. C. and Bappu, M. K.: 1957, *Astrophys. J.* **125**, 661.
Zirin, H.: 1966, *The Solar Atmosphere*, Blaisdell Publishing Co., London, pp. 228–230.
Zirker, J. B.: 1968, *Solar Phys.* **3**, 164.

Discussion

Deutsch: Can one estimate the magnitude of the differential velocities that would be required to reproduce the asymmetry observed in K_2?

Hearn: The paper by Miyamoto gives a calculation of the effect of velocities on the asymmetries of the Ca, He, K lines and this leads to an estimate of the magnitude of the velocities required.

ON THE CONTRIBUTION OF SOLAR ACTIVITY TO THE
ULTRAVIOLET SPECTRUM OF THE SUN

A. V. BRUNS, V. K. PROKOFIEV, and A. B. SEVERNY

Crimean Astrophysical Observatory, U.S.S.R.

Abstract. As measured from space, the contribution of one moderate flare to the emission spectrum of the Sun in the far-ultraviolet (304 Å, Lyman continuum, etc.) is comparable with the emission of the whole undisturbed solar disc.

The grazing incidence spectrometer of the Crimean astrophysical observatory installed on board Kosmos 166 (flown on June 16, 1967) recorded the solar spectrum in the region 800–950 Å from June 16, 1967 until July 5, 1967. The spectrometer used a concave grating (300 lines mm^{-1}) whose radius of curvature was 500 mm, and a scanning system permitting photoelectric recording of the region 800–950 Å (each cycle took 1.5 min) with the aid of a copper-beryllium photomultiplier and a suitable counting-rate technique. The guiding system kept the solar beam fixed on the slit of the spectrometer with an accuracy of $\pm 10'$ (Bruns *et al.*, 1968).

Kosmos 166 also carried X-ray counters installed by Prof. S. L. Mandelstam, who kindly supplied us with observational data of X-ray emissions during the periods of our observations.

About 50% of all records have been measured and reduced to give the data on the *variations* of ultraviolet radiation of the Sun. Three types of these variations were found:

(1) Short-term variations connected with solar flares. Figure 1 is an example showing variations connected with a flare of class 1 N, which appeared at $18^h 06^m$ UT, on June 18, 1967 at the Eastern limb (18.06–19.28) (maximum at 18.30). At the same time the X-ray flare was recorded in the corona at a height of about 20000 km (according to the data of Prof. Mandelstam). We started recording the UV-radiation at 18.45 UT when the intensity was increasing, although the X-ray intensity was decreasing. The maximum of UV radiation was attained at $19^h 10^m$, about 30 min *after* the onset of the flare. Such diversity in the run of different line-emissions of flares is probably a quite normal phenomenon (see Hall and Hinteregger, 1968), and it is well known that the yellow coronal line frequently appears around the flare sometimes half an hour after the onset and then remains for a long time.

(2) Moderately slow variations in a matter of a few hours are shown on Figure 2 and are connected with a series of solar flares leading sometimes to intense flares, or to enhanced activity. In the first example (on June 22), the observations covered a period of 9 hours and several brightenings and minor flares were followed by two more intense flares of classes 1 N and 1 B. The relative increase of intensity in the ultraviolet reaches 25–30%. In the other case (June 17) the recordings covered 6 hours. They began during small flares (1 N and faint brightenings) and ended with two important flares, one of which is a 2B-flare. We also had an appreciable increase of intensity (by 15–30%) connected with the increase of flare activity.

Houziaux and Butler (eds.), Ultraviolet Stellar Spectra and Ground-Based Observations, 256–259.
All Rights Reserved. Copyright © 1970 by the IAU.

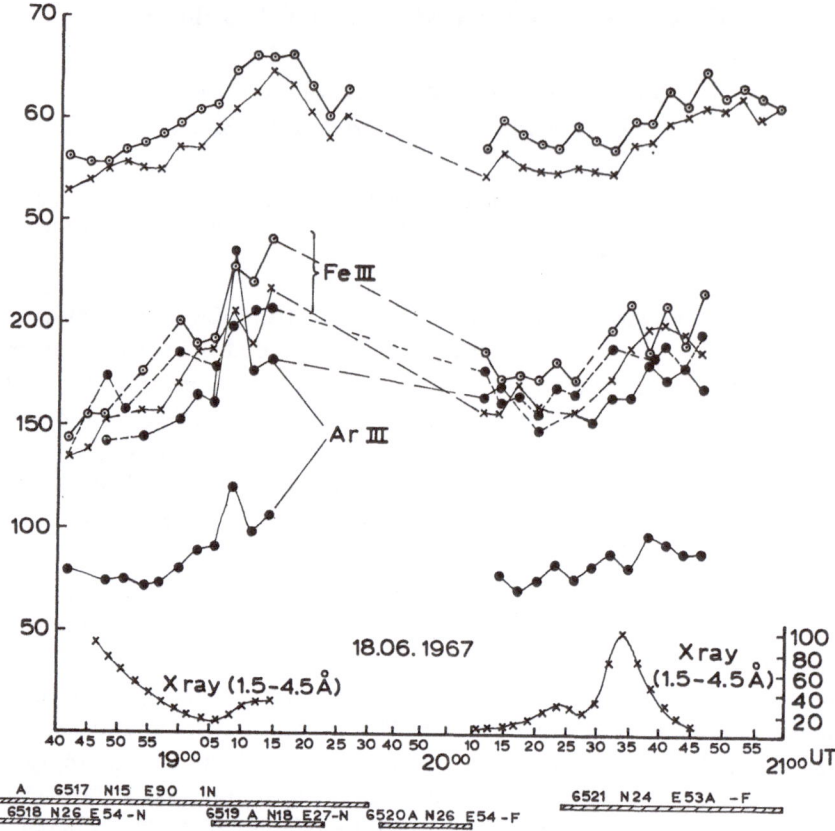

Fig. 1. The intensities (arbitrary units) in the UV regions during the solar flares June 19, 1967. – Above: two different parts near the limit of the Lyman series (920–950 Å): × = the short wave part, and ○ = the minimum at the middle wavelength. Fe III: three different parts in the region 870–840 Å: × = the long-wave part, ● = the short-wave part, and ○ = at the middle-wavelength. Ar III: the region 875–890 Å: the upper curve = the short wave part, and the lower curve = the long wave part. – Bottom: the durations of solar flares, the moments of their maxima, numbers of group, coordinates of flares and their importance (according to Boulder bulletin) are noted.

(3) Slow day to day variations of ultraviolet radiation of the Sun accompanying the variations in solar active regions. On Figure 3, we show mean values of the intensity (average for each revolution of the satellite around the Earth) for the period from June 16 until July 2, 1967. On the same figure the total area S_w of sunspots is also plotted (crosses). There, we have also a relative increase, by 25–40% the total intensity of the solar disc accompanying the increase of solar activity, provided that the total area of sunspots is adopted as a measure of this activity.

Therefore, we see that solar activity and flares in particular can contribute appreciably (up to 50%) to the total flux of ultraviolet emission from the whole disc of the Sun, and it is not impossible that the same kind of activity in stars can produce important changes in the ultraviolet stellar spectra.

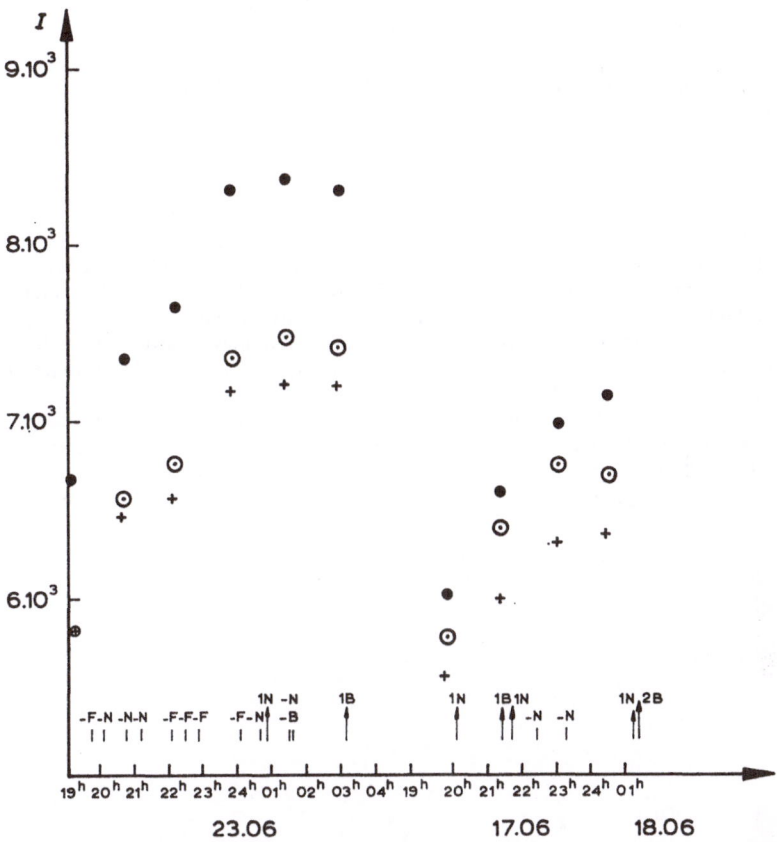

Fig. 2. The intensities in the region of the limit of the Lyman series (920–950 Å) during the series of solar flares of June 17 and June 23, 1967. ● = the short wave maximum in this region, ○ = the long wave maximum in this region, and + = the minimum of intensity in this region. – Bottom: the moments of onset of solar flares and their importance (from *Boulder Bulletin*) are noted.

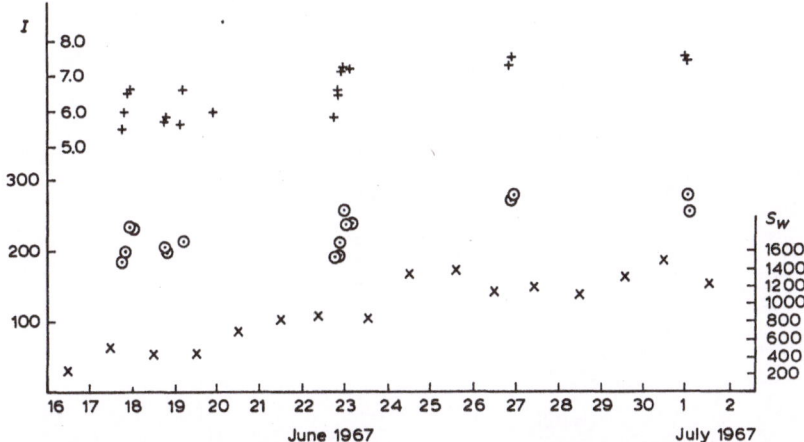

Fig. 3. The intensities in the UV region (arbitrary units) and the area of the sunspots (S_w). + = the region near the limit of the Lyman series (920–950 Å) ○ = the region of the Fe III lines (840–870 Å), and × = the area of the sunspots (S_w) according to Bulletin *Solnechnye Dannye*.

References

Bruns, A. V., Prokofiev, V. K., and Severny, A. B.: 1968, XI COSPAR, Tokyo, May.
Hall, L. A. and Hinteregger, H. E.: 1968, XI COSPAR, Tokyo, May.

Discussion

Jordan: I would like to ask which lines are observed to increase during the flare?

Severny: These lines presumably (not finally identified due to low resolution of our records) are 834 Å (blend of O II, O III, S III), Fe xv 284 Å (from the 3rd order), 845 Å, 850 Å Fe III and 878 Å Ar III.

Carruthers: Did you see any enhancement of the Lyman continuum in flares?

Severny: Yes, we suspected such enhancements and very preliminary consideration showed that the increase of electron temperature (estimated by the slope of the density curve in the region Lyman α to continuum) can reach 2000–3000 K.

Burton: Is the spectrum observed in the region 800–950 Å complicated by overlapping spectral lines formed by higher grating orders ($m = 3, 4$)? The high intensity of He II 304 Å and ionized iron emission at about 200 Å is a possible source of blending in the wavelength region which you observe.

Severny: Yes, of course, and we use this blending for rough estimates of intensity changes in 304 Å and 284 Å Fe xv.

INTENSITY DISTRIBUTION IN THE LYMAN-α LINE
AT THE SOLAR LIMB

J. C. VIAL

Laboratoire de physique stellaire et planétaire, Verrières-le-Buisson, France

Abstract. The distribution of the solar intensity in the Lyman α line has been measured close to the visible limb. It is compared to a computation including LTE departures (as evaluated by Y. Cuny). As far as the profile of the line and the intensity (integrated over the line) are concerned, the interspicular model of Coates is the only one which seems to agree with the observations.

1. Introduction

To date, models of the solar chromosphere have been deduced from eclipse flux measurements in the solar visible continuum, radio wavelength measurements and, more recently, far UV spectra of the Sun. Brightness measurements at the limb have been made in the Lyman α line during the eclipse of November 1966. The results are compared to theoretical computations including several models of the chromosphere. As a result, the agreement between computations and observations can be improved if the temperature gradient in the chromosphere is modified.

2. Experimental Data

The measurements were made from a rocket during the eclipse of November 1966, visible from the Southern Hemisphere. The rocket was launched from Argentina. The Service d'Aéronomie du C.N.R.S. (France) was responsible for the scientific payload.

The experimental results have been reported in [3]. The detector was a LiF window ionisation chamber filled with CS_2. The field of view was 2° and the flux was received from the whole disc. The bandpass was 1050–1250 Å. The measured flux ϕ is related to the phase of the eclipse through:

$$\phi(t) = \int_{\Sigma(t)} I(P)\, dS$$

where $I(P)$ is the intensity over the band of the detector at point P on the disc. The integration is made over the whole uncovered area $\Sigma(t)$ of the disc at time t. Through inversion of this integral, one can obtain the value of $I(P)$.

A description of the inversion method used is given in [3] together with the different assumptions introduced in the mathematical processing of the data.

The net result is shown in Figure 1 where the emerging intensity I is plotted versus the position r on the solar disk, measured in arc sec. The origin of position is deduced from simultaneous flux measurements made during the same rocket flight in the in-

Houziaux and Butler (eds.), Ultraviolet Stellar Spectra and Ground-Based Observations, 260–270.

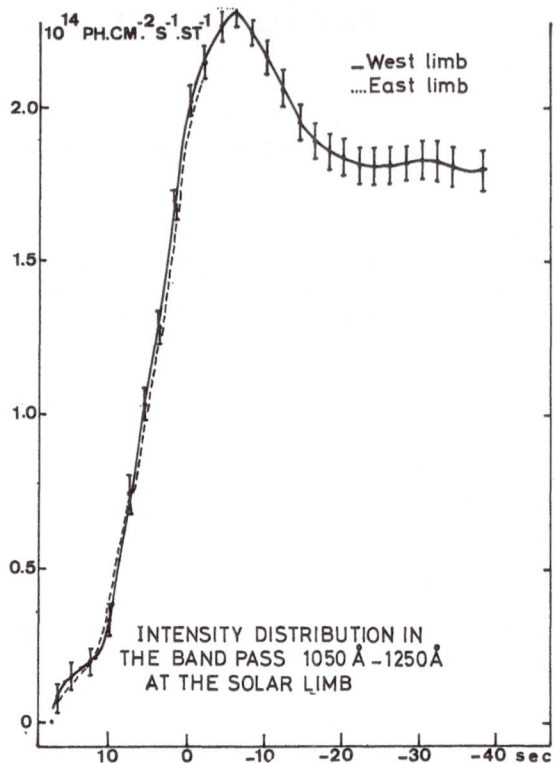

Fig. 1.   Computed intensity integrated over frequency as a function of the distance to the visible limb in seconds of arc (from Blamont and Malique).

frared, with a photomultiplier. Figure 1 has been communicated to me by Blamont and Malique, before publication.

The curve of Figure 1 shows two important characteristic features:

(i) the position of the point of inflexion in the curve of variation of intensity at the limb is located 3″ of arc outside the visible limb, within an accuracy of ±½″.

This position will hereafter be called 'Lyman α solar limb'.

(ii) the Sun is limb-brightened, the maximum being 6″ inside the visible limb.

This curve has been used to check the results of a computation using three solar models of the chromosphere and taking into account departures from LTE.

A description of these computations is given below.

3. Computation of the Emerging Intensity

A. METHOD OF COMPUTATION

The actual transfer equation in a spherically symmetrical medium is:

$$\frac{dI}{ds} = \mu_0 \frac{\delta I}{\delta r} + \frac{1 - \mu_0^2}{r} \frac{\delta I}{\delta \mu_0} = k(I - S) \quad \text{(see Kourganoff [8])}$$

where k is the absorption coefficient.

Close to the limb, the second term of the right-hand side of this equation cannot be considered as negligible in comparison to the first one.

The method we have used was also applied by Ambartsumyan [1] to the computation of the intensity emitted by prominences and by Pagel [10] for the computation of the continuous spectrum emitted in an optically thin medium.

The emergent intensity I at a given point in the line profile is

$$I_\lambda = \int\limits_{-y_0}^{+y_0} k_\lambda S(y) \exp\left[- \int\limits_{-y_0}^{+y} k_\lambda \, dy'\right] dy$$

where y measures the geometrical path along a line of sight at a given distance from the limb.

The absorption coefficient at wavelength λ is given by: $k_\lambda = k_{\lambda 0} H(a, v)$ where the subscript 0 refers to the line centre; $H(a, v)$ is the Hjerting function:

$$H(a, v) = \frac{a}{\pi} \int\limits_{-\infty}^{+\infty} \frac{e^{-z^2} \, dz}{a^2 + (v - z)^2}$$

with $a = \delta/\Delta v_D$, where δ is the natural width; Δv_D is the Doppler width,

$$\Delta v_D = \frac{v_0}{c} \sqrt{(2KT/m) + \xi^2}$$

and

$$v = \Delta\lambda/\Delta\lambda_D .$$

T is the temperature and ξ the microturbulent velocity.

B. MODEL ATMOSPHERES

Three models have been used (Figure 2) to compute the source function and the optical depth.

1. *H.A.O. model* [2]

This model has been obtained from the continuum emission at 4760 Å and 3640 Å. It describes the atmosphere for a single set of parameters T_e, n_e, p (electron temperature, electron density and total pressure).

2. *Coates' model* [4]

This model is deduced both from eclipse measurements at 8.6 mm and from a scan of the disk at 4.3 mm.

The model attempts to describe the atmosphere by two sets of parameters T_e, n_e, p. For both measurements, the spatial resolution was not high enough to separate spicules from interspicular matter. One can just argue that the model thus established represents correctly the observations. In the interspicular matter, the model is characterized by the existence of a plateau between 3000 and 4000 km and then, a rapid increase in temperature.

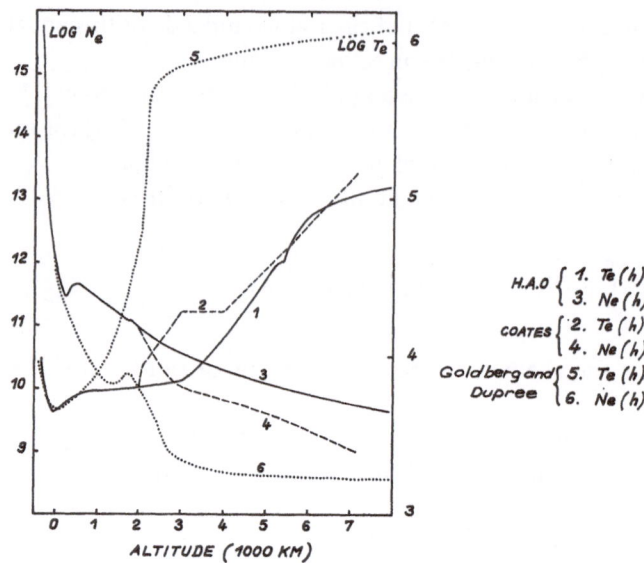

Fig. 2.   1–3 H.A.O. model; 2–4 Coates' model; 5–6 Goldberg and Dupree model; 1–2–5 electronic temperatures plotted vs. the altitude in the chromosphere; 3–4–6 electron densities versus the altitude in the chromosphere.

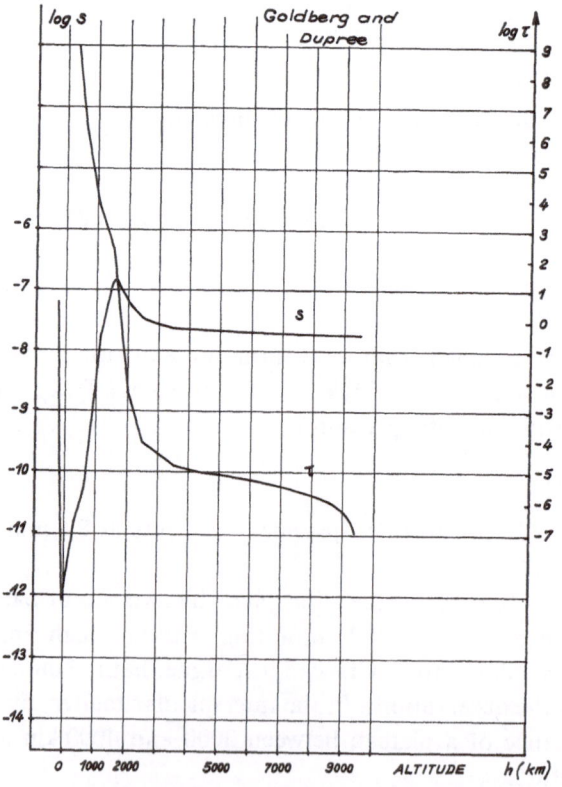

Fig. 3.   For Goldberg and Dupree model, the source function S and the optical depth τ are plotted on logarithmic scale vs. altitude.

Spicules are very cold (6400 K) and dense.
We have used, here, the interspicular model.

3. *Goldberg and Dupree model* [7]

This model was deduced from the recent OSO IV measurements of the solar UV intensity which is emitted in the chromospheric layers and the corona. It presents a sudden rise in temperature at about 2000 km that leads to coronal temperatures without any transition zone. Electron densities are consequently very low.

C. COMPUTATION OF THE SOURCE FUNCTION AND THE OPTICAL DEPTH

Y. Cuny's computing program has been used for the simultaneous resolution of the radiation transfer and statistical equilibrium equations which leads to the source function in the Lyman α line.

The source function S and the optical depth τ in the centre of the line are represented as functions of height in the chromosphere, on Figure 3, for the Goldberg and Dupree model.

4. Results of the Computation

A. INTENSITY RECEIVED AT 1 AU INTEGRATED OVER FREQUENCY

The calculation has been made for the first two models by Cuny [6]. The three computed values (Table I) lie higher than the value measured by Tousey of 6 ergs sec^{-1} cm^{-2} at the Earth, but they depend on the values of the electron collision cross-sections adopted for the computation and the width adopted for the line.

Coates' model leads to the best result with a value of 9.5 ergs sec^{-1} cm^{-2} at the Earth.

Table II indicates for Coates' model how the value of the flux does depend upon the wavelength interval of integration.

TABLE I

Intensity integrated over the disc and over the line profile within a bandpass of 5 Å in ergs sec^{-1} cm^{-2} at 1 AU

Model	H.A.O.	Coates	Goldberg and Dupree
Flux	24	9.5	9.95

TABLE II

Coates' model
Variation of the intensity integrated over the disc and in the line with the width of the bandpass $\Delta\lambda$

$\Delta\lambda$ (Å)	I ergs sec^{-1} cm^{-2} at the earth
1.6	6.75
3	8.4
5	9.5
10	10.9

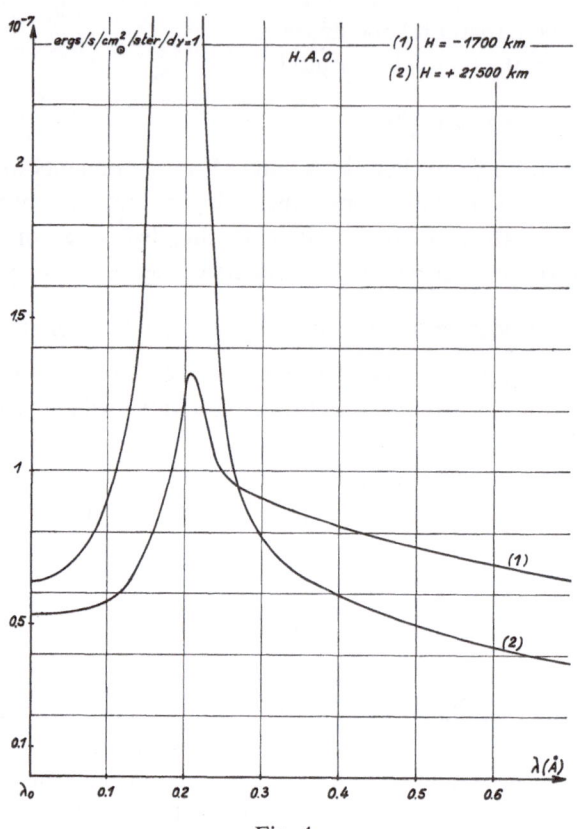

Fig. 4.

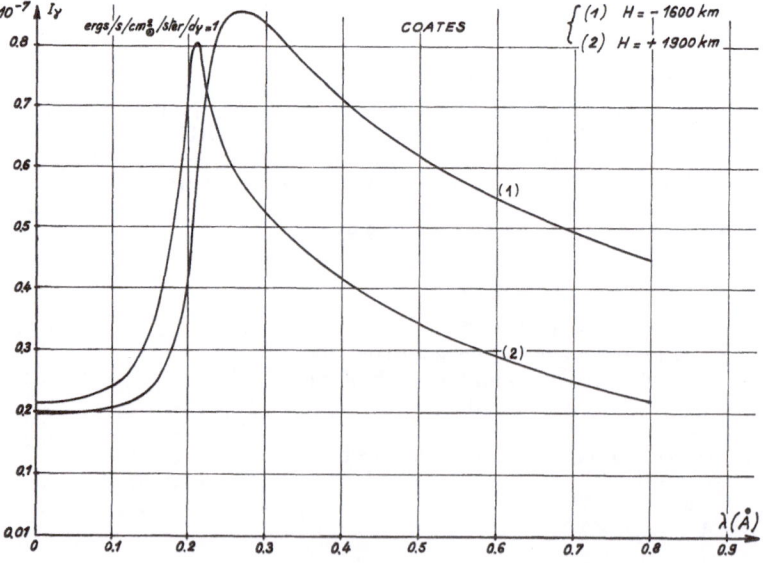

Fig. 5.

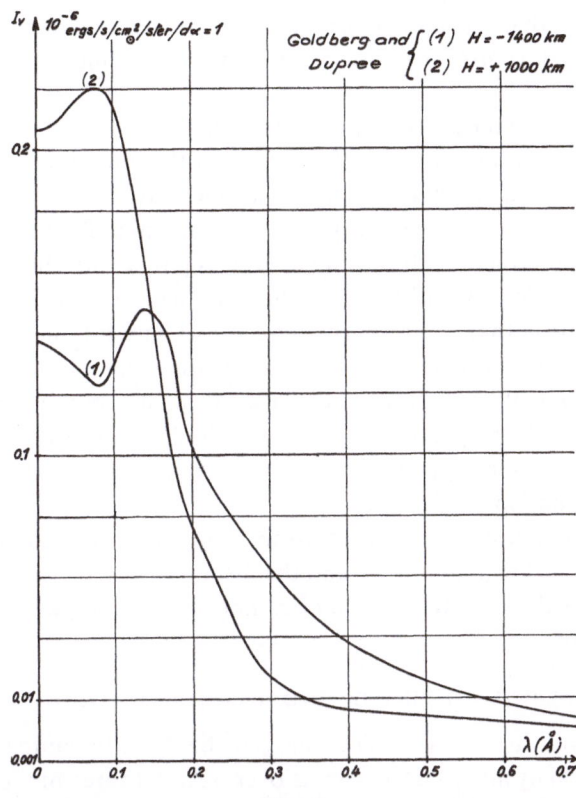

Fig. 6.

Figs. 4–6. Monochromatic intensity versus the distance from the centre of the line, at two positions on the disc close to the limb; (1) outside the visible limb; (2) inside the visible limb; (4) H.A.O. model; (5) Coates' model; (6) Goldberg and Dupree model.

The high values given by the H.A.O. model which are in complete disagreement with Tousey's measurements, are likely to be due to the high electron densities of this model. The Coates and Goldberg and Dupree models give values in better agreement with Tousey's measurements.

B. LINE PROFILE FOR VARIOUS POSITIONS ON THE DISC

The results are shown in Figures 4, 5 and 6.

TABLE III

Ratio of the intensities at the peak and at the centre of the line for the three models at the disc centre

Model	H.A.O.	Coates	Goldberg and Dupree
I_P/I_0	2.1	3.2	1.0

It can be noticed that all three models lead to broader and deeper profiles close to the limb, than at the centre of the disc, which is in agreement with Tousey's observations.

Let us take I_p as the intensity at a peak of the line.

In Table III we indicate the value of the ratio of the intensity emitted at the centre of the disc in the peak (I_p) to that emitted in the line centre.

It must be noticed that Tousey [11] gives a mean value for this ratio of 1.6. This measured value is intermediate between values computed with the H.A.O. and Goldberg and Dupree models but differs strongly from that given by Coates' model. This might be an indication that, as formerly noticed by Cuny [6], the steeper the temperature gradient for a given temperature, the lower the value of I_p/I_0.

On the other hand, the two first models lead to a distance between the peaks at the disc centre of 0.4 Å, in agreement with Tousey's observations [11].

That distance corresponds approximately to $\tau \sim 1$ for the altitude in the atmosphere where the source function is a maximum.

The Goldberg and Dupree model leads to special results: close to the limb, the line is reversed with a central emission peak and the distance between the peaks is about 0.3 Å. At the centre of the disc, the profile does not present any reversal, which contradicts the observations.

C. VARIATION OF THE INTENSITY INTEGRATED OVER THE LINE WIDTH AT THE LIMB

The results are given in Figure 7. They have to be directly compared to the observations. The interval of integration extends over 5 Å. All three models lead to a limb-brightening but none is able to represent the position of the maximum intensity according to observations.

H.A.O. Model: The 'Lyman α limb' is 4600 km above the minimum temperature or 4900 km above the visible limb, which corresponds to an angular distance of 7″, a value of the order of 2 greater than what is deduced from observations.

The brightening is 20% of the intensity at the disc centre.

Coates' model limb is 2900 km above the visible limb, i.e. $\simeq 4″$, in better agreement with observations. The brightening is very considerable and sharp (about 100% of the value at the centre of the disc).

It has also to be noticed that the intensity decreases continuously as one goes towards the centre.

Goldberg and Dupree model's 'Lyman α limb' is 2000 km above the visible limb and in good agreement also with observations. The model leads to oscillations in the intensity at the limb.

We have checked whether the results depend:

(1) on the interval of integration. If the intensity at the limb depends on $\Delta\lambda$ the position of the 'Lyman α limb' and of the maximum intensity remains fairly constant; so does the value of the intensity at the centre of the disc.

In figure 7, the curves 1, 2 and 3 correspond to an interval of integration of $\Delta\lambda \simeq$ 1.6 Å whereas curve 2′ refers to Coates' model with $\Delta\lambda = 5$ Å. Here we can see how

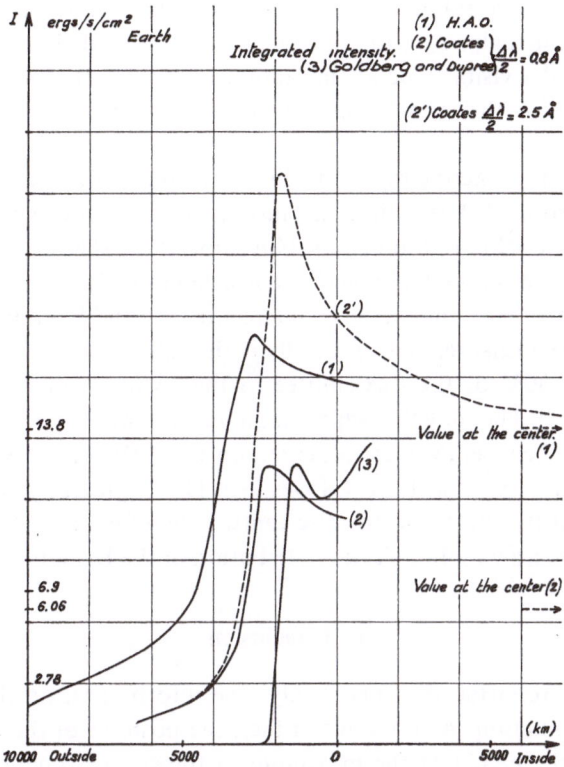

Fig. 7.   Integrated intensity over frequency vs. the distance on the disc (on a band pass of 1.6 Å)
(1) H.A.O. model; (2) Coates' model; (3) Goldberg and Dupree model;
(2') integrated intensity over 5 Å.

important would be the knowledge of measured values of the absolute intensity emitted by the Sun in order to discriminate the models. Unfortunately this value could not be obtained during the flight of Blamont and Malique.

(2) on the choice of the microturbulent velocity distribution. For this purpose we have tried out a model with a constant value of $\xi = 4$ km sec^{-1} all across the chromosphere, and another one for which $\xi = 20$ km sec^{-1} for $h > 2000$ km and $\xi = h/100$ for $h < 2000$ km.

We did not notice any important differences between the two results given by the two kinds of models.

5. Discussion

The comparison between the distributions of computed integrated intensities and Blamont and Malique's observations shows that the Goldberg and Dupree model would be satisfactory, for the position of the 'Lyman α limb'. Within the experimental error limits (about $\pm$ 350 km) Coates' limb agrees also with observations.

The Goldberg and Dupree model leads to a profile of the line at the limb which does not present any reversal at all. We see here the importance of detailed measured profiles for different regions on the disc. However, none of the models leads to any brightening inside the visible disc as that indicated by the observations.

Before incriminating the models, several sources of experimental error have to be discussed:

(1) the presence of an active region at the limb would affect the distribution of intensity and the position of the maximum. The only feature noticed is a facula appearing on the K line spectroheliogram taken on November 12, 1966. This feature was taken into account by Blamont and Malique who corrected their flux curve.

(2) The effect of other emission lines intercepted by the bandpass of the detector might affect the flux measurements, specially if they show a very intense limb-brightening, since about 80% of the flux emitted by the Lyman α line is concentrated in a band of 2 to 3 Å [12]. The only line which might be of some importance is that emitted by Si III at 1206 Å. The recent values measured on OSO-IV for the flux emitted in this line is 0.059 ergs cm^{-2} sec^{-1} at 1 AU (Noyes [9]). The brightening at 0.9 R is only 1.75 in units of the intensity emitted at the disc centre. Therefore, such a brightening is not large enough appreciably to modify the distribution of the Lyman α line intensity close to the limb.

6. Conclusion

The measurements reported in [3] can only give information on the position of the maximum source function. As a matter of fact, the position of the 'Lyman α limb' is closely related to the height of the maximum of the source function. As noticed by Cuny, the maximum takes place at heights where the electron temperature reaches a value of some 20000 K, which conditions the distance between the two peaks of the self reversal profile. The Goldberg and Dupree model fits this condition. However, it does not lead to any reversal except close to the limb, because at the altitude where the source function S is maximum, the optical depth is very low.

Coates' model might agree with measurements, as far as absolute intensity, distance between peaks a position of the Lyman α limb, are concerned. Furthermore, due to the plateau in temperature between 3 and 4×10^3 km of this model, the central reversal in the Lyman β line can be accounted for (Cuny [6]).

The value of the depression of the profile ($I_P/I_0 \sim 3$) is larger than that which was formerly measured by Tousey [11] in a quiet region of the active Sun ($I_P/I_0 \sim 1.6$), but agrees with the value measured, on August 22, 1962, in a quiet region of the quiet Sun ($I_P/I_0 \sim 3$) [11]. Anyway, a higher gradient of temperature above the plateau of Coates' model might reduce the importance of this depression.

However, one must keep in mind that the chromosphere is completely heterogeneous and that the experimental distribution of intensity reported by Blamont and Malique gives us information only in terms of an average model.

Therefore, further improvements have to be achieved in the direction of simultaneous observations and computations of the Lyman α and β profiles at a given position on the disc.

Acknowledgements

I am much indebted to Y. Cuny for having provided me her computer program and to her and R. M. Bonnet for helpful discussions. I wish to thank C. Malique for communicating his most recent results before publication.

References

[1] Ambartsumyan, V. A.: 1958, *Theoretical Astrophysics*, Pergamon Press, p. 317.
[2] Athay, R. G., Menzel, D., Pecker, J. C., and Thomas, R. N.: 1955, *Astrophys. J. Suppl. Ser.* **1**, 505.
[3] Blamont, J. E. and Malique, C.: 1969, *Astron. Astrophys.* **3**, 135.
[4] Coates, R. F.: 1958, *Astrophys. J.* **128**, 83.
[5] Cuny, Y.: 1967, *Ann. Astrophys.* **30**, 143.
[6] Cuny, Y.: 1968, *Solar Phys.* **3**, 204.
[7] Goldberg, L., Noyes, R. W., Parkinson, W. H., Reeves, E. M., and Withbroe, G. L.: 1968, *Science* **162**, 95.
[8] Kourganoff, V.: 1967, *Introduction à la théorie générale du transfert des particules*, Gordon, and Breach, New York.
[9] Noyes, R.: private communication.
[10] Pagel, B. E. J.: 1956, *Monthly Notices Roy. Astron. Soc.* **116**, 608.
[11] Tousey, R.: 1963, *Space Sci. Rev.* **2**, 57.
[12] Zirin, H., Hall, L. A., and Hinteregger, H. E.: 1963, *Space Res.* **3**, 760.

Discussion

Deutsch: Does the geocoronal Lyman α interfere with the determination of the central intensity of the chromospheric line?

Vial: The geocoronal absorption is too narrow to have any influence on the intensity integrated over frequency.

A HIGH-RESOLUTION SOLAR SPECTRUM 2000 Å–2200 Å

B. B. JONES, B. C. BOLAND and R. WILSON

Science Research Council, Astrophysics Research Unit, Culham Laboratory, Berks., England,

and

S. T. F. ENGSTROM

Stockholm Observatory, Saltsjöbaden, Sweden

A high-resolution solar spectrum in the range 2000–2200 Å was obtained in a recent flight of a sunpointing Skylark rocket. This was launched at 04.21 hr UT on April 22, 1969 from Woomera and reached an apogee of 178 km. An optical alignment system operating on the main vehicle pointing system gave a net stabilisation of ± 3 arc sec in the position of the solar image relative to the spectrograph slit. The slit, of length 1.0 mm, was set in the north-east quadrant parallel to and 5 arc min from the north/south axis, its lower edge being 1 arc min from the equator. The roll control of $\pm 2.5°$ was provided entirely by the standard Elliott Bros. type of vehicle stabilisation.

The spectrograph is an all-reflecting echelle system incorporating (in order of the light progression) a 1-m concave collimator, a plain echelle of 73.25 groove mm^{-1} and a 1-m concave grating acting as a camera mirror and cross-disperser. The system corresponds to an Ebert configuration in the direction of the echelle dispersion, and to a Wadsworth configuration in the direction of the grating dispersion. The two-dimensional spectral format is recorded on Kodak 101-01 photographic emulsion. The collector mirror and collimator are both coated with $Ge + ZnS$ multiple coatings to give a high discrimination against longer wavelengths and thereby solve the scattered light problem. The echelle and concave grating were both coated with $Al + MgF_2$. The linear dispersion is about 1 Å mm^{-1} and the spectral resolution (pre-flight) of the instrument was measured as 0.02 Å in the laboratory for the operating slit width of 0.018 mm.

During the flight, four exposures of 10, 25, 100 and 50 sec were obtained in that order from 130 km through apogee to 140 km. Solar spectra were recorded over the range 2000–2200 Å and the background fog density was 0.1 on the longest exposure.

The spectrum obtained from the 50-sec exposure is reproduced in Figure 1. An examination of the detail indicates a flight resolution of 0.03 Å, and extends the high resolution solar spectrum to wavelengths below that obtained by Purcell *et al.* (1963). The solar spectrum is still particularly rich in this region and contains nearly 500 detectable absorption lines. Identification of the lines has just started and many are due to Fe I and Fe II. An interesting feature is a broad (~ 1.5 Å) shallow absorption line which can be seen crossed by a number of sharp lines in the fifth echelle cycle near 2124 Å. The possible source of this feature has not been identified at the present time.

The echellogram embraces the absorption edge present in the solar spectrum near

Houziaux and Butler (eds.), Ultraviolet Stellar Spectra and Ground-Based Observations. 271–273.

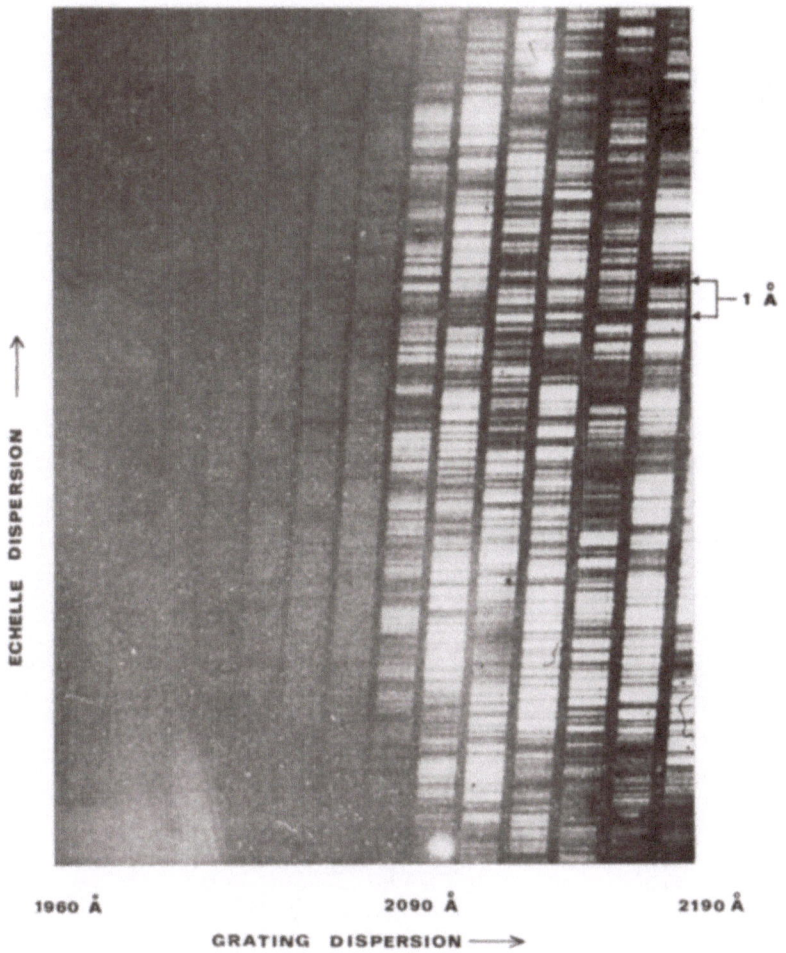

ECHELLE DISPERSION ⟶

1960 Å   2090 Å   2190 Å

GRATING DISPERSION ⟶

Fig. 1.   High-resolution solar spectrum.

2090 Å, suspected to be due to AlI (Tousey, 1963; Bonnet *et al.*, 1967). In Figure 1 this is more apparent in the frame of the grating dispersion and appears as a sudden drop in the intensity of the eighth visible echelle cycle. The photometric reduction of the echellogram, which will take out the echelle 'ripples' by reference to a laboratory continuum source, should reveal an accurate location of the edge and the high resolution will enable checks to be made on the element responsible by a search for the lead-in members of the series.

References

Bonnet, R. M., Blamont, J. E., and Gildwarg, P.: 1967, *Astrophys. J.* **148**, L115.
Purcell, J. D., Garrett, D. L., and Tousey, R.: 1963, in *Space Research*, vol. III, p. 781.
Tousey, R.: 1963, *Space Sci. Rev.* **2**, 3.

Discussion

Underhill: Could the broad shallow line be an auto-ionization line?

Wilson: That is one possibility, but we have not been able to find such a line at that wavelength in the published data on autoionization spectra.

FABRY-PÉROT INTERFEROGRAMS OF THE SOLAR
Mg II RESONANCE LINES

B. BATES, D. J. BRADLEY, C. D. McKEITH, and N. E. McKEITH

*Department of Pure and Applied Physics, The Queen's University of Belfast,
Belfast, N. Ireland*

and

W. M. BURTON, H. J. B. PAXTON, D. B. SHENTON, and R. WILSON

Astrophysics Research Unit, Culham Laboratory, Abingdon, Berks., England

The resonance lines of Mg II occur at wavelengths (2802.7 Å, 2795.5 Å) just beyond
the extinction limit of the Earth's atmosphere. At such wavelengths sophisticated
optical techniques can now be employed and this fact, together with the high cosmic
abundance of magnesium, makes these lines particularly important for study in
UV Astronomy. In the case of the Sun, the lines consist of a broad absorption with a
pronounced emission core.

The first high resolution observations (~ 0.03 Å) were obtained by Purcell *et al.*
(1962) using an echelle spectrometer in a Sun-pointing rocket. The resulting profiles
were composite ones averaged over one-third of the solar disk, but since then spatially
resolved (~ 10 arc sec), high resolution (0.045 Å) spectra have been obtained with
balloon borne spectrographs by Lemaire and Blamont (1967) and Lemaire (1969).
These give, for the first time, information on the centre-to-limb variation of the line
profiles.

The purpose of this note is to indicate some new results obtained with an optically
contacted Fabry-Pérot interferometer, internally mounted in an echelle spectrograph,
which was flown on a stabilised Skylark Rocket launched from the Woomera Range,
South Australia at 05.35 UT on December 4, 1968. The experiment was designed to
give a spectral resolution of 0.03 Å with spatial resolution of 6 arc sec along a solar
diameter. Forty-five interferograms were recorded producing 500 useful spectral
profiles of both the Mg II lines, at 15 different positions on the solar disk. The payload
also included a pinhole camera assembly to record monochromatic solar images at
different wavelengths, including the extreme ultraviolet.

Figure 1 shows a section of one interferogram near the Mg II doublet, together with
an Hα photograph taken by the Carnarvon Station, Western Australia at approximately the same time. The exact location of the slit has not yet been established from the
telemetry and analysis but its approximate position is indicated. The broad absorption
is particularly evident, especially towards the limb and the emission components
stand out clearly. The Fraunhofer absorption lines show up as breaks in the channels.

Densitometer traces of one set of Mg II profiles covering the emission cores and
obtained by scanning along the different channels are shown in Figure 2. In general
they are similar to those reported by Lemaire (1969). The central intensities of the

Houziaux and Butler (eds.), Ultraviolet Stellar Spectra and Ground-Based Observations, 274–276.

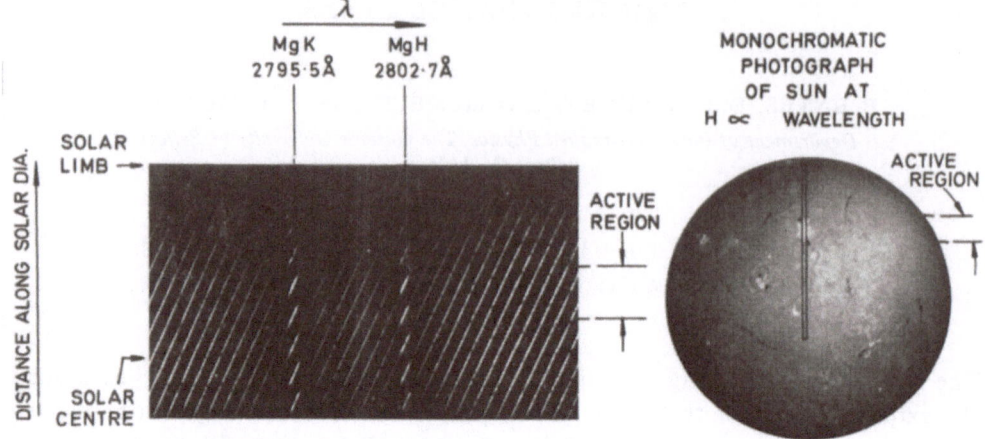

Fig. 1. Solar interferogram covering the Mg II H and K lines at 2802.7 Å and 2795.5 Å. The lower limit of the fringes is set by the end of the slit which covers the solar disk and the upper limit is set by the solar limb. Also shown is a monochromatic photograph of the Sun recorded at the Hα wavelength.

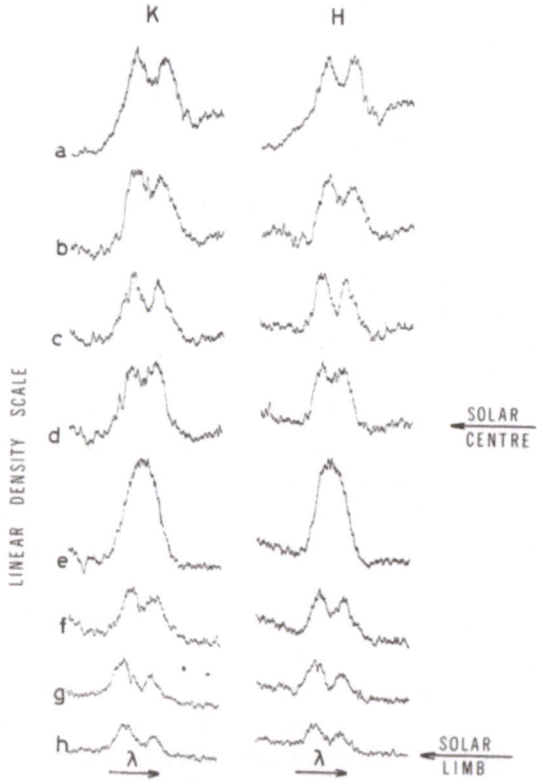

Fig. 2. A selection of microdensitometer tracings of the Mg II H and K emission cores at various points on the solar disk.

emission peaks decrease whereas their separation increases from centre to limb. The marked feature of the interferograms is the wide variation in profiles over the disk. In some cases, strong H_2 and K_2 emission lines are crossed by two absorption cores whereas in others, usually near active regions, absorption is entirely absent as in (e) of Figure 2. However, most of the observed profiles show a double emission in the core, usually unsymmetrical. The preliminary analysis of H line emission profiles (H_2) in all the interferograms shows that about 50% are asymmetrical with the stronger component to the violet, 10% are asymmetrical with the stronger component to the red, 30% are symmetrical and 10% show the single enhanced emission feature. The more detailed analysis including the Fraunhofer lines is now in progress.

References

Purcell, J. D., Garrett, D. L., and Tousey, R.: 1962, in *Space Research*, vol. III, p. 781.
Lemaire, P.: 1969, *Astrophys. Letters* 3, 43.
Lemaire, P. and Blamont, J. E.: 1967, *Astrophys. J.* 150, L129.

PART III

INTERSTELLAR ABSORPTION
AND EMISSION

A. ABSORPTION LINES

OBSERVATIONS OF INTERSTELLAR LYMAN-α ABSORPTION

EDWARD B. JENKINS

Princeton University Observatory, Princeton, N.J., U.S.A.

Abstract. Absorption at the Lyman-α transition from interstellar neutral hydrogen has been observed in the ultraviolet spectra of 18 nearby O and B stars. Radiation damping is the dominant cause of line broadening, which makes the derived line-of-sight column densities proportional to the square of the observed equivalent widths. An average hydrogen density on the order of 0.1 atom cm^{-3} has been found for most of the stars observed so far. This is in contrast to the findings from surveys of 21-cm radio emission, which suggest 0.7 atom cm^{-3} exists in the local region of the Galaxy. Several effects which might introduce uncertainties into the Lyman-α measurements are considered, but none seems to be able to produce enough error to explain the disagreement with the 21-cm data. The possibility that small-scale irregularities in the interstellar gas could give significantly lower values at Lyman-α is explored. However, a quantitative treatment of the factor of ten discrepancy in Orion indicates the only reasonable explanation requires the 21-cm flux to come primarily from small, dense, hot clouds which are well separated from each other. The existence of such clouds, however, poses serious theoretical difficulties.

1. Introduction

Stellar spectroscopy in the far ultraviolet has the potential of substantially increasing our knowledge on the distribution and composition of interstellar matter. A rich variety of resonance lines for various elements are observable between the Lyman limit at 912 Å and wavelengths where conventional ground-based equipment can operate (Spitzer and Zabriskie, 1959), and this prospect provides a strong incentive for further development of ultraviolet astronomical instrumentation. As of now, however, resolutions ranging from 1 to 10 Å have limited us to collecting significant data only for the strongest transition of the most abundant element: the 1216 Å Lyman-α line of atomic hydrogen. In interstellar space the absorption from this transition is strong enough to produce line widths on the order of 10 Å over typical distances to the brightest and most easily observed O and B stars. A measurement of the absorption's equivalent width allows a direct determination of the neutral hydrogen abundance along the line of sight.

Presently, the total number of observations of the Lyman-α absorption in various directions is small, and the coverage of the sky is still rather spotty (see Figure 1). We cannot yet really class these measurements in the usual sense as a 'general survey' of H I near the Sun, and even in the near future we cannot hope to measure up to the overall scope and detail of information on the distribution of hydrogen provided by the 21-cm radio observations. In view of this, the ultraviolet spectroscopic data would therefore seemingly represent a paltry and unimportant contribution to our understanding of interstellar hydrogen. Nonetheless, it is important to realize that the Lyman-α observations represent a truly independent means of collecting neutral hydrogen data, and of particular interest is the fact that many of the measurements available so far have strongly disagreed with the 21-cm results.

Houziaux and Butler (eds.), Ultraviolet Stellar Spectra and Ground-Based Observations, 281–301.

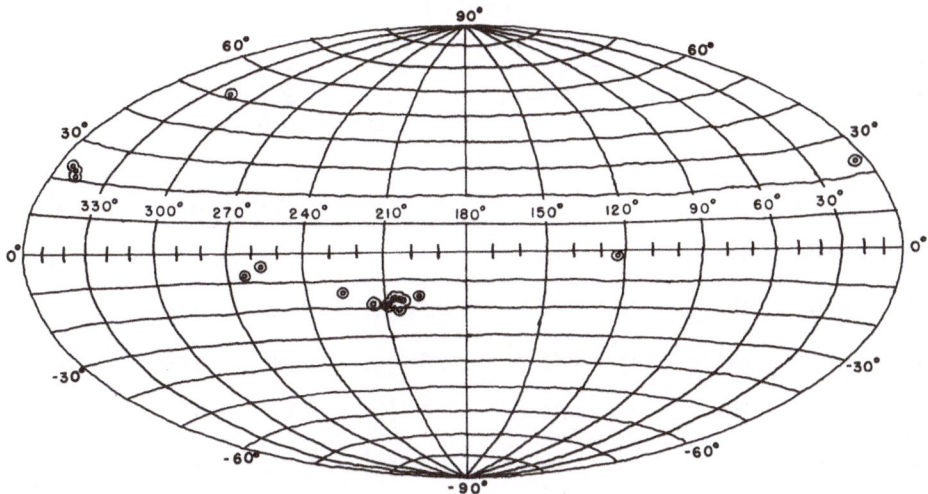

Fig. 1.   The distribution in l^{II} and b^{II} of stars whose Lyman-α lines have been observed to date (see Table I). Stars near the galactic anticenter in the middle of the diagram show the deficiency of H I, while π, δ, and β^1 Sco (far left) and ζ Oph (far right) are in better agreement with the 21-cm data.

With the exception of the results for ζ Oph and three stars in Scorpius, the strengths of the Lyman-α absorptions has indicated the presence of roughly one-tenth the column densities found by 21-cm emission measurements in the same direction. Or, from a slightly different viewpoint, the average volume densities to the stars (all less than 0.5 kpc away) are of the order of 0.1 atom cm^{-3}, in contrast to a description by Kerr and Westerhout (1965) of 21-cm data which indicates the density is more nearly 0.7 atom cm^{-3} in the solar neighborhood.

Table I is a compilation of data presently available from the Lyman-α measurements for 18 stars, all of which have been observed on rocket flights. The stars are listed in ascending order of l^{II} and are grouped according to mutual proximity in the sky. For stars which have been observed more than once, the column density N_H is computed from a weighted average of the separate measurements of the equivalent width W_λ, with the weights being proportional to the inverse squares of the respective error estimates σ_i for each observation. The quoted error, when specified by an author, is used to compute the uncertainty in N_H, and multiple observations are assigned a net uncertainty in W_λ equal to $(\sum_i \sigma_i^{-2})^{-1/2}$. The estimated distance d to each star is used for deriving the average volume density n_H along the line of sight. The value shown for α Vir represents an upper limit since a good fraction of the Lyman-α absorption is probably stellar.

The far greater capability of satellite telescopes for providing a large volume of measurements should soon allow us to contemplate a much longer list of stars. The tabulation is already on the verge of being out of date as spectroscopic data will shortly be available from the Wisconsin Experiment Package aboard the first Orbiting Astronomical Observatory.

TABLE I

Interstellar hydrogen measurements

Star	MK	l^{II}	b^{II}	E(B–V)	N_H (10^{20} atom cm^{-2})	d (pc)	n_H (atom cm^{-3})	Investigators
ζ Oph	O9.5 V	6°	+24°	0.32	4.2	170	0.8	4
γ Cas	B0IVe	124	− 3	0.08	1.2	200	0.2	10
γ Ori	B2III	197	− 16	0.02	$0.39^{+0.19}_{-0.15}$	95	0.135	3
δ Ori	O9.5II	204	− 18	0.08	$1.25^{+0.33}_{-0.28}$	460	0.088	1, 2
η Ori	B0.5 V	205	− 20	0.11	$2.7^{+3.3}_{-2.0}$	380	0.23	2
ε Ori	B0Ia	205	− 17	0.09	$1.34^{+0.34}_{-0.29}$	460	0.094	2
ζ Ori	O9.5Ib	206	− 17	0.09	$1.30^{+0.21}_{-0.19}$	460	0.092	1, 2, 3, 8
σ Ori	O9.5 V	207	− 17	0.06	$2.7^{+1.5}_{-1.2}$	460	0.19	2
θ¹ Ori C	O6p	209	− 19	0.31	$4.8^{+2.7}_{-2.1}$	500	0.31	8
ι Ori	O9III	210	− 20	0.07	$0.99^{+0.26}_{-0.24}$	500	0.064	2, 3, 8
κ Ori	B0.5Ia	215	− 19	0.04	$0.89^{+0.33}_{-0.28}$	460	0.071	3
β CMa	B1II–III	226	− 14	0.01	$0.57^{+0.48}_{-0.34}$	210	0.088	3
ζ Pup	O5f	256	− 5	0.05	$0.61^{+0.12}_{-0.11}$	390a	0.051	3, 6, 9
γ Vel	WC7+O7	263	− 8	0.03–0.06	$0.31^{+0.21}_{-0.16}$	390a	0.026	3, 6
α Vir	B1 V	316	+ 51	0.03	< 0.13	100	< 0.042	9
π Sco	B1 V	347	+ 20	0.06	7.5^{+8}_{-4}	170	1.4	7
δ Sco	B0 V	350	+ 23	0.18	12^{+6}_{-3}	170	2.4	7
β¹ Sco	B0.5 V	353	+ 24	0.20	16^{+16}_{-8}	170	3.0	7

a Estimated distance to the near edge of the Gum Nebula, which probably surrounds both stars.

Key to Investigators:

1. Morton (1967a) (not included in average since no errors were quoted).
2. Jenkins and Morton (1967).
3. Carruthers (1968).
4. Stecher (1968).
5. Morton *et al.* (1967).
6. Morton, Jenkins and Brook (1969).
7. Jenkins *et al.* (1969).
8. Carruthers (1969).
9. Smith (1969).
10. Morton, Jenkins and Bohlin (1970).

2. Profile Structure

Before exploring the implications of the Lyman-α results in more detail, it would be appropriate to review the theory which applies to the observed profiles and follow with a discussion of the various effects which could lead to errors in the measurements. The large abundance of interstellar hydrogen and the strength of the transition dictate that to even the nearest stars the absorption is very strongly saturated. An optical depth at the line center is typically about 10^7, which is sufficient to place an equivalent width measurement far up on the damping portion of the curve of growth, and at interstellar densities radiation damping is the dominant cause of broadening. Any turbulent or thermal Doppler broadening is completely insignificant. We are fortunate to have the other forms of broadening of no importance here, since in the end we can obtain a very straightforward relation for the column density N_H in terms of the equivalent width W_λ without having to worry about any assumptions regarding temperatures, densities, or velocities.

The basic equation for a transition's optical depth τ as a function of frequency v

$$\tau(v) = \frac{e^2}{mc} f_{12} \frac{\gamma}{4\pi} \frac{N_H}{(v - v_0)^2 + (\gamma/4\pi)^2} \tag{1}$$

may be immediately simplified by dropping the $(\gamma/4\pi)^2$ term in the denominator since $|v-v_0| \gg \gamma/4\pi$ at any point where τ is not unreasonably large. After substituting in the expression

$$\gamma = \frac{g_1}{g_2} f_{12} \frac{8\pi^2 e^2 v^2}{mc^3}, \tag{2}$$

and re-expressing the equation in terms of wavelength λ, we obtain a more convenient representation

$$\tau(\lambda) = 2\pi N_H \left(\frac{e^2}{mc^2}\right)^2 \lambda_0^2 \left(f_{12,\,1/2}^2 \frac{g_1}{g_{2,\,1/2}} + f_{12,\,3/2}^2 \frac{g_1}{g_{2,\,3/2}}\right)(\lambda - \lambda_0)^{-2}. \tag{3}$$

In this formula the f and g values for the two terms in the doublet are shown separately. The integration of $1 - \exp\left[-\tau(\lambda)\right]$ is straightforward and we obtain

$$N_H = (1.865 \times 10^{18} \text{ atom cm}^{-2} \text{ Å}^{-2}) W_\lambda^2 \tag{4}$$

after substituting in the numerical values for the various constants which are all well known.

3. Possible Misinterpretations from the Lyman-α Data

From Equation (4) it is evident that the factor of ten discrepancy in density results from equivalent width measurements which are typically a factor of 3 too low for agreement with the radio data. This amount seems to be well outside any reasonable experimental errors in the actual Lyman-α measurements, and one can be confident that the observed 21-cm brightness temperatures should be even more accurate. Allowances for some of the complications which commonly arise in optical interstellar line investigations tend only to aggravate the discrepancy. For instance, we assume that stars earlier than about B3 should have their own Lyman-α absorption line well within the saturated region of the interstellar absorption. If the stellar component were wide enough to be of consequence, we would be forced to conclude the actual interstellar absorption is narrower than the observed line which is already too narrow. An excessively large velocity for some very small fraction of the interstellar hydrogen would similarly worsen the disagreement with the radio results. On the whole, the existence of a large number of weak absorptions on top of the stellar continuum should neither increase nor decrease the observed equivalent width, unless these absorptions were very oddly distributed near the Lyman-α wavelength.

In the category of possible errors in the actual processing of the recorded data, one should include incorrect conversions of densities recorded on film into relative intensities, through the application of a badly determined H and D curve. Although this could lead to some inaccuracies in the equivalent widths, from the author's experience any reasonable variations in the shape of the H and D curve do not seem to have an appreciable effect on the derived values. Another point to consider is that since the wings of the theoretical profile extend fairly well beyond the main core, have the measurements really included all of the absorption? We might question whether the published equivalent widths were based on an assumed continuum level which had been drawn in too low. This might be a possibility if a star's continuum were rather confusing, and it could be a particularly tricky question when the resolution of a recording is comparable to W_λ and the profile is badly washed out.

Some insight on the validity of a few of the W_λ derivations is available from spectra having a resolution of wavelength intervals considerably smaller than W_λ. In such cases the actual shape of the profiles may be discerned, and in a consistency check we can make use of this extra recorded detail plus our *a priori* knowledge of the theoretical line shape. The bottom tracings in Figure 2 are a sampling of such spectra obtained by Princeton investigators. In each case the observations were recorded on film by objective spectrographs in a rocket, and the resolutions were on the order of 1 Å. To gain a perspective on the acceptability of various values of N_H, we may consider the shape of the original tracings after they have been multiplied by $e^{+\tau(\lambda)}$. These profiles are basically reconstructions (except near the core) of how the stellar output should have appeared in the absence of interstellar hydrogen. The profiles shown above the original data use $\tau(\lambda)$ functions based on one, two, and four times the column densities derived from the W_λ estimates. Whether or not the larger values of N_H are plausible should depend upon our judgement on how reasonably the reconstructed stellar flux behaves near 1216 Å. The drawings seem to reveal that four times the originally quoted densities are out of the question, and even where only twice these densities are assumed, the tracing turns upward reasonably far from the line center, i.e. at a wavelength where a measurable intensity had been recorded. With the actual assumed densities (N_H times one), however, it appears that the flux falls below the continuum at

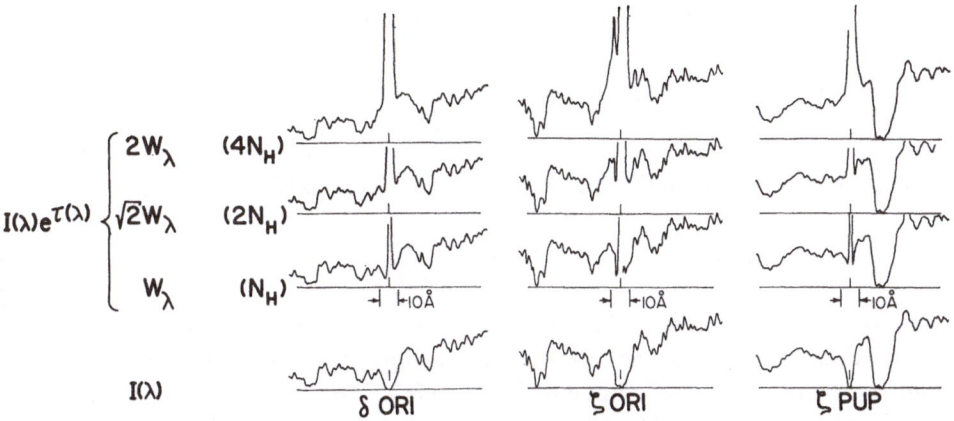

Fig. 2. A plausibility test for different interstellar hydrogen densities. Above the originally recorded spectra of δ Ori, ζ Ori and ζ Pup are three reconstructions of how the stellar output would have appeared in the absence of interstellar Lyman-α absorption. The examples show the original tracing multiplied by $e^{+\tau(\lambda)}$, where $\tau(\lambda)$ is derived from Equation (3) in the text, assuming 1, 2 and 4 times the column densities quoted by Jenkins and Morton (1967) for the Orion stars and Morton *et al.* (1969) for ζ Pup. These cases respectively correspond to 1, $\sqrt{2}$ and 2 times the measured equivalent widths.

some distance from the line core, although the upturn is nicely confined to wavelengths where the recorded signal loses significance. The fact that the theoretical absorption profile is not reproduced perfectly by the data is evidenced by our inability to obtain a fairly smooth continuum near both the core and the wings, but the fidelity seems to be as good as one could expect from these observations. Generally speaking, for spectra of moderately good resolution, a reasonably placed continuum level should not result in a quoted column density being more than a factor of 2 lower than the real value.

After looking at the top curves in Figure 2, we might be tempted to ask whether material in the immediate vicinity of the star (or for that matter, the entire H II region) is not actually emitting a strong and quite broad Lyman-α line which could pull up the wings of the observed profile and reduce the apparent W_λ. For this effect to be important, it would be necessary to have an emission strength comparable to the stellar continuum extending at least several Å from the line center. Ordinary thermal and turbulent Doppler broadening from a 10^4 K gas having an rms bulk velocity of the order of 10 km sec^{-1} would lead to a Gaussian profile with a dispersion $\sigma = 0.06$ Å. Since this profile approaches zero very rapidly as $\Delta\lambda$ exceeds several σ, random velocities cannot be responsible for pushing any of the Lyman-α emission into a wavelength region where the interstellar absorption is not overwhelming.

On the other hand, the more gradual fall-off of the Lorentz profile might render broadening from radiation damping an important consideration if the total emission flux were strong enough. For each stellar photon more energetic than the Lyman limit we would expect the eventual production of one Lyman-α photon from either within or not far beyond the edge of the surrounding H II region. If we neglect the other forms of broadening, the emission profile $N_{\alpha e}(v)$ will equal the number of photons

N_L having $\lambda < 912$ Å times the expression in Equation (1) normalized in frequency,

$$N_{ae}(v) = N_L \frac{\gamma/4\pi^2}{\Delta v^2 + (\gamma/4\pi)^2},$$ (5)

where $\Delta v = v - v_0$. If we now solve for the value of $\Delta v/v_0$ which gives an emission flux $N_{ae}(v)$ equal to the star's continuum radiation $N_{ac}(v)$ near 1216 Å, we obtain

$$\frac{\Delta v}{v_0} = (1.615 \times 10^{-12} \text{ Hz}^{-1/2}) \left(\frac{N_L}{N_{ac}(v)}\right)^{1/2}.$$ (6)

(Again we have ignored the $(\gamma/4\pi)^2$ term.) For representative values of N_L and $N_{ac}(v)$ we consider as a worst case the output from the hottest type star observed and adopt the fluxes $N_{ac}(v) = 2.4 \times 10^9$ phot cm^{-2} sec^{-1} Hz^{-1} and $N_L = 3.5 \times 10^{24}$ phot cm^{-2} sec^{-1} given by Hickok and Morton (1968) and Morton (1969a) for a model O6.5 star ($T_e = 37450$ K). Substituting these values into Equation (6) gives $\Delta v/v_0 = 6.2 \times 10^{-5}$, which corresponds to 0.075 Å from the line's center. Or to put it differently, at 1 Å away from the center the emission would be only 0.0057 times the stellar continuum flux. Thus our rough calculation appears to demonstrate that, unless some violent nonequilibrium processes are important, the total Lyman-α emission is insufficient to generate enough photons at wavelengths which are relevant to the interstellar hydrogen measurements.

In view of the evidence for mass loss from the supergiant stars observed by Morton (1967a), Carruthers (1968), Morton et al. (1968), and Morton et al. (1969), we should explore yet another possibility for a significant effect from Lyman-α emission. The observations have shown that prominent stellar absorption lines exhibit negative velocity shifts corresponding to about 10 Å. We therefore must recognize a possible source of Lyman-α photons which unquestionably could be shifted well out of the region of heavy interstellar absorption. The ejected material from the star is highly ionized, and we should explore the possibility that hydrogen recombinations could generate enough Lyman-α photons to be of concern to us. Morton (1967b) has developed a model for the mass loss from some Orion supergiant stars, and in spite of some simplifying assumptions in his analysis, the model should be sufficiently close to reality for us to determine whether or not the emission would be troublesome.

In Morton's model, the velocity of the gas did not change with increasing distance from the star's surface, and hence the electron density n_e followed the relationship

$$n_e = n_{e*} \frac{r_*^2}{r^2},$$ (7)

where r_* was the stellar radius and n_{e*} was the density at the surface. The total Lyman-α emission N_{ae} is the volume integral of the number of recombinations $n_e^2 \alpha_2$ above the star's surface

$$N_{ae} = \int_{r_*}^{\infty} \alpha_2 \frac{n_{e*}^2 r_*^4}{r^4} (4\pi r^2 \, dr) = 4\pi \alpha_2 n_{e*}^2 r_*^3.$$ (8)

The recombination coefficient α_2 is assumed not to vary over the volume, which is in effect saying that $T = 10^4$ K everywhere. Next, we again wish to compare the derived $N_{\alpha e}$ with the continuum flux from the star, and in this case we want the total flux $4\pi r_*^2 N_{\alpha c}(\lambda)$. For a B0I–II star, it is appropriate to use Hickok and Morton's (1968) B0V model flux reduced by a factor of 2.7 (Morton, 1969b), which gives $N_{\alpha c}(\lambda) = 8.7 \times 10^{20}$ phot sec^{-1} cm^{-2} Å^{-1}. Assuming $n_{e_*} = 5 \times 10^9$ elec cm^{-3} and $r_* = 2 \times 10^{12}$ cm, we find the quotient $N_{\alpha e}/N_{\alpha c}(\lambda)$ equals 0.017 Å, which is to say that the Lyman-α recombination emission is equivalent to the stellar continuum flux within a 0.017 Å passband and hence is insignificant.

As a final consideration, we might briefly review the consequences of having extremely small scale irregularities existing in the distribution of interstellar hydrogen. If these fluctuations were sufficiently strong and small to cause substantial variations in $N_{\rm H}$ across the apparent disk of the star, then the observed W_λ would be too narrow for the corresponding average $N_{\rm H}$ over the area. This prospect is untenable, however, because a supergiant's radius is about 2×10^{12} cm, and we would not expect to find the scale length for the density gradients of interstellar hydrogen to be much less than the mean free path, which is of the order of 10^{16} cm.

4. Comparison with 21-cm Data

As the introductory discussion has indicated, there is a need to resolve an apparent disagreement between the available Lyman-α data and the findings from surveys of 21-cm emission in the Galaxy. A really precise comparison of the two is difficult to achieve since there are several distinct differences in the methods of sampling. These factors which set the Lyman-α and 21-cm results apart, however, should be a key to the explanation of the apparent differences in density, and in turn we may gain an insight on some otherwise obscure attributes of the interstellar hydrogen.

To some extent, we are unsure of how much of the observed 21-cm emission came from beyond the stars, which generally are not too distant since they must have a relatively large apparent brightness to have been observed to now. For stars which are reasonably far off the galactic plane, little of the radio energy is likely to have come from beyond, and these stars offer the most secure comparisons with the 21-cm data in the same direction. Conversely, at small Galactic latitudes it is difficult to tell how much of the radiation is background from the galaxy, and the stars are not far enough away to make velocity ranging from galactic rotation a useful means for separating foreground and background 21-cm components. Unfortunately, a majority of the bright O and B stars have a small $b^{\rm II}$, and hence for many observations we must fall back on a more indirect comparison with an overall estimate of the local density from the 21-cm data.

Together with the Orion association, which will be treated in some detail in a later section, the stars listed in Table II have a high enough galactic latitude to permit meaningful comparisons with the radio data. The results of McGee *et al.* (1966) are listed for the 21-cm readings which were made closest to each star. If we refer to

Table I, it should be evident that both β CMa and α Vir show considerably lower column densities from the Lyman-α equivalent widths than from the strength of the 21-cm emission. Near $l^{II}=0$, on the other hand, the Lyman-α data for the Scorpius stars and ζ Oph indicate N_H values slightly greater than those listed in Table II.

TABLE II

21-cm column densities

Star	N_H[a] $(10^{20}$ atom cm$^{-2})$
ζ Oph	9.5
β CMa	5.3
α Vir	6.9
π Sco	5.9
δ Sco	10.3
β^1 Sco	8.8

[a] Derived from the $\int T_b\, dv$ measurements by McGee *et al.* (1966) using Equation (9) in the text.

In addition to the emission measurements, 21-cm absorption has been observed in the radiation from two different continuum sources, both of which are bright H II regions not far from stars seen at Lyman-α. The Orion Nebula, which surrounds θ^1 Ori C, has been observed by Clark (1965), who obtained 16 km sec^{-1} for the integral of the 21-cm optical depth τ over velocity. Earlier measurements by Muller (1959) and Clark *et al.* (1962) have ranged from 8.3 to 17 km sec^{-1}, but we shall adopt Clark's (1965) value for the purposes of discussion. Similarly, Menon (1969) has observed 4.9 km sec^{-1} for NGC 2024, which is very near ζ Ori. For the absorption studies, the well defined end-point in the path eliminates the uncertainty on how much of a contribution may come from beyond, but since the optical depth is proportional to the column density N_H divided by the hydrogen's spin temperature T_s, we now are troubled by an uncertainty of a different nature, namely, the actual value of T_s. If we adopt the value of 125 K (Schmidt, 1957), which has been a popular estimate in the past, the resulting column densities of 3.7×10^{21} atom cm^{-2} for the Orion Nebula and 1.1×10^{21} atom cm^{-2} for NGC 2024 are higher than the Lyman-α measurements for θ^1 Ori C and ζ Ori by factors of 7.7 and 8.5, respectively. Although we could again say that the radio data show the presence of considerably more hydrogen than the Lyman-α measurements, we could equally well propose that T_s is more on the order of 15 K. In view of the observed brightness of 21-cm radiation over the sky, a 15 K T_s may seem unusually low, but close comparisons of 21-cm emission and absorption in certain directions have demonstrated that T_s may vary over a wide range (Clark, 1965; Radhakrishnan and Murray, 1969). It may not be too unreasonable to find lower than usual temperatures selectively located near the radio sources, which might be associated with dense H I clouds where strong cooling may occur. For the Orion

Nebula we should be cautious about our conclusions from the comparison, since Clark (1965) has shown that the 21-cm absorption is produced by a small, opaque cloud located in front of the relatively large region of continuum emission, instead of a uniform, weaker absorption distributed evenly over the apparent area of the source. Thus it is conceivable that the line of sight to θ^1 Ori C could be missing the main absorbing clump.

Returning to the 21-cm emission measurements, it should be clear that we cannot reconcile the disagreement with the Lyman-α data by raising the spin temperature, since the formula customarily used for deriving the column density,

$$N_H = 1.835 \times 10^{18} \text{ atom cm}^{-2} K^{-1} (\text{km sec}^{-1})^{-1} \int T_b \, dv \qquad (9)$$

already assumes that $\tau \ll 1$, which is to say that the observed $T_b \ll T_s$. It follows that Equation (9) really represents a lower limit for the amount of hydrogen, regardless of the actual spin temperature. This assertion, however, does not take into account the possible existence of negative spin temperatures, which could be responsible for a strong maser emission at 21 cm. In a treatment of the evidence for high velocity clouds in the galactic corona, Shklovsky (1967) proposed that blue-shifted Lyman-α radiation from HII regions could, according to the theoretical criteria of Varshalovich (1967), lead to an optical pumping of the hydrogen. The consequent population inversion of the hyperfine levels could produce a 21-cm signal which was overrepresentative of the hydrogen actually present (using Equation (9)). Fischel and Stecher (1967) have recognized this possibility might also apply to the emission from hydrogen near the stars in Orion, which could explain the difference between the Lyman-α and 21-cm data. The analyses of Van Bueren and Oort (1968) and Storer and Sciama (1968), on the other hand, point out that the Lyman-α radiation becomes rapidly thermalized to the local velocity field as it scatters through many optical depths, and the required frequency gradient in the radiation is lost for virtually all of the neutral hydrogen which is not immediately outside the boundary of the HII region. It therefore appears difficult for us to reason that a significant 21-cm contribution could come from a maser amplification in the interstellar medium.

5. Interstellar Gas Clumps

We have not yet discussed one particularly significant difference in the sampling geometry of the 21-cm and Lyman-α measurements: the radio antenna beam averages the brightness distribution over a finite solid angle, usually of the order of 1° in diameter, whereas the Lyman-α measurements subtend a virtually infinitesimal sampling area. One could expect therefore to find some differences in the two results if there were any substantial irregularity or bunching in the distribution of hydrogen over a scale much smaller than the radio beam size.

The possibility that this concept may be of importance is illustrated by some sample comparisons in Figure 3 of the velocity profiles for interstellar matter observed both

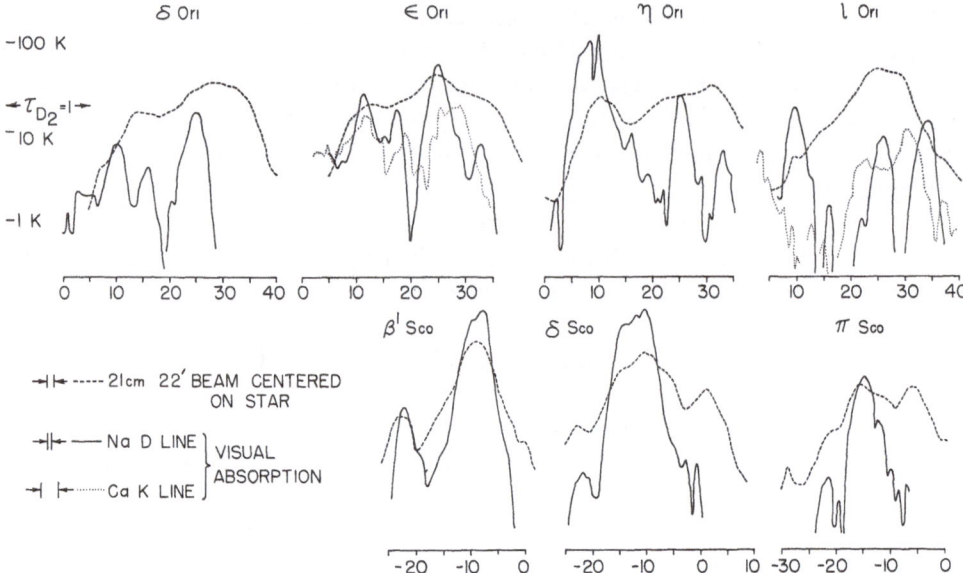

Fig. 3.   A comparison of velocity profiles for optical line absorptions and 21-cm emission. Logarithms of the absorption optical depths and the 21-cm brightness temperatures are plotted against heliocentric velocities. The 21-cm, Na I D and Ca II K line observations are from the results of Goldstein (1968), Hobbs (1969) and Takakubo (1963), respectively, and their velocity resolutions are shown in the lower left corner.

in 21-cm emission and optical absorption in the visible. In each case we have a pair of observations which are closely analogous to the direct comparisons of 21-cm and Lyman-α results. The dashed lines show 21-cm emission as seen in a radio beam 22′ in diameter centered on a star (Goldstein, 1968). The solid lines depict the velocity profiles for each star's interstellar sodium D absorption line (Hobbs, 1969) and in some cases data were available from Takakubo's (1963) study of the Ca II K line absorptions (dotted lines). Except for the K line observations, the velocity resolutions of the visual and radio determinations are comparable. Since it is difficult to assign a multiplying factor to normalize one observation to the other, the 21-cm T_b and the visual line τ values have been plotted logarithmically with an arbitrary vertical placement of one curve with respect to the other. The greater detail in the velocity structure of the D line absorption would suggest that the interstellar gas is not homogeneous over an apparent angular length corresponding to the beam size. It is not difficult to imagine that the optical absorption path could cut through only a few small clouds which have narrow, well-defined velocities, while the 21-cm emission is observed to come from a large collection of neighboring clumps which together exhibit a blended spectrum of velocities. To be sure, this comparison may be further complicated by possibly large fluctuations in the relative sodium abundances or the ionization equilibria at various places. Without question, however, the contrast between these optical and radio observations demonstrates that it is still difficult to say with much conviction that one is seeing truly equivalent samples of interstellar gas in the two cases.

It may also be interesting to note that a simple interpretation of the observed galactic absorption of the soft, intergalactic X-ray background shows a deficiency of hydrogen similar to the Lyman-α results. From this, Bowyer and Field (1969) have on their own concluded that the material is unevenly distributed and that each clump is optically thick for the X-rays. The overall attenuation of X-rays would be therefore less than if the gas distribution were uniform.

With this encouragement, we might now ask whether we can invoke the existence of small-scale irregularities to explain the apparent deficiency at Lyman-α. What picture could we construct for an interstellar gas distribution which could satisfy both observations and yet remain a statistically reasonable explanation? The basic approach will be to presume that the voids between the concentrations occupy a large enough area in the sky to make all but a very few of the Lyman-α observations show an abundance significantly lower than the overall average hydrogen density. We could then suppose that up to now no observations have penetrated any of the very dense clumps which are responsible for most of the radio emission. There should be little doubt that the conclusions derived from the arguments do not constitute a sound observational premise on the actual properties of the medium, since far too few observations are at hand. It is better to say that the spirit of the analysis will be to gain a perspective on the minimum quantitative extremes which are necessary to make the explanation viable.

For a case study, we shall simplify the situation by focusing our attention on just one region of the sky. The Orion region is a particularly appropriate choice since, relatively speaking, it has been intensively studied by both the radio and rocket astronomers. There is a total of 8 stars in close proximity to each other which have been observed at Lyman-α. All of the stars have shown substantially less hydrogen than one would expect from the 21-cm emission in the same general direction observed by Menon (1958) and more extensively by Schwartz and Van Woerden (1967). Both of the 21-cm surveys agree that the column densities range from 1.5 to 2.5×10^{21} atom cm^{-2} which is about 10 times the Lyman-α N_H. Again, we should consider how much of the hydrogen could be beyond the stars, but with $b^{II} \approx 18°$ and $d \approx 460$ pc (see Table I), the stars should be approximately 140 pc off the plane of the Galaxy. Thus it seems difficult to propose that roughly nine-tenths of the hydrogen is beyond the stars, since in our neigborhood of the Galaxy the total thickness between the half-density points is about 220 pc (Schmidt, 1957).

In all fairness, however, we should be aware that significant interstellar reddening does occur beyond the Orion association. The Orion stars have a B-V color excess of the order of 0.1, while out to 1 kpc in the same direction an additional reddening of around 0.1 and 0.2 has been observed (FitzGerald, 1968). If the gas to dust ratio were constant, it might be reasonable to suppose that as much as one-half to two-thirds of the gas may lie beyond the stars. Shane and Wirtanen's (1967) galaxy counts also indicate that the Orion stars lie within a spur of absorption projecting out of the main plane of galactic obscuration. On the other hand, γ Ori at 95 pc is much closer than the other Orion stars, and yet it shows an average density n_H not much different from the others. This would suggest the hydrogen seen at Lyman-α is somewhat evenly distributed between here and the more distant stars and is not just concentrated around the association. In the analysis that follows, we shall assume that one-half of the 21-cm emission comes from beyond the stars. Even with this fairly liberal allowance, a substantial disagreement with the Lyman-α column densities is still evident.

Starting with a uniform distribution of hydrogen, we can envisage the successively greater degrees of condensation which are shown in the sequence in Figure 4. Whether

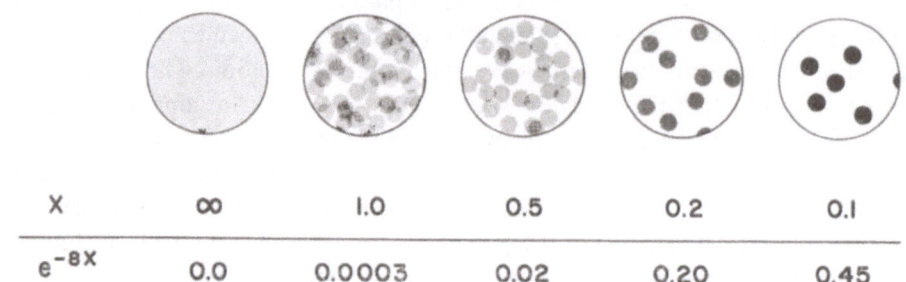

x	∞	1.0	0.5	0.2	0.1
e^{-8x}	0.0	0.0003	0.02	0.20	0.45

Fig. 4. An illustration of the appearances at 21-cm for hypothetical, small clouds within the solid angle of a radio beam. Each example represents a different value for x, but all cases would give the same average brightness temperature.

these condensations are actually filamentary or globular is of minor importance for this study; for simplicity in the analysis here, the hydrogen is assumed to be contained entirely within small spheres of uniform density n_c and radius a (this is a common practice in discussions on interstellar clouds). If we let the parameter x represent the average number of clouds penetrated along various random lines of sight, the probability of missing a cloud in one observation is e^{-x}. To have missed the clouds in all eight observations, we obviously need to have x very small since the joint probability equals e^{-8x}. In a portrayal of the sizes and distributions of clouds, a specification for x is basically an expression of how extreme the clumping must be to satisfy the condition that there is now still an inadequate number of Lyman-α observations to give a representative sample of what hydrogen is really present.

We should next consider the constraints on the problem imposed by the radio observations and require that the clouds radiate a sufficient amount of energy within the radio beam to maintain the observed 21-cm flux. As the clouds become fewer, their 21-cm surface brightness must correspondingly increase to compensate for the absence of emission in the voids. It would normally follow that the average radiation would remain the same if the overall density of hydrogen did not change as the condensation into well-separated, dense clouds took place. However, the clouds soon become optically deep to 21-cm radiation if $T_s \approx 100$ K, and naturally, self absorption prevents the surface brightness of a cloud from ever exceeding T_s. It should be clear that as x decreases, we must substantially raise T_s.

A rectangle 50 K high and 20 km sec^{-1} wide is a simple but adequate approximation for the shape of the observed 21-cm radiation profile in the Orion region. A reduction of both the height T_b and width Δv by a factor of $\sqrt{2}$ allows for our estimate that one-half of the emission may come from beyond, and we obtain the profile shown in the corner of Figure 5, which will be adopted in the analysis that follows.

Within any of the radio beams illustrated in Figure 4, the expression

$$T_b = \sum_{I=0}^{\infty} P_x(I)\, T_s\left[1 - \exp\left(-I\sigma_c/k\Delta v T_s\right)\right] \tag{10}$$

is a reasonably accurate prediction for the average brightness temperature, where the Poisson distribution term $P_x(I)$ represents the fraction of the beam's area covered by I

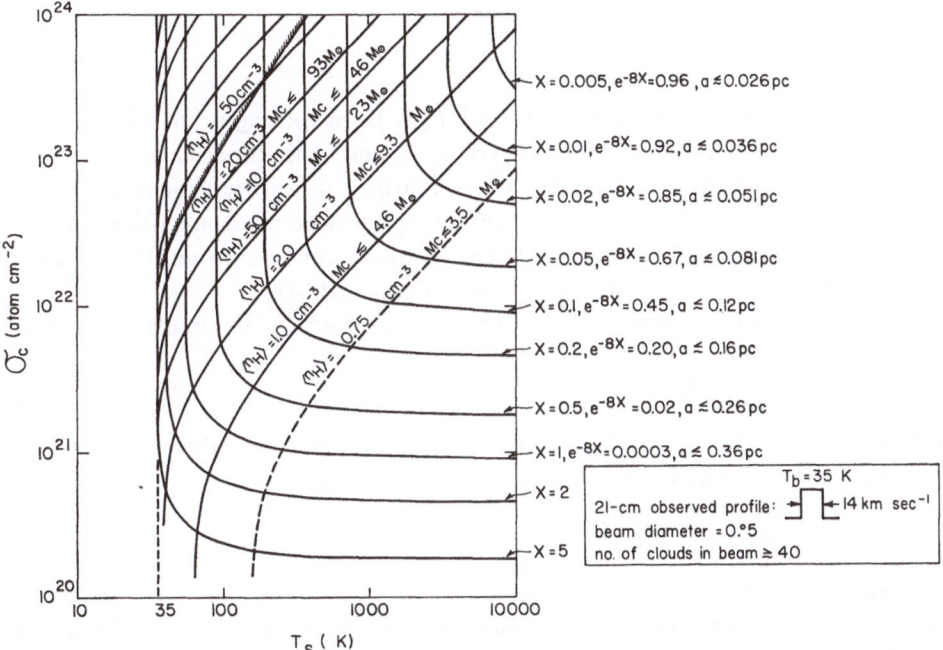

Fig. 5. Possible combinations of cloud parameters (see text) subject to the idealized 21-cm observations shown in the box. Gravitational binding for individual clouds is important only above the feathered line in the diagram.

clouds in line. Each cloud has a projected mean surface density specified by σ_c (which equals $4an_c/3$), and k is the constant shown in Equation (9). A straightforward evaluation of the sum gives

$$T_b = T_s\left(1 - \exp\left\{-x\left[1 - \exp\left(-\sigma_c/k\Delta v T_s\right)\right]\right\}\right). \tag{11}$$

Our main objective will be to try to arrive at a description of the clouds and their distribution which avoids any unreasonable values for the physical parameters x, σ_c and T_s, which we are free to vary. In so doing, it is helpful graphically to display these parameters and their behavior in Equation (11). If Equation (11) is rewritten in the form

$$\sigma_c = - k\Delta v T_s \ln \left[1 + \frac{\ln(1 - T_b/T_s)}{x} \right], \tag{12}$$

with $T_b = 35\,\mathrm{K}$ and $\Delta v = 14\,\mathrm{km\ sec}^{-1}$, we obtain the family of curves for different values of x shown in the plot of $\log \sigma_c$ vs. $\log T_s$ in Figure 5. The vertical portions of the x curves correspond to clouds which are quite opaque at 21 cm, where Equation (11) converges to

$$T_b = T_s(1 - e^{-x}) \tag{13}$$

as σ_c becomes much larger than $k\,\Delta v\,T_s$. The horizontal sections represent optically thin clouds where Equation (12) approaches

$$\sigma_c = k\Delta v T_b/x, \tag{14}$$

when T_s is much greater than both T_b and T_b/x. In proposing a model, we wish to avoid unnecessarily large values for either σ_c or T_s, and hence the intermediate sections of the lines of constant x (where they curve sharply) are of most interest to us. We should, in fact, hesitate to consider values for T_s much greater than 1000 K, since in the Orion region many of the observed velocity dispersions σ for the 21-cm Gaussian components are as small as 3 km sec^{-1} (Van Woerden, 1967).

The average volume density of hydrogen over all space (both clouds and voids) is given by

$$\langle n_H \rangle = \tfrac{4}{3}\pi a^3 R n_c, \tag{15}$$

where R is the number of clouds per unit volume. If the stars are situated at a distance d, we may use x to eliminate R in Equation (15) and obtain

$$\langle n_H \rangle = \tfrac{4}{3}axn_c/d = \sigma_c x/d. \tag{16}$$

The diagonal lines in Figure 5 depict the different values for $\langle n_H \rangle$ with $d = 460$ pc. It is evident that over a wide range of x values, the average density does not differ much from 1 atom cm^{-3} in the region where the constant x lines have their lowest σ_c and T_s.

Even after we have decided upon a favorable location in the diagram, the quantities a and R remain undetermined. In principle we are unable to distinguish between a radio beam which sees many small clouds or just a few much larger clouds, either of which cover the same relative area in the sky. A very approximate lower limit for the number of clouds inside the beam may be found by examining how constant the 21-cm emission is for a number of adjacent, independent measurements. Within the limits $171° < l^I < 179°$ and $-18° < b^I < -13°$, 28 profiles exhibited by Schwarz and Van Woerden (1967) have an rms fluctuation in $\int T_b\, dv$ equal to one-ninth its mean value. This would appear to preclude our having less than about 81 clouds in the beam, since a smaller number of clouds would produce statistical fluctuations greater than the observed variation. Since we are assuming one-half of the hydrogen emission may have come from beyond the stars, there should be more than 40 clouds out to $d =$

460 pc within a beam whose diameter is approximately 0.01 radians. Assuming, at worst, all of the clouds were actually at the distance d, we may specify an upper limit for the clouds' radius

$$a \lesssim \frac{1}{2}\left(\frac{x}{40}\right)^{1/2} 0.01 \, d = 0.36 \, x^{1/2} \, \text{pc}, \tag{17}$$

which is tabulated alongside the values in Figure 5. From Equations (16) and (17) we find each cloud's mass M_c should satisfy the relation

$$M_c = \frac{4\pi}{3} n_c a^3 = \pi d a^2 \langle n_H \rangle / x \leqslant (4.6 \, M_\odot \, \text{atom}^{-1} \, \text{cm}^3) \langle n_H \rangle. \tag{18}$$

The choice of a 'most conservative' model is governed by one's willingness to tolerate the ever diminishing measures of plausibility, e^{-8x}, which are associated with increasing values of x. How large x can be, then, is to a large degree a matter of personal judgement. Let us say, however, that it would be reasonable to adopt a value of 0.1 for x, and along the curved portion of the $x=0.1$ line we arrive at $\sigma_c = 1.5 \times 10^{22}$ atom cm^{-2} and $T_s = 600$ K. Our desire for moderation might also favor the consideration of the a and M_c upper limits as actual equalities, which would lead us to 4.6 $M_\odot$ clouds which are 0.12 pc in diameter.

As we contemplate the existence of such clouds in interstellar space, we must face the difficult question of what holds the clouds together. An exceptionally large external pressure would be required to prevent the clouds from expanding, since $n_c T = 1.5 \times 10^7$ atom cm^{-3} K. Gravitational binding is relatively unimportant for regions in the $\log \sigma_c$ $-\log T_s$ diagram below the feathered line in the upper left corner, if we assume the kinetic temperature equals T_s throughout the cloud. We might also be hard pressed to explain what dynamical conditions might have led to this state for the interstellar matter. Unless some favorable theoretical conclusions or more compelling observational material can be presented to uphold the picture that neutral hydrogen can exist in the form we have just considered, it would appear that the main accomplishment of the foregoing analysis has been to reveal the inadequacy of the explanation that the lack of agreement between the 21-cm and Lyman-α observations in Orion is chiefly a result of the differing responses of the two sampling modes to a very inhomogeneous distribution of gas. Also, this example has illustrated how in the future one might investigate a possible repetition of the problem in Orion, when more Lyman-α data are available elsewhere in the sky.

6. Correlations of Dust and Hydrogen

Early comparisons of 21-cm fluxes with the optical absorption from dust grains have generally shown a positive correlation of the two for various regions of the sky, with a gas to dust density ratio of the order of 100 (Lilley, 1955; Lambrecht and Schmidt, 1957). Significant variations in the ratio have been observed, however, and the relative deficiency of 21-cm emission from regions rich in dust has been interpreted as indirect evidence that some of the hydrogen is in molecular, rather than atomic, form (Van de Hulst *et al.*, 1954; Heeschen, 1955; Bok *et al.*, 1955; Varsavsky, 1968). Several recent studies of specific areas having a very strong optical obscuration have shown the deficiency to be rather pronounced (Garzoli and Varsavsky, 1966; Kerr and Garzoli, 1968; Mészáros, 1968). Such observations support the contention that a particularly favorable environment for the formation of H_2 could exist within dense dust clouds, where there is a strong attenuation of starlight which would normally heat the grains above a critical temperature for molecule formation (Knaap *et al.*, 1966; Solomon and Wickramasinghe, 1969).

Although the correlations of the optical and radio data have shown convincing evidence for anomalies in the gas to dust ratio, they are subject to the same ambiguities arising from sampling differences that we have considered in connection with the Lyman-α and 21-cm data. On the other hand, a comparison of the Lyman-α equivalent widths and B-V color excesses for various stars could provide a more secure basis for comparison, since the path lengths and solid angles are precisely equal. In addition, it is not entirely clear how much of the 21-cm deficiency in the clouds could be attributable to lower temperatures inside, unless, of course, the 21-cm profile is measured in absorption (Gosachinskii, 1966). Lyman-α measurements could provide a value for the atomic hydrogen density which is independent of temperature. Unfortunately, before probing the dense dust clouds, we must await the development of ultraviolet spectrographs which are considerably more sensitive than present day instrumentation,

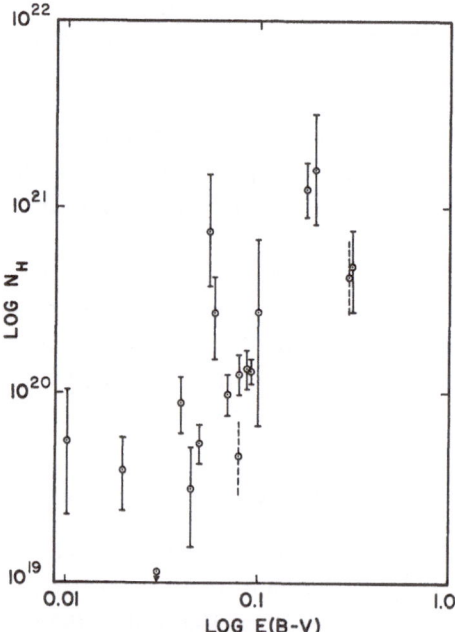

Fig. 6. A plot of hydrogen column densities (atom cm⁻²) vs. B-V color excesses for the stars listed in Table I.

since absorption in the ultraviolet is far stronger than the extinction at visible wavelengths.

Figure 6 shows a plot of the color excesses and Lyman-α column densities for the stars listed in Table I. In view of the fact that the Lyman-α data generally show less hydrogen than the 21-cm measurements, it is not surprising that the gas to dust ratio indicated by a best fit (with slope $= 1$) through the points in Figure 6 is lower than the previous estimates. However, the ratio here is less by only a factor of 3, as opposed to the factor of 7 between the Lyman-α and 21-cm local density values. This difference is probably a consequence of selecting stars which generally have a less than average reddening per unit distance. The most reddened stars observed so far, ζ Oph and θ^1 Ori C ($E_{B-V} = 0.3$), do not seem to show a significant departure from the overall trend. The other studies have shown, however, that somewhat more absorption is necessary before one can expect to notice a decrease in the ratio.

7. Conclusion

The most noteworthy aspect of the results on interstellar Lyman-α absorption is the apparent contrast with the extensive information on the distribution of hydrogen provided by surveys of line radiation at 21 cm. We have explored a number of possible effects, related either to the experimental method or to the interstellar medium, for explaining the much lower column densities deduced from the Lyman-α equivalent widths. A satisfactory explanation for the difference seems to be lacking, although we cannot deny the possibility that a combination of several of the previously considered factors may play a large enough role. It is also conceivable that the overall discrepancy may reflect the fact that the Sun may be actually situated in a hole where the density of hydrogen is considerably less than normal for this region of the Galaxy. One might also argue that some unforeseen selection effects play an important role when we observe hot O and B stars.

In the end, a better insight on the Lyman-α measurements will undoubtedly arise when more data are available. In some respects, a more diverse collection of observations is required before many meaningful correlations and generalizations can be made. For instance, it is not yet clear whether the four stars ζ Oph, π, δ and β^1 Sco on their own exhibit a large interstellar hydrogen density because (a) the resolutions were about 10 Å, as opposed to around 1 or 2 Å for all the other stars, (b) these stars are, on the whole, less luminous than many of the others, or (c) they are in a markedly different region of the sky. The answers to these and similar questions are not far from being realized, since a rapid accumulation of many more Lyman-α observations should be assured by our swiftly advancing technology in ultraviolet stellar spectroscopy.

Acknowledgements

It is a pleasure to thank D. C. Morton and J. P. Ostriker for their stimulating discussions on the material presented here. S. J. Goldstein, T. K. Menon, L. M. Hobbs, and A. M. Smith kindly provided their results in advance of publication. Support for travel and the preparation of this paper was provided by contract NSr-31-001-901 of the U.S. National Aeronautics and Space Administration.

References

Bok, B. J., Lawrence, R. S., and Menon, T. K.: 1955, *Publ. Astron. Soc. Pacific* **67**, 108.
Bowyer, C. S. and Field, G. B.: 1969, preprint.
Carruthers, G. R.: 1968, *Astrophys. J.* **151**, 269.
Carruthers, G. R.: 1969, preprint.
Clark, B. G.: 1965, *Astrophys. J.* **142**, 1398.
Clark, B. G., Radhakrishnan, V., and Wilson, R. W.: 1962, *Astrophys. J.* **135**, 151.
Fischel, D. and Stecher, T. P.: 1967, *Astrophys. J.* **150**, L51.
FitzGerald, M. P.: 1968, preprint.
Garzoli, S. L. and Varsavsky, C. M.: 1966, *Astrophys. J.* **145**, 79.
Goldstein, S. J.: 1968, private communication.
Gosachinskii, I. V.: 1966, *Soviet Astron. – AJ* **9**, 714.
Heeschen, D. S.: 1955, *Astrophys. J.* **121**, 569.
Hickok, F. R. and Morton, D. C.: 1968, *Astrophys. J.* **152**, 203.
Hobbs, L. M.: 1969, *Astrophys. J.* **157**, 135.
Jenkins, E. B. and Morton, D. C.: 1967, *Nature* **215**, 1257.
Jenkins, E. B., Morton, D. C., and Matilsky, T. A.: 1969, *Astrophys. J.* **158**, 473.
Kerr, F. J. and Garzoli, S.: 1968, *Astrophys. J.* **152**, 51.
Kerr, F. J. and Westerhout, G.: 1965, in *Stars and Stellar Systems*, Vol V; *Galactic Structure* (ed. by A. Blaauw and M. Schmidt), University of Chicago Press, Chicago, p. 167.
Knaap, H. F. P., Van den Meijdenberg, C. J. N., Beenakker, J. J. M., and Van de Hulst, H. C.: 1966, *Bull. Astron. Inst. Netherl.* **18**, 256.
Lambrecht, H. and Schmidt, K. H.: 1957, *Astron. Nachr.* **284**, 71.
Lilley, A. E.: 1955, *Astrophys. J.* **121**, 559.
McGee, R. X., Milton, J. A., and Wolfe, W.: 1966, *Australian J. Phys., Astrophys. Suppl.* no. 1.
Menon, T. K.: 1958, *Astrophys. J.* **127**, 28.
Menon, T. K.: 1969, private communication.
Mészáros, P.: 1968, *Astrophys. and Space Sci.* **2**, 510.
Morton, D. C.: 1967a, *Astrophys. J.* **147**, 1017.
Morton, D. C.: 1967b, *Astrophys. J.* **150**, 535.
Morton, D. C.: 1969a, *Astrophys. J.* **158**, 629.
Morton, D. C.: 1969b, private communication.
Morton, D. C., Jenkins, E. B., and Brooks, N. H.: 1969, *Astrophys. J.* **155**, 875.
Morton, D. C., Jenkins, E. B., and Bohlin, R. C.: 1968, *Astrophys. J.* **154**, 661.
Morton, D. C., Jenkins, E. B., and Bohlin, R. C.: 1970, in *Ultraviolet Stellar Spectra and Ground-Based Observations* (ed. by L. Houziaux and H. Butler), Reidel Pub. Co., p. 138.
Muller, C. A.: 1959, in *Paris Symposium on Radio Astron.* (ed. by R. N. Bracewell), Stanford University Press, Stanford, Calif., p. 370.
Radhakrishnan, V. and Murray, J. D.: 1969, *Proc. Astron. Soc. Australia* **1**, 215.
Schmidt, M.: 1957, *Bull. Astron. Inst. Netherl.* **13**, 247.
Schwarz, U. J. and Van Woerden, H.: 1967, preprint (see also Van Woerden, 1967).
Shane, C. D. and Wirtanen, C. A.: 1967, *Publ. Lick Obs.* **22**, part 1.
Shklovsky, I. S.: 1967, *Soviet Astron. – AJ* **11**, 240.
Smith, A. M.: 1969, private communication.
Solomon, P. M. and Wickramasinghe, N. C.: 1969, preprint.
Spitzer, L. and Zabriskie, F. R.: 1959, *Publ. Astron. Soc. Pacific* **71**, 412.
Stecher, T. P.: 1968, private communication.
Storer, S. H. and Sciama, D. W.: 1968, *Nature* **217**, 1237.
Takakubo, K.: 1963, *Sendai Astron. Rap.* No. 86.
Van Bueren, H. G. and Oort, J. H.: 1968, *Bull. Astron. Inst. Netherl.* **19**, 414.

Van de Hulst, H. C., Muller, C. A., and Oort, J. H.: 1954, *Bull. Astron. Inst. Netherl.* **12**, 117.
Van Woerden, H.: 1967, in *Radio Astronomy and the Galactic System, IAU Symposium No. 31* (ed. by H. Van Woerden), Academic Press, New York, p. 3.
Varsavsky, C. M.: 1968, *Astrophys. J.* **153**, 627.
Varshalovich, D. A.: 1967, *Soviet Phys. JETP* **25**, 157.

Discussion

Van de Hulst: In your suggested explanation the Lyman-α strength is low because the line of sight to most stars accidentally misses the small hot clouds. Should not other stars be observed in which the line of sight accidentally goes through these clouds?

Jenkins: Yes, one would expect on occasion to find a star having a very wide Lyman-α line, and in the situation I have discussed, it is suggested that this happens rarely enough that we have not yet come across just such a star with the few observations at hand. (For instance, if $x = 0.1$, then roughly one time out of ten we would expect to find an absorption line with an equivalent width of about 90 Å.) It would be interesting if in the future, when several times as many stars have been observed, some very wide Lyman-α profiles are discovered.

Oort: Would not the dense hot clumps that you have proposed cool off very fast, and would it not then be very difficult to imagine the bulk of the interstellar hydrogen to be condensed in such objects, so that it would presumably disappear from the interstellar medium?

Jenkins: The cooling of the clouds, as well as the other problems I mentioned, all present serious difficulties if one were to advocate seriously the existence of these concentrations. Of course, one means of answering the objection on cooling rate would be to propose some means of heating the clouds. Again, I should stress that we are far short of having any really convincing evidence for these hot clouds; I present this picture more to demonstrate the quantitative extremes which must be faced in considering the possibility that the difference in measurements arises solely from the effect of inhomogeneities in the two sampling methods.

Walker: Is there not in fact a selection effect in the stars so far observed for Lyman-α? They are all only slightly reddened. If the normal ratio of gas to dust holds for the dense clouds, you suggest a star seen through one, would show a large color excess. There is a very heavily reddened star (NGC 2024 in Orion) which could be such a star.

Jenkins: It is true that the observations are biased toward unreddened stars, since the absorption is especially strong in the ultraviolet, and the limiting magnitude is an important factor in the target selection for a rocket flight. Nevertheless, aside from a tendency to avoid the more distant stars at a low galactic latitude, I do not think that variations in reddening could be responsible for prejudicing the samples in favor of clear patches in any given region. Let us again consider the Orion region as an example. I doubt that you could find very many stars belonging to the association which have an absolute magnitude comparable to the observed stars, but which have been avoided because of a substantial reddening loss. This is not to say that selection effects do not exist. The simple fact that we are looking toward hot stars may influence the sample considerably. Verschuur has obtained evidence at 21-cm which indicates that hot stars may locally deplete the neutral hydrogen, and he has discussed the physical means by which this could occur.

Van de Hulst: Would you care to comment on the remote possibility that something is wrong with the theory of natural line broadening, which is used here but has never been confirmed by laboratory measurements?

Jenkins: This possibility has occurred to me in the past, but my enthusiasm for questioning the damping theory has waned since the relatively normal widths have been observed for ζ Oph and the Scorpius stars. Nevertheless, there is some chance these observations may later be shown to be in error, and thus I think this idea merits some discussion, especially since a departure from the theory might lead to rather exciting implications in basic physics. As you suggest, pure natural line broadening several Angströms from the core has probably never been observed in the laboratory owing to the difficulty in achieving the necessary column density at a low enough pressure, and thus the Lyman-α observations represent a unique situation.

Basically, we want to re-examine the validity of the prediction that the absorption cross section for the transition should continue to follow the Lorentz distribution at a frequency some 10^5 times the 'Natural line width' from resonance. Consider the simple, semiclassical picture that the time dependence of the excited state may be represented by a sinusoidal wave of frequency v_0 modulated by a function equal to 0 for $t \leqslant 0$ and $e^{-\gamma t}$ for $t < 0$, where γ is the reciprocal of the state's lifetime. The Fourier transform of this function gives us the Lorentz distribution in frequency, and the $(v - v_0)^{-2}$ dependence far from resonance arises from the finite discontinuity which occurs when the state is formed at $t = 0$. The asymptotic behavior at arbitrarily large frequency shifts from the line core is dependent upon the discontinuity in time being infinitely sharp. If, however, the discontinuity is not perfect – let us say it is rounded or sloppy over some short time interval Δt – then the frequency function will begin to approach zero more rapidly than $(v - v_0)^{-2}$ for $|v - v_0| > \Delta t^{-1}$. It should be evident that the present theory for the curve of growth of the observed Lyman-α absorptions would be in trouble if Δt were as long as 10^{-13} sec (which is still much less than γ^{-1}), but offhand I see no reasonable justification for saying the formation of the state occurs gradually, or that somehow some uncertainty principle limits our ability for specifying the exact time the electron jumped to the excited level.

This line of reasoning is mostly intuitive conjecture on my part, but it may be an interesting point to reconsider if the Lyman-α widths continue to be much narrower than expected as more stars are surveyed.

OBSERVATIONS OF INTERSTELLAR LYMAN-α
WITH THE ORBITING ASTRONOMICAL OBSERVATORY

B. D. SAVAGE and A. D. CODE

Space Astronomy Laboratory, Washburn Observatory,
University of Wisconsin, Madison, Wis., U.S.A.

Abstract. The equivalent width of the blended line at Lyman α is given for 48 stars measured with the OAO-A2 scanning spectrometer. This provides an upper limit to the neutral hydrogen column density. In the Orion association these upper limits are significantly lower than the column densities determined from 21-cm emission line measurements. The determination of the Lyman α equivalent width for θ Ori by Carruthers is rediscussed and agreement between 21-cm absorption measures and Lyman α absorption is obtained for a spin temperature in the range of 40–70 K. It is suggested that the most likely explanation for the discrepancies found for the other Orion stars is that the 21-cm emission primarily occurs beyond the Belt stars.

The correlation between the OAO blended equivalent widths and color excess, 4430 Å absorption, and interstellar sodium absorption are examined. Excellent correlation between sodium and hydrogen column densities is found.

1. Introduction

The short wavelength photoelectric scanning spectrometer on board OAO-A2 has scanned approximately 60 early-type stars (as of June 1969) from 1050 Å to 1800 Å. The instrumental resolution is 15 Å full width at half maximum. Some 50 of these stars have sufficient flux at 1200 Å to permit measurements of the equivalent width of the interstellar Lyman α line. A single observation consists of two scans and many objects were observed several times. Comparison of scans of the same stars separated by time intervals of several months agree to within 2%. Several representative scans in the region of Lyman α are shown in Figure 1, where the measured digital counts are plotted as a function of wavelength. The counting statistics on a single scan are quite good. The relative weakness of Lyman α in ε Per and the great strength in θ Ori is almost entirely due to the interstellar contribution. The variation of the strongest stellar line features with spectral type and luminosity has been described by Code and Bless (1969). A small background count of the order of ten percent of the continuum intensity near Lyman α has been subtracted from the scans in Figure 1. This background is primarily due to scattered light and has been determined from measurements made shortward of 1050 Å where only scattered light and dark counts are recorded. The sky background does not produce a significant contribution on the sensitivity range on which these spectra have been recorded. Any variation in the details of weak spectral features on repeated scans of the same object is due to the discrete nature of the spectrometer steps. The grating is rotated by a stepping motor in 10 Å steps. Since the spectrometer has an objective grating, a variation in absolute pointing of the spacecraft from one set of observations to another of 30 sec of arc may occur corresponding to a displacement of 2.5 Å. This results in a modification of the line shapes but has little effect on the equivalent widths. It is unlikely that photometric

Houziaux and Butler (eds.), Ultraviolet Stellar Spectra and Ground-Based Observations, 302–314.

errors, background counts, and spacecraft motions introduce an error in the equiva-lent widths of strong lines larger than 10%. However, the uncertainty of the continuum level is likely to be the main source of error in the equivalent width measurements.

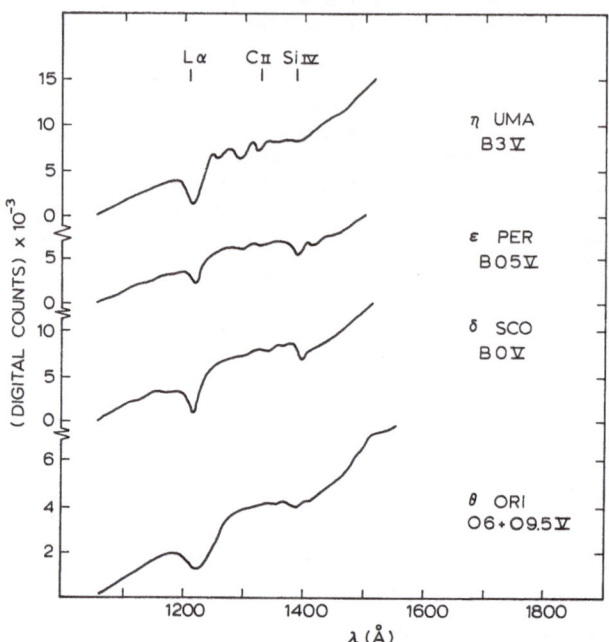

Fig. 1.   Spectrometer scans in the region of Lyman α of 4 early-type stars made with the OAO-A2 short wavelength scanner. Digital counts are plotted vs. wavelength.

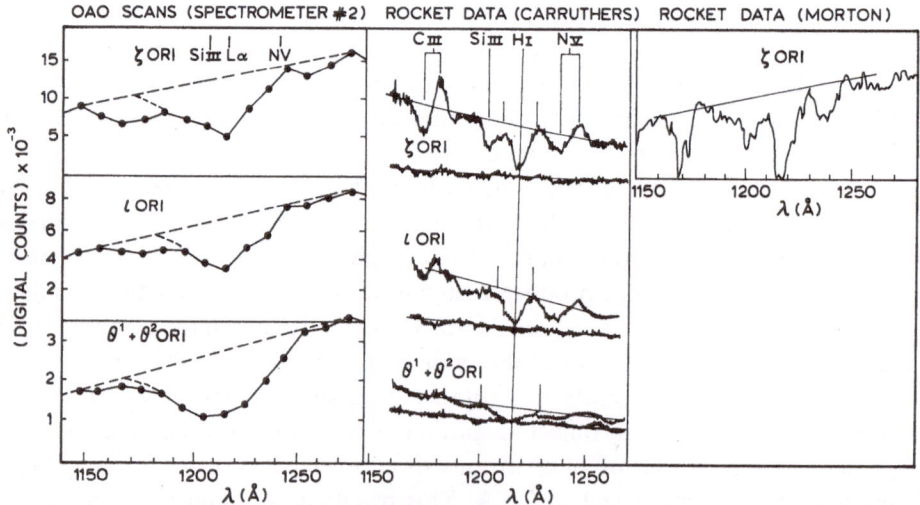

Fig. 2.   A comparison of OAO-A2 scanner observations with rocket observations of Carruthers (1969) and Morton (1969). Dashed lines on OAO scans indicate adopted continuum level and portion of Lyman α blend which is measured. For detailed information about the stars see Table I.

TABLE I

1 HD	2 Name	3 l^{II}	4 b^{II}	5 Sp. T.	6 V	7 B-V	8 E(B-V)	9 Sodium EW (Å) D_1	D_2	10 4430 Å Ac (%)	11 r (pc)	12 Blend EW OAO (Å)	13 EW Lα Rocket (Å)	14 OAO N_{Hl} Upper limit 10^{20} cm^{-2}	15 21 cm N_{Hl} 10^{20} cm^{-2}
11415	ε Cas	129.84	+ 1.65	B3IVp	3.37	−.16	+.04				141	28 ± 5		15.	9.9*(k, .6)
24398	ζ Per	162.28	− 16.70	B1Ib	2.86	+.12	+.31	.16	.22 (o)	4.1 (m)	331	25 ± 7		12.	
24760	ε Per	157.35	− 10.10	B0.5V	2.89	−.18	+.10	.06	.11 (o)	3.8 (m)	207	14 ± 3		3.7	13.0*(k, 4.7)
29763	τ Tau	176.64	− 15.08	B3V	4.28	−.13	+.07				142	29 ± 6		16.	
32630	η Aur	165.35	+ 0.27	B3V	3.17	−.18	+.02			.2 (n) 1.0 (m)	92	25 ± 6		12.	
34816	λ Lep	214.83	− 26.23	B0.5IV	4.29	−.27	+.01				552	14 ± 3		3.7	{ 6.2 (a) 8.6 (b) 12.5*(k, 5.4) 11.7 (a)
35411	η Ori	204.86	− 20.40	B0.5V	3.35	−.19	+.09				259	20 ± 4	12 ± 6 (c)	7.5	{ 18 (d) 11.3*(k, 2.7)
35439	25 Ori	200.95	− 18.30	B1Vpe	4.94	−.21	+.05				475	19 ± 5		6.7	17.7*(k, 2.3)
35468	γ Ori	196.92	− 15.94	B2III	1.64	−.23	+.01				110	15 ± 4	4.6 ± 1 (e)	4.2	20.8*(k, 4.7)
35497	β Tau	178.00	− 3.74	B7III	1.65	−.13	−.01				45	35 ± 5		23.	
36485−86	δ Ori	203.85	− 17.74	O9.5II	2.20	−.21	+.09			.2 (n)	334	20 ± 4	8.2 ± 1 (c)	7.5	{ 18 (d) 17.7*(k, 1.9)
36512	υ Ori	210.44	− 20.98	B0V	4.61	−.27	+.03				608	13 ± 3		3.2	{ 10 (d) 12.5*(k, 2.1)
36822	φ' Ori	195.40	− 12.30	B0IV	4.41	−.17	+.13	.20	.29 (o)		578	31 ± 6		18.	24.1*(k, .9)
37022−41	θ Ori	209.01	− 19.37	O6p O9.5Vp	5.13 5.07	+.04 −.09	+.36 +.21	.16	.26 (o)	1.3 (l) 1.8 (n) 2.5 (m)	799 637	33 ± 5	16 ± 4 (f)	20.	19 (g) 29 (p)
37043	ι Ori	209.52	− 19.60	O9III	2.76	−.24	+.07	.07	.14 (o)	0.0 (m)	445	17 ± 4	9 ± 3 (c) 8.8 ± 1.5 (f)	5.4	12.5*(k, 3.5)

HD	Name	l	b	Sp	V										
37128	ε Ori	205.21	−17.25	B0Ia	1.70	−.19	+.05	.13	.21 (o)	−.1 (n) / 0.0 (m)	354	24 ± 5	8.5 ± 1 (c)	11.	18 (d) / 17.7* (k, 3.0)
37202	ζ Tau	185.69	− 5.63	B2IVp	2.98	−.18	+.06				165	21 ± 5		8.2	
37468	σ Ori	206.80	−17.34	O9.5V	3.76	−.24	+.06			0.0 (m)	430	16 ± 5		4.8	18 (d) / 17.7* (k, 4.6)
37742–43	ζ Ori	206.45	−16.59	O9.5Ib	1.76	−.21	+.06	.17	.27 (o)	.5 (n) / 1.0 (m)	327	25 ± 5	9.3 ± 1 (c) / 7.7 ± 1.5 (f)	12.	18* (d) / 17.7* (k, 4.1)
38771	κ Ori	214.51	−18.50	B0.5Ia	2.06	−.18	+.04	.12	.20 (o)	2.0 (m)	425	22 ± 4	6.9 ± 1.2 (e)	9.0	16* (d) / 20.0* (k, 2.8) / 23.8* (k, 3.0)
44743	β CMa	226.05	−14.26	B1II	1.98	−.23	+.02			.4 (n)	211	10 ± 3	5.5 ± 2 (e)	1.9	
52089	ε CMa	239.83	−11.33	B2II	1.50	−.22	.00			1.0 (n)	182	12 ± 3		2.7	
87901	α Leo	226.43	+48.93	B7V	1.35	−.12	.00				22	43 ± 7			
116658	α Vir	316.12	+50.86	B1V	0.97	−.24	+.02			.7 (l)	80	11 ± 3		2.3	1.64 (a)
120315	η Uma	100.69	+65.32	B3V	1.88	−.19	+.01				51	22 ± 5		9.0	
122451	β Cen	311.77	+ 1.26	B1III	0.61	−.23	+.03				127	12 ± 3		2.7	
127381	σ Lup	318.93	+ 9.24	B2V	4.41	−.20	+.04				228	20 ± 4		7.5	
127972	η Cen	322.78	+16.67	B1.5Vne	2.31	−.21	+.04				109	15 ± 3		4.2	
128345	ρ Lup	320.14	+ 9.86	B5V	4.04	−.15	+.01				100	36 ± 7		24.	
129056	α Lup	321.61	+11.43	B1V	2.31	−.21	+.05				141	14 ± 4		3.7	
132200	κ Cen	326.86	+14.76	B2V	3.12	−.21	+.03				127	17 ± 4		5.4	
133242–43	π Lup	325.32	+ 9.92	B5IV	3.88	−.15	+.01				135	26 ± 5		13.	
133955	λ Lup	326.80	+11.12	B3V	4.05	−.18	+.02				137	22 ± 5		9.0	
136298	δ Lup	331.32	+13.82	B2IV	3.21	−.23	+.01				198	14 ± 3		3.7	
136664	φ² Lup	333.84	+16.74	B5V	4.53	−.16	.00				128	26 ± 5		13.	
139365	τ Lib	341.08	+20.44	B2.5V	3.65	−.17	+.05				131	22 ± 4		9.0	
141637	1 Sco	346.10	+21.70	B2.5Vn	4.63	−.05	+.17			3.4 (m)	174	30 ± 6		17.	8.75 (a)
142983	48 Lib	356.39	+28.63	Bpe	4.86	−.09						12 ± 6		2.7	
143018	π Sco	347.22	+20.23	B1V	2.90	−.19	+.07				180	18 ± 4	20 (h)	6.0	5.84 (a)
143275	δ Sco	350.10	+22.50	B0V	2.32	−.11	+.19	.12	.18 (o)	1.8 (l) / 1.8 (n)	168	24 ± 5	25 (h)	11.	10.2 (a) / 10.7 (i)
144470	ω¹ Sco	352.76	+22.76	B1V	3.97	−.04	+.22	.19	.25 (o)	5.6 (m)	238	33 ± 7		20.	10.2 (a)
147165	σ Sco	351.31	+17.00	B1III	2.89	+.14	+.40	.15	.21 (o)	5.7 (m)	162	31 ± 6		18.	15 (i)
149757	ζ Oph	006.28	+23.58	O9.5V	2.57	+.02	+.32	.18	.26 (o)	1.3 (l) / 1.6 (n) / 8.8 (m)	172	26 ± 5	15 (j)	13.	10.6 (a) / 13.5 (i)

Table I (continued)

1 HD	2 Name	3 l^{II}	4 b^{II}	5 Sp. T.	6 V	7 B-V	8 E(B-V)	9 Sodium EW (Å) D_1 D_2	10 4430 Å Ac (%)	11 r (pc)	12 Blend EW OAO (Å)	13 EW Lα Rocket (Å)	14 OAO N_{HI} Upper limit 10^{20} cm^{-2}	15 21 cm N_{HI} 10^{20} cm^{-2}
151890	μ' Sco	346.20	+ 3.80	B1.5V	3.02	− .23	+ .02			156	20 ± 5		7.5	
158926	λ' Sco	351.74	−02.21	B1V	1.62	− .20	+ .06			101	12 ± 3		2.7	
160578	κ Sco	351.04	−04.73	B2IV	2.41	− .21	+ .03			133	16 ± 4		4.8	
189103	θ' Sgr	5.°52	−28.°46	B3IV	4.35	− .15	+ .05			218	29 ± 6		16.	
209952	α Gru	350.00	−52.47	B5V	1.73	− .15	+ .01			35	33 ± 6		20.	

* Nearest radio region chosen.
a) Goldstein and MacDonald (1969).
b) Habing (1968).
c) Jenkins and Morton (1967).
d) Menon (1958).
e) Carruthers (1968).
f) Carruthers (1969).
g) Muller (1959) ($T_{spin} = 125$ K).
h) Jenkins et al. (1969).
i) Howard et al. (1963).
j) Herbig (1968).
k) Takakubo and Van Woerden (1966). (The number in parentheses is the angular distance in degrees of the radio region to the star.)
l) Wampler (1966).
m) Duke (1951).
n) Stoeckly and Dressler (1964).
o) Merrill et al. (1937).
p) Clark (1965) ($T_{spin} = 100$K).

The OAO resolution of 15 Å blends the interstellar Lyman α line with the stellar lines of Si III 1206.5 Å and N v 1238.5 Å, 1242.8 Å and for stars earlier than O9.5 a stellar feature at approximately 1195 Å. A number of higher resolution spectra have been obtained by Jenkins and Morton (1967) and Carruthers (1969) which show these stellar contributions. Figure 2 shows a comparison of the OAO data with rocket data of Carruthers (1969), which is plotted on a density scale and of Morton (1969) on an intensity scale. It should be possible to separate out the effects of the stellar contribution to the blended Lyman α line by means of high resolution scans of brighter stars along with the large number of OAO scans of diverse types and amounts of interstellar absorption. In this preliminary report, however, we merely present measurements of the blend equivalent widths at Lyman α. It will be seen that interesting results can be obtained from such measurements. Such measurements provide an upper limit to the column density of neutral hydrogen, which is interesting because the discussions to date (Carruthers 1968, 1969; Jenkins *et al.*, 1967, 1969) have indicated that the hydrogen densities determined from Lyman α measurements were in general considerably smaller than has been inferred from 21-cm data.

Table I lists the equivalent widths for the OAO line blend at Lyman α for 48 stars, along with other pertinent data. We estimate the possible error to be approximately $\pm 20\%$, the main uncertainty being due to the continuum level estimate. Column 9 gives the equivalent widths of Na D_1 and D_2 lines from Merrill *et al.* (1937), while column 10 lists the central depths of the diffuse 4430 Å band from several sources. The distances given in column 11 are spectrophotometric distances derived from the spectral types and photometric data given in columns 5, 6, and 7. Column 12 gives the OAO equivalent widths of Lyman α and column 13 the published equivalent widths of Lyman α from rocket data. Column 14 gives the upper limit to the neutral hydrogen column density from the OAO observations. These upper limits were obtained assuming radiation damping and therefore we used the relation $Nl = 1.865 \times 10^{18} W_\lambda^2$ atoms cm^{-2}, with W_λ expressed in Å (Morton, 1967). The final column gives the 21-cm column densities in the direction of these stars. Except for θ Ori the 21-cm data refer to emission measurements. In the case of the emission measurements no attempt has been made to separate out emission in front and behind the star (see Jenkins *et al.*, 1969).

2. Comparison of OAO Data with Rocket Data

An examination of Table I shows that for all the stars for which rocket data exist, except the Scorpius stars, the OAO blend equivalent width exceeds the rocket Lyman α equivalent widths by a factor of 2 to 3. In the case of the Scorpius observations of Jenkins *et al.* (1969), however, the OAO blend equivalent width is approximately equal to their measured Lyman α equivalent width. These latter observations were of resolution comparable to the OAO resolution, and it appears that their measurements may include some blending with Si III 1206.5 Å.

It appears unlikely to us that the Si III and N v contributions to the OAO blend could cause a factor of 2 to 3 increase in the OAO blend equivalent width over the rocket Lyman α measurements at higher resolutions. Indeed for all of these objects the equivalent widths of the stellar lines account for only about one third of the discrepancy, as judged from the higher resolution spectra. We believe that a more likely explanation of the difference is that in the rocket measurements the continuum level has been systematically underestimated and furthermore inadequate allowance has been made for the broad damping wings of the interstellar Lyman α.

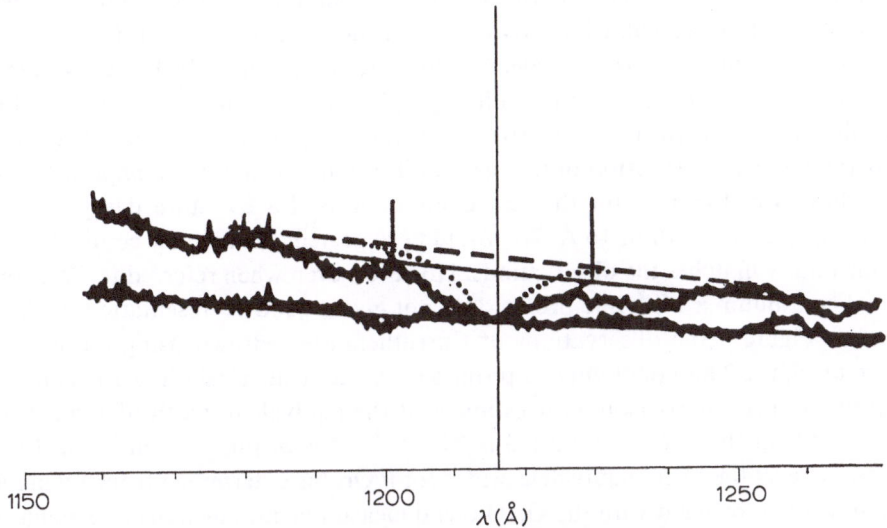

Fig. 3.   The θ Ori observation of Carruthers (1969). Solid line is Carruthers' estimate of the continuum level. We have drawn in a new continuum level (dashed line). The dotted line is the line profile for Lyman α having an equivalent width of 16 Å when referred to the new continuum level.

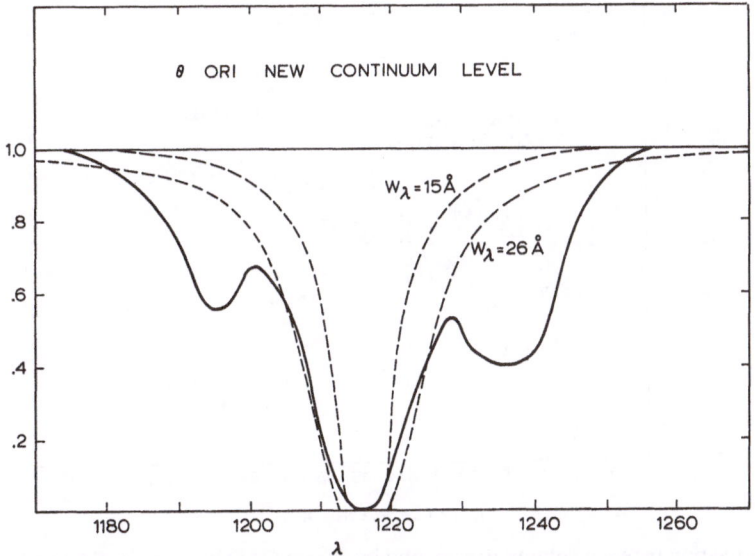

Fig. 4.   θ Ori observation of Carruthers (1969) redrawn using the new continuum level of Figure 3. Lyman α lines with equivalent widths of 15 Å and 26 Å are drawn in for comparison.

The case of θ Ori is particularly important since 21-cm absorption measurements exist for the Orion nebula. Carruthers (1969) measured an equivalent width of 16 ± 4 Å from his θ Ori observations. This yields a hydrogen column density of about $\frac{1}{3}$ the value obtained from 21-cm absorption measures assuming a spin temperature of 100 K. He suggests that the discrepancy could be removed if the spin temperature was about 20 K. The OAO blend equivalent width is 33 ± 5 Å. Figure 3 shows the observations of Carruthers (1969) on which a new continuum has been indicated by the dashed line. The three solid lines (two vertical and one horizontal) define the area Carruthers measured in his equivalent width determination of 16±4 Å. We assume in this discussion that, for the low photographic densities recorded in Carruthers' θ Ori observation, intensity is directly proportional to photographic density (unfortunately there was no discussion of intensity calibration in Carruthers' paper). We have also indicated in Figure 3 for the new continuum level a Lyman α damping profile with an equivalent width of 16 Å. As would be expected, for the new continuum level this line poorly matches the observations. However, even when referred to Carruthers' original continuum a 16 Å line profile does not give a good representation of the line shape. In Figure 4 the observations of Carruthers are redrawn using the new continuum of Figure 3 and the Lyman α profiles for an equivalent width of 15 Å and 26 Å are drawn in for comparison. Our estimate of the equivalent width of Lyman α for θ Ori based on the data in Figure 3 is 24 ± 8 Å. It is of interest that with the new continuum level the blend equivalent width for θ Ori (as determined from Figure 4) is 34 Å, in good agreement with the OAO blend measurement. Since the interstellar line

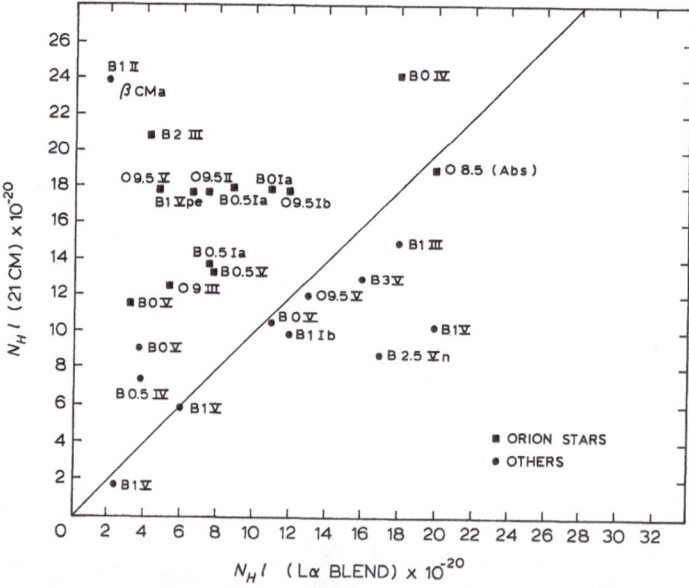

Fig. 5.   Correlation between column density obtained from OAO blend equivalent widths (an upper limit to the column density) and the 21-cm radio emission and absorption column densities. Orion stars are indicated with a ■, other stars are indicated with a ●.

is on the square-root portion of the curve of growth, the 50% increase in the equivalent width of Lyman α for θ Ori results in an increase in the column density of more than a factor of 2. The column density for an equivalent width of 24 Å is $N_H l = 11 \times 10^{20}$ atoms per cm^2. Muller (1958) gives a value of $N_H l = 19 \times 10^{20}$ per cm^2 from the 21-cm absorption line based on a spin temperature of 125 K. If we equate the Lyman α and 21-cm densities we derive a spin temperature of approximately 70 K. Carrying out the same calculation but using the radio absorption data of Clark (1965) ($N_H l = 2.9 \times 10^{21}$ for $T_{spin} = 100$ K) we obtain a spin temperature of approximately 40 K.

In Figure 5 the column density determined from 21-cm measures is plotted against the upper limit to the column density given by the OAO blend. Thus correcting the OAO Lyman α measurements for the effects of blending would move points to the left. The Orion stars are plotted as squares, the other stars as circles. Excluding the Orion stars, the 21-cm densities are usually somewhat less than the OAO upper limit

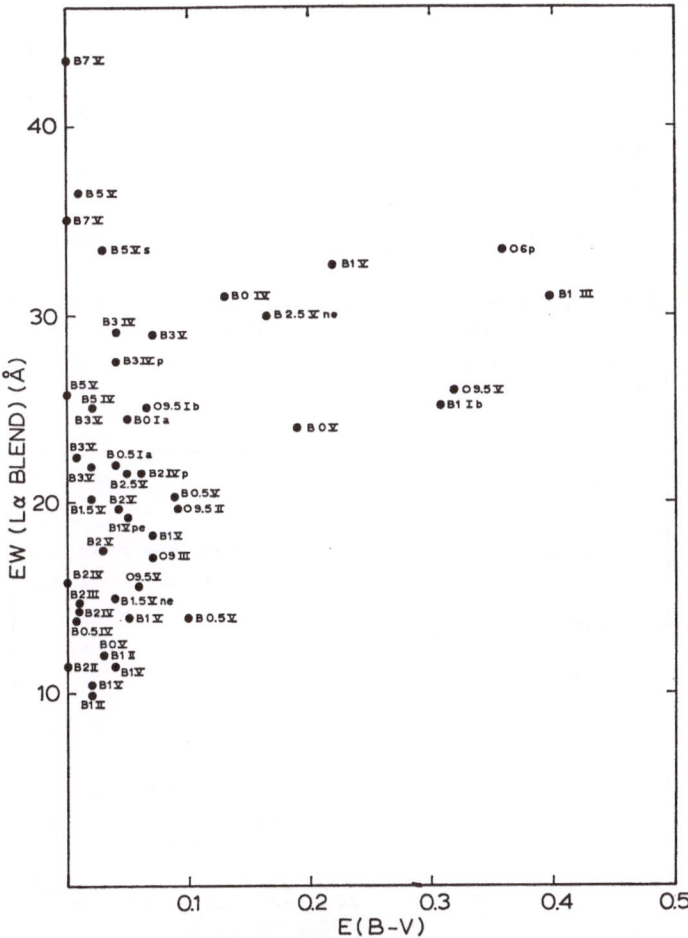

Fig. 6.   Correlation between Lyman α blend equivalent width and color excess. The spectral type of each star is indicated on the figure.

as would be expected if most of the 21-cm emission is not background emission. In Orion the 21-cm column densities range from 1.5 to 5 times the OAO upper limit. In view of the reasonable agreement found for the 21-cm absorption measurement in front of the Orion Nebula and the Lyman α absorption for θ Ori it is reasonable to believe that most of the 21-cm emission in the direction of the Orion association occurs at distances greater than the bright Orion stars.

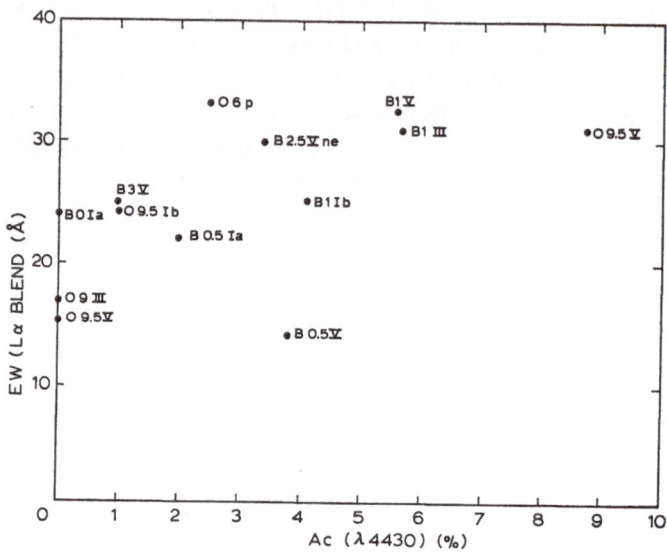

Fig. 7.   Correlation of the equivalent width of the Lyman α blend with the 4430 Å central absorption measurements of Duke (1951).

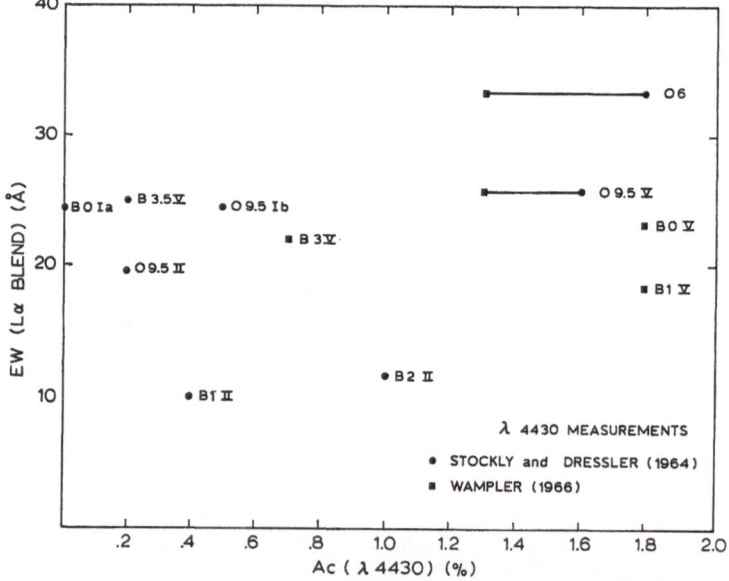

Fig. 8.   Correlation of the equivalent width of the Lyman α blend with the 4430 Å central absorption measurements of Stoeckly and Dressler (1964), and Wampler (1966).

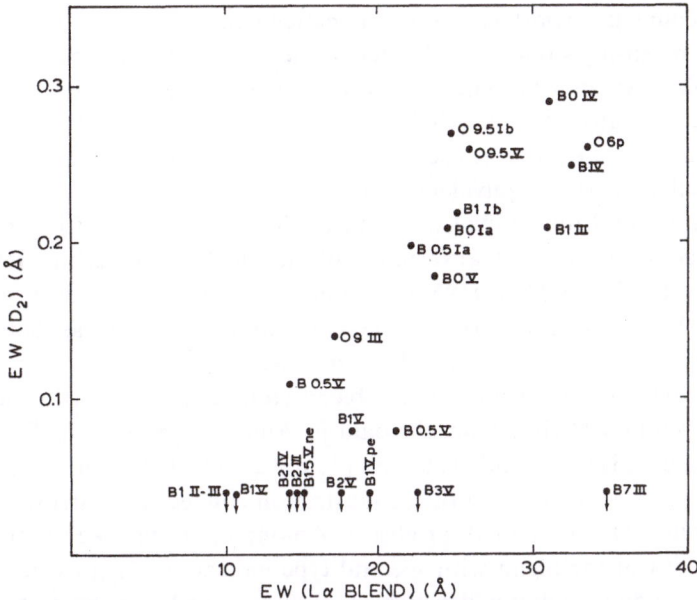

Fig. 9. Correlation of the equivalent width of the Lyman α blend with the equivalent width of the sodium D₂-line. Points connected to arrows indicate the upper limit on D₂ equivalent width. Sodium data is from Merrill *et al.* (1937).

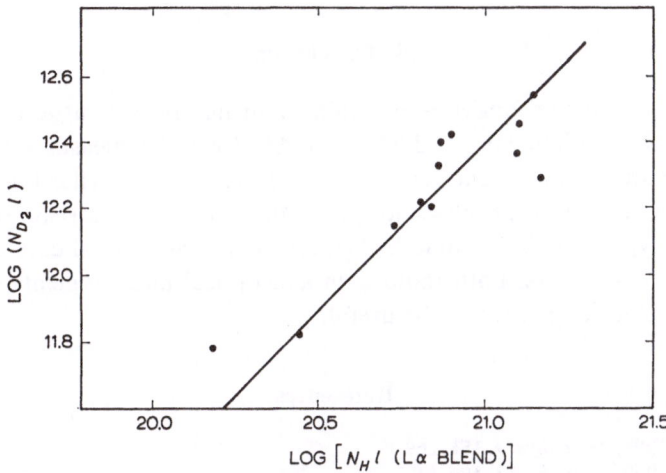

Fig. 10. Column density of atomic hydrogen (obtained from the Lyman α blend) vs. column density of neutral sodium atoms.

3. Correlations of Blend with Other Interstellar Features

In Figure 6 the equivalent width of the OAO Lyman α blend is plotted against the $B-V$ color excess of the stars. The early-type stars do show a general correlation of increasing equivalent width with increasing color excess. For stars with small color excesses the equivalent width shows a systematic increase with spectral type. This is a result of the increasing strength of the stellar lines included in the blend. In particular Si III and stellar Lyman α become quite strong in stars later than B3. For B0 V stars the stellar lines contribute 5 to 10 Å to the measured equivalent widths. A simple subtraction, however, cannot be used to go from the Lyman α blend equivalent width to the interstellar Lyman α equivalent width.

In Figures 7 and 8 the equivalent width of the Lyman α blend is plotted against measurements of the central absorption of the 4430 Å diffuse interstellar band. Figure 7 shows the data of Duke (1951) and Figure 8 the measurements by Stoeckly and Dressler (1964), and Wampler (1966). The correlation is poor, just as that between the 4430 Å band and the sodium D-line is known to be.

Figure 9 shows the correlation of the blend equivalent width with the equivalent width of the sodium D_2 line from the data of Merrill et al. (1937). The correlation between these data is very good. The stars plotted are located in Orion, Scorpius, and Perseus. The stars at the bottom of the illustration connected to arrows are stars for which interstellar D is very weak or absent. Among these stars we see the increasing equivalent width of the blend with spectral type indicating the growth of the stellar contribution. Sodium column densities were determined for these stars from the doublet ratio and the results are plotted in Figure 10. The agreement between the D_2 column densities and the hydrogen densities is good and in reasonable agreement with abundance ratios and ionization estimates.

4. Conclusion

The discussion presented above is preliminary in nature but suggests the direction which future work might take. Additional OAO Lyman α measurements are being made. We are attempting to extract the effect of the blended stellar lines both on the basis of higher resolution profiles and on the differential comparison of stars of the same spectral type. Clearly it would be helpful to have additional data on interstellar absorption for these stars, both radio data and optical measurements of interstellar Ca, Na, CH, 4430 Å, etc., would be useful.

References

Carruthers, G.: 1968, Astrophys. J. 151, 269.
Carruthers, G.: 1969, Astrophys. J. 156, L97.
Clark, B. G.: 1965, Astrophys. J. 142, 1398.
Code, A. D. and Bless, R. C.: 1969, this volume, p. 173–177.
Duke, D.: 1951, Astrophys. J. 113, 100.
Goldstein, S. J. and MacDonald, D.: 1969, Astrophys. J., in press.
Habing, H. J.: 1968, Bull. Astron. Inst. Netherl. 20, 120.
Herbig, G. H.: 1968, Z. Astrophys. 68, 243.
Howard, W. E., Wentzel, D. G., and McGee, R. X.: 1963, Astrophys. J. 138, 988.
Jenkins, E. B. and Morton, D. C.: 1967, Nature 215, 1257.
Jenkins, E. B., Morton, D. C., and Matilsky, T. A.: 1969, in press.
Menon, T. K.: 1959, Astrophys. J. 127, 28.

Merrill, P. W., Sanford, R. F., Wilson, O. C., and Burwell, C. G.: 1937, *Astrophys. J.* **86**, 274.
Morton, D. C.: 1967, *Astrophys. J.* **147**, 1017.
Muller, C. A.: 1959, *Proc. IAU Symp.* **9**, Stanford Univ. Press, Stanford, Calif., p. 360.
Stoeckly, R. and Dressler, K.: 1964, *Astrophys. J.* **139**, 240.
Takakubo, K. and Van Woerden, H.: 1966, *Bull. Astron. Inst. Netherl.* **18**, 488.
Wampler, E. J.: 1966, *Astrophys. J.* **144**, 921.

Discussion

Carruthers: You mentioned that the equivalent width of your 'blend' is about a factor of two greater than our equivalent width of the Lyman α line for θ Ori. I would like to point out that the N v line is very strong in an O6 star, in the case of θ Ori this line is comparable to or possibly even stronger than the Lyman α line, hence the combined strength of both lines would be in good agreement with your measurements.

Bless: N v is indeed strong in θ Ori but the factor of 2–3 refers to all the OAO-rocket comparisons, except those in Scorpius recently observed by Princeton. This includes stars in which N v is not very strong.

Carruthers: Would you agree that it is possible to raise the level of your continuum as I suggested earlier?

Bless: Yes, I agree that there is room for raising the continuum. Our spectrum of θ Ori is rather weak in this wavelength range and there is an appreciable uncertainty in the continuum level.

Greenberg: If we believe in a generally good correlation of 4430 Å with extinction, how do you reconcile your observation of lack of correlation of Lyman α and 4430 Å depth with the result which Jenkins presented showing an almost surprisingly excellent correlation of Lyman α with extinction?

Bless: I don't know. The correlation – if it exists – might be masked by our measuring a blend, rather than Lyman α alone. Also, we have observed relatively few stars for which 4430 Å measurements exist.

Morton: Is it possible to reduce the problems of the blends by concentrating on stars of types B1 to B3 where the contributions of N v and Si III may be rather small without too serious a contribution from the photospheric Lyman α line?

Bless: Yes, this is one of the things we intend to do to sort out the effects of the blend.

Morton: Since the OAO Lyman α data tend to agree with the rocket observations toward Orion, would the radio astronomers object to a real hole in the H I density in this direction?

Oort: It is perhaps not implausible to think that the interstellar gas in the Orion region would form some sort of a relatively thin sheet. If this is supposed to be related to the Orion association one could easily imagine that there would be holes in the neutral hydrogen around the high-temperature stars while the rest of the sheet would remain neutral. In such case the Lyman α absorption results for the H I density might well be much lower than those derived from the 21-cm emission measurements.

Bless: Since OAO observations suggest that for some stars in other parts of the Galaxy there is reasonable agreement between Lyman α and 21 cm measurements, such a local density anomaly in the Orion region seems a little more plausible than looking for fundamental difficulties with 21-cm observations, etc.

INTERSTELLAR LINES OTHER THAN HYDROGEN

G. H. HERBIG

Lick Observatory, University of California, Santa Cruz, Calif., U.S.A.

I would like to extrapolate beyond the present state of interstellar line spectroscopy toward the next generation of satellite-borne equipment, but with due recognition of the fact that more work still has to be done from the Earth's surface.

Experience with narrow interstellar absorption lines shows that the faintest detectable lines have equivalent widths of about 0.1 times the resolution. Since the best far-ultraviolet spectrograms now available have a resolution of about 1 Å, so the detection limit is about 100 mÅ. Most of what I have to say is directed toward the advantages and problems of the era when lines having one-tenth of this strength are observable from outside the atmosphere.

One of the most central problems of this subject is that of the chemical composition of the interstellar gas, and the possibility that it varies from place to place in the interstellar medium. From the Earth's surface, the total abundance of an element in the gas has to be inferred from the abundance of one or two of its ions. This calculation involves the competition between the ionization rate (a radiation field and an energetic-particle spectrum operating upon wavelength-dependent photoionization and collisional-ionization probabilities) and the recombination rate (dependent upon the electron-capture cross-sections for all the levels, and on the electron temperature). Unfortunately, these rates have to be multiplied by the local concentrations of the various species, $n(r)$ (in cm^{-3}), while the observations give the total number of ions in the column between Earth and star, N(in cm^{-2})$=\int n(r)$ dr. Therefore, to correct for ionization one has either to make some assumption about the distribution of matter along the line of sight, or to minimize the effect by some manipulation of the data.

There is evidence from terrestrial observations, interpreted in this way, that the interstellar abundances of Na and K are approximately 'cosmic' with respect to atomic H, while Ca and Ti are abnormally low. Upper limits that are not in conflict with chondritic data can be set on Li and Be (Herbig, 1968). But the results are no stronger than the weakest link in the rather extended calculations that lead to the corrections for ionization.

Surely one of the most attractive prospects of space spectroscopy is that it will provide direct information on more ions so that, in some elements, column abundances can be obtained directly without these very difficult allowances for ionization. The elements listed in Table I are particularly favorable in this respect. Listed also in the table are the wavelengths of the K absorption edges for these same elements, as a reminder that there is in principle a means of determining the total column abundance of an element that does not involve the state of ionization or whether the atom is free or bound, although information on the chemical association of the atom may be contained in the fine structure of the edge. But the result, however accurate or however

Houziaux and Butler (eds.), Ultraviolet Stellar Spectra and Ground-Based Observations, 315–319.
All Rights Reserved. Copyright © 1970 by the IAU.

obtained, is still N, and the determination of the local concentration $n(r)$ must depend upon knowledge of the radiation field, upon the availability of the necessary laboratory data, and upon some model for the intervening interstellar medium. Let us consider these last two problems in more detail.

TABLE I

Abundant atoms having ions with
resonance lines $\lambda > 911$ Å

Ion	Resonance Line	Ion	Resonance Line
CI	1656 Å	NI	1200 Å
CII	1334	NII	1083
CIII	977	NIII	989
CIV	1549	(NIV	765)
K edge C	43.77	NV	1240
		K edge N	31.05
AlI	3944	SiI	2514
AlII	1670	SiII	1808
AlIII	1854	SiIII	1206
K edge Al	7.95	SiIV	1393
		K edge Si	6.74
PI	1774	SI	1807
PII	1532	SII	1259
PIII	1334	SIII	1190
PIV	950	SIV	1062
K edge P	5.79	K edge S	5.02

A pressing need exists for more detailed information on absolute photo-ionization cross-sections as a function of wavelength, from threshold to about 504 Å (below which there is ordinarily too little stellar radiation to contribute significantly). The experimental data are most complete for the ground states of the alkali metals (LiI, NaI, KI) and for CaI, but there is urgent need for more work, especially on MgI and AlI. But special priority should be attached to studies of the atoms listed in Table I. For ionized atoms, however, the technical limitations of absorption-tube work are such that photoionization probabilities for ions and for excited states will apparently have to come from theory. The laboratory absorption data include the autoionization structure if any, so the photoionization rate includes the contribution of autoionization. But the contribution of dielectronic recombination to the recombination rate under interstellar conditions has not been taken into account.

The increasing strength of interstellar lines with distance means that the line of sight is passing through an increasing number of regions with increasingly heterogeneous properties. The problem of recovering n from N thus becomes progressively more difficult. It seems obvious that more will be learned from a concentrated attack upon the interstellar spectra of a few, relatively nearby stars in which the line structure is not complex, and for which some independent estimate can be made of the radiation field and the density profile in the line of sight. Therefore I wish to recommend for particular attention by optical, radio, and ultraviolet observers the four stars listed in Table II. These stars show especially simple line structure in Hobbs' (1969) atlas of high-resolution Na I profiles.

TABLE II

Stars recommended for detailed attention

Star	b^{II}	Spectral type	E(B-V)	N(Na I)
δ Cyg AB	$+10°$	B9.5 III	0.00	2×10^{11} cm^{-2}
δ Sco	$+22$	B0 V	$+0.19$	42×10^{11}
o Per	-18	B1 III	$+0.32$	–
ζ Oph	$+24$	O9.5 V	$+0.32$	490×10^{11}

The long-standing Ca/Na abundance anomaly is a good example of the abundance problems that await solution. The apparent deficiency of Ca in the interstellar gas is still unresolved, despite recurrent optimism that improvement of some aspect of the ionization correction would provide an explanation. The amount of the Ca deficiency with respect to Na seems to vary in different directions (e.g., it is particularly large in front of ζ Oph), but one must be cautious with N(Ca II) values determined from K/H ratios, which are susceptible to serious systematic error. The next members of the 4^2S-$n^2P°$ series of Ca II lie at 1649, 1652 Å ($n=5$) and 1341, 1342 Å ($n=6$). High priority should be given with space spectrographs to the determination of accurate equivalent widths for these higher members in a few interstellar spectra *, to be sure at least that the curve-of-growth formalism is correct. This same problem will probably arise when observations become available of the interstellar lines of Mg II at 2795, 2802 Å. In the case of Na I, the second members of the series, at 3302 Å, fortunately are accessible from the Earth's surface.

If this apparent underabundance of Ca in the gas persists, then the following more interesting hypotheses will have to be considered as possible explanations:

(1) A superionization phenomenon: the missing Ca I and Ca II may have been transferred to Ca III. This will be a difficult matter to confirm even from space, since the strongest resonance lines of Ca III lie at 490 Å and shortward (Borgström, 1968).

(2) A chemical phenomenon: the missing Ca may have been bound up in the interstellar dust. This would be quite understandable in the context of the recent hypothesis that the interstellar particles are composed of refractory Mg, Al, and Ca silicates plus metallic iron (Larimer, 1967; Herbig, 1969). One would then expect that the concentrations of free Mg, Al, etc. in the gas would also show the effects of such selective depletion.

---

* One expects these lines to be weak, because Ca II is isoelectronic with K I, in which the f-values of the higher series members are very small.

(3) A molecular process: the missing Ca may have been tied up in some tightly bound, spectroscopically undetectable molecule in the gas. A candidate might be CaC, whose spectrum has not yet been observed in the laboratory.

Whatever the mechanism, it is clear that Ca must be present in the interstellar medium in some form and in approximately the solar or chondritic proportions, as demonstrated by the Ca contents of stars formed from that same interstellar material (although there is a hint that the Ca/H ratio may vary slightly with age; Spite, 1968).

References

Borgström, A.: 1968, *Ark. Fys.* **38**, 243.
Herbig, G. H.: 1968, *Z. Astrophys.* **68**, 243.
Herbig, G. H.: 1969, 'Pre-Main Sequence Stellar Evolution', *Mem. Soc. Roy. Sci. Liège* (in press).
Hobbs, L. M.: 1969, *Astrophys. J.* **157**, 135.
Larimer, J. W.: 1967, *Geochim. Cosmochim. Acta* **31**, 1215.
Spite, M.: 1968, *Ann. Astrophys.* **31**, 269.

Discussion

Spitzer: There is one central feature relevant to the sodium-calcium ratio which was pointed out years ago and which should be stressed again here. Observations indicate that for the high-velocity components of interstellar lines the ratio of sodium atoms to calcium atoms appears to be nearly normal.

If the low calcium-sodium ratio in normal clouds is attributed to the calcium being locked up either in grains or in molecules, than one must find some mechanism by which the calcium can be released back into the gas as atoms when the interstellar medium is being accelerated. Quite possibly acceleration of clouds by newly born O stars or by supernova explosions can provide such a mechanism.

Wickramasinghe: It is probably worth noticing that the three elements Ca, Be, Ti share two properties in common: (1) they form strongly bound gaseous oxides, and (2) these gaseous oxides in turn can form highly refractory solids. It may be argued that much of the Ca, Be, Ti reaching the interstellar medium from stellar sources is locked away in the form of these highly refractory solids.

Herbig: There is no longer any evidence that Be is underabundant (see Herbig, *Z. Astrophys.* **68**, 243, 1968). As regards free gaseous TiO (and possibly CaO as well), resonance lines of these molecules are accessible, but despite specific search, have not been found in interstellar absorption. If ejection of solids by cool oxygen stars were common, I would think that some of these diatomic constituents of the atmospheres would be carried off into space along with the solids, and am therefore surprised that none is found.

Wickramasinghe: This would not be so if all the ambient oxide molecules were deposited as solid particles before they reached the interstellar medium. It may also be possible that these metallic oxides combined with SiO_2 and actually condensed as silicates.

Greenberg: I should just like to mention another correlation (or anti-correlation) between interstellar Ca abundance and 4430 Å. It appears that where Ca abundance is relatively normal the 4430 Å is weak and this would seem to indicate that 4430 Å (and perhaps the other diffuse lines) is due to Ca in some (as yet unknown) form either as bound in molecules or in the grains.

Herbig: Of course, it would be immensely satisfying if the carrier of the diffuse bands could also account for the Ca deficiency: e.g., if the CaC molecule could explain both.

Praderie: The ratio of heavy elements to calcium is known to be very different in normal stars as compared to Ap or Am stars. Concerning the origin of the peculiar stars and their relation with the neighbouring interstellar matter, have interstellar lines of heavy elements (e.g., rare earths) been ever detected?

Herbig: I have looked for interstellar lines of several rare earths in ζ Oph, without success (see *Z. Astrophys.* **68**, 243, 1968).

Note added November 1969: Since my remark at Lunteren regarding the CaC molecule as a constituent of the interstellar medium, I noted a paper by Espenhain *et al.* (*Z. Astrophys.* **61**, 77, 1965), in which observations were reported of Ca vapor in contact with carbon in a King furnace at relatively high H_2 pressures. They noted 11 unidentified bands in the 6256–6304 Å region, but neither their tracings nor those of Wurm and Meister (*Z. Astrophys.* **13**, 25, 1936) show any hint of an absorption feature near 4430 Å. There is a diffuse feature at 4550 Å, found originally by Wurm and Meister, which is regarded as unidentified although lines of the 4^3P^0–4^3D multiplet of Ca I occur near that wavelength.

INTERSTELLAR MOLECULAR HYDROGEN

PHILIP M. SOLOMON

Columbia University, New York, N.Y., U.S.A.

Abstract. The dominant photodissociation process for H_2 under interstellar conditions is absorption in the lines of the ultraviolet Lyman bands at $912 \text{ Å} < \lambda < 1108 \text{ Å}$ and re-emission into the continuum at $\lambda \sim 1600 \text{ Å}$. It is shown that 40% of all upward transitions in this band lead to emission into the continuum of the ground state and therefore dissociation of the molecule. The photodissociation rate is calculated and the theory is developed for the abundance of interstellar H_2. It is shown that interstellar space will be divided into atomic and molecular regions in a manner analogous to the division between H_{II} and H_I regions. The molecular regions will exist only where the rate of formation of H_2 is sufficiently great to build up a self-shielding layer. This is expected in cold clouds with number densities greater than 100 atoms cm^{-3}. Normal regions of interstellar space will have an extremely small abundance of molecular hydrogen.

Houziaux and Butler (eds.), Ultraviolet Stellar Spectra and Ground-Based Observations, 320.
All Rights Reserved. Copyright © 1970 by the IAU.

GENERAL DISCUSSION

Stecher: How does the variation of the dipole moment with intermolecular separation affect the photodestruction of the molecules?

Solomon: It increases the photodissociation slightly but the main effect is the very large overlap between the vibrational wave functions of the high V^1 states and the vibrational continuum in the ground electronic states.

Stecher: What grain model did you use in obtaining the grain temperature?

Solomon: Graphite core, ice mantle models were used but the grain temperature is not a sensitive function of the model unless one assumes a grain with a very low optical absorption efficiency such as for silicates, in which case the grain temperature would be substantially lower.

Carruthers: Perhaps the best chance to detect molecular hydrogen in the far ultraviolet would be to observe the resonance fluorescence in the Lyman bands at the interface between a hot star and a dense dust cloud. Such an opportunity is presented by the Orion nebula, in which Werner and Harwitt feel that they have observed the vibration-rotation infrared bands which follow in cascade the Lyman emission to the ground state.

Solomon: This might be possible but one of the difficulties is that the emission in the ultraviolet will be shared amongst over a hundred lines and no single line will be particularly strong. In addition, the detection by Werner and Harwitt still seems to be uncertain.

Houziaux and Butler (eds.), Ultraviolet Stellar Spectra and Ground-Based Observations, 321.
All Rights Reserved. Copyright © 1970 by the IAU.

B. EMISSION

THE NIGHT SKY BRIGHTNESS MEASURED FROM
SATELLITES KOSMOS 51 AND 213

N. A. DIMOV, A. B. SEVERNY and A. M. ZVEREVA

Crimean Astrophysical Observatory

Abstract. (1) The minimum measured brightness of the night sky (in the visual region) is about 100 stars of the 10th magnitude per square degree near the galactic poles.

(2) The ratio of fluxes in the ultraviolet (2300–3000 Å) and visual regions is approximately in agreement with expected theoretical data based on the models of stellar atmospheres for spectral classes B0 to G5 and on the distribution of stars of the different spectral classes over the sky.

(3) The nature of possible deviations (theory minus observation) is discussed.

The survey of night sky brightness was designed more than 5 years ago to obtain from space observations the following: (1) Data about the limiting magnitudes of night sky brightness, which are important not only for the choice of cosmological models of the Universe, but mainly for having an idea about the penetrating power of space telescopes. (2) Information about the ultraviolet radiation and possible sources in a sky which is free of the influence of the Earth's atmosphere. The preliminary communications were made in [1].

The measurements of night sky brightness were made twice with the aid of photoelectric photometers installed on spacecrafts Kosmos 51 (December 1964) and Kosmos 213 (April 1968). The necessary data about these satellites and the character of the observations are presented in Table I. The device consists essentially of two similar tubes especially designed to reduce the scattered light to minimum (10^{-5} in the case of the solar beam making the angles larger than 70° with the axis of the photometer) and having a photometric half-width of the field of view equal to 18° (see Figure 1). Behind these tubes there is a disk, common for both photometers with two circular diaphragms with the ratio 7.6 for the transmitted flux. For absolute calibration of brightness we have also the small circular area covered by radioactive luminofor (Carbon-14). Each cycle of observation consists of successive settings of the disk in

TABLE I

Data about experiments

Data	K 51	K 213/212
Date	December 10, 1964	April 15, 1968
Period of revolution	92^m5	89^m16
Max. distance	554 km	291 km
Min. distance	264 km	205 km
Inclination	$48°8$	$51°4$
Stabilization	No	around sun-satellite axis
Period of observation	Dec. 10–31, 1964	April 17–22, 1968
Type of information	Direct telemetry	Tape

Houziaux and Butler (eds.), Ultraviolet Stellar Spectra and Ground-Based Observations, 325–333.

four positions: first diaphragm ('large' sky), the second one with reduced flux ('small' sky), the radioactive standard, and 'darkness' (no diaphragm). All this is done by motor; the duration of the cycle is 15 sec.

To stabilize the electronic system we have a rotating disk with four equal diaphragms producing fast modulation of the light (120 cycles per second). Just behind these disks we have two photomultipliers, the first one sensitive to the ultraviolet (2000–3000 Å) and the second to the visual part of the spectrum. The general view of the photometer can be seen in Figure 2.

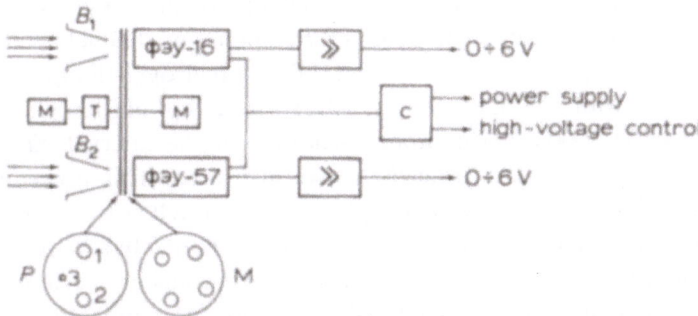

Fig. 1. Scheme of the photometer: B_1, B_2, visual and ultraviolet photometers resp., M = motors driving the programme-disk P and modulating disk M. C = electronics for B_1 and B_2-photometers.

Fig. 2. General view of the photometer.

The most difficult and laborious task in the whole experiment is the calibration of the photometers. In the first experiment with Kosmos 51 the following procedure was adopted (see Figure 3). For the U photometer we measured first the spectral characteristic of a hydrogen lamp and of the double-monochromator with the aid of sodium salicylate, and green-blue filter transmitting the luminescent radiation of sodium salicylate (the quantum yield of this material is constant for $\lambda \leqslant 3200$ Å). With the aid of the same monochromator the response of the U photometer was recorded in the same spectral region. Comparing responses in channels I and II we obtain the spectral sensitivity $\varphi_1(\lambda)$ of the U photometer.

The same procedure was used for the visual photometer. We used a tungsten-filament lamp. Besides, we simply measured the response of channel II and compared it with the known spectral characteristics of the hydrogen lamp in the region 3000–4000 Å and with the measured spectral sensitivity of the photomultipliers (in the region 4000–7000 Å). Additional measures in the interval 3000–3200 Å of the visual channel with sodium salicylate were also used. This procedure gives us the spectral sensitivity $\varphi_2(\lambda)$ of the visual photometer as well as the ratio of the maximum transmissions b_u and b_v in the U- and V-regions (about 0.08). With these data the telemetry reading (in volts) and the ratio U/V can be transformed into the ratio of fluxes. A somewhat different procedure of calibration was adopted for the second experiment with Kosmos 213. The spectral characteristics for both experiments are presented in Figure 4.

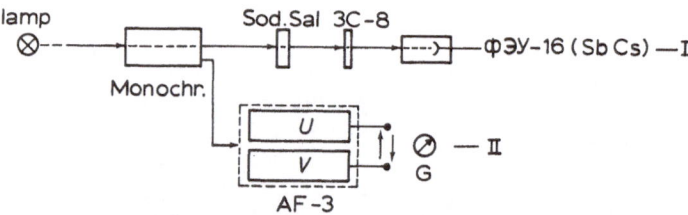

Fig. 3.   Scheme of calibration (see text).

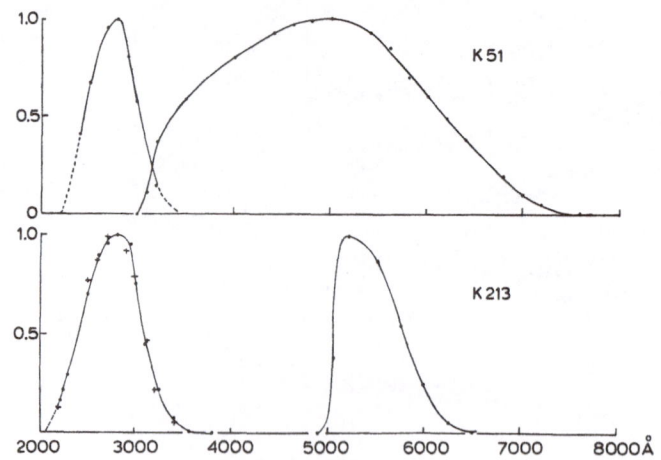

Fig. 4.   The response curves of the photometers for both experiments.

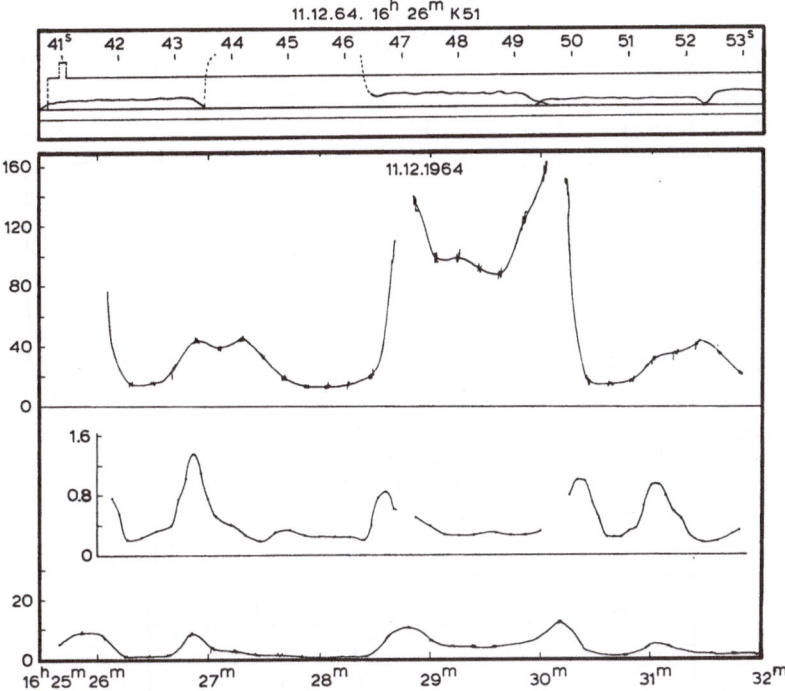

Fig. 5.   Example of telemetry above, and below the reduced readings in visual (*V*), ultraviolet (*U*) regions, with the ratio *U*/*V* in the middle.

Luckily we had a chance to calibrate the ratio U/V also on board the spacecrafts, because in the first experiment we saw Jupiter, and in the second Kosmos 212 receding after junction. As the albedos of both objects are known in the region 2000–3000 Å we have no difficulty in calibrating U/V from actual observed values of U/V and the distribution of intensity in the solar spectrum. In the first case we get the value b_u/b_v which is not quite reliable (because of the absence of reliable values for the albedo of zodiacal light in the region 2000–3000 Å), but still this value is only 30% lower than the experimental one. For the second case we found a value which is in very good agreement with the laboratory calibration.

In Figures 5 and 6 we show examples of telemetry records for both experiments, as the Milky Way at the boundaries of the constellations Lyrae-Cygnus passes across the field of view of our photometers and the corresponding increase of the ratio U/V in these regions. In the first experiment a great amount of recordings was accumulated, but as we did not have any stabilization the identification is extremely difficult and laborious. It is still not completed. In the second case when partial stabilization existed (the only rotation was around the Sun–satellite axis) we could easily identify the band on the sky crossed by the field of view, and this band is shown in Figure 7. It includes also the Milky Way, the constellations Lyra, Cygnus and Aquila and other southern parts of the sky.

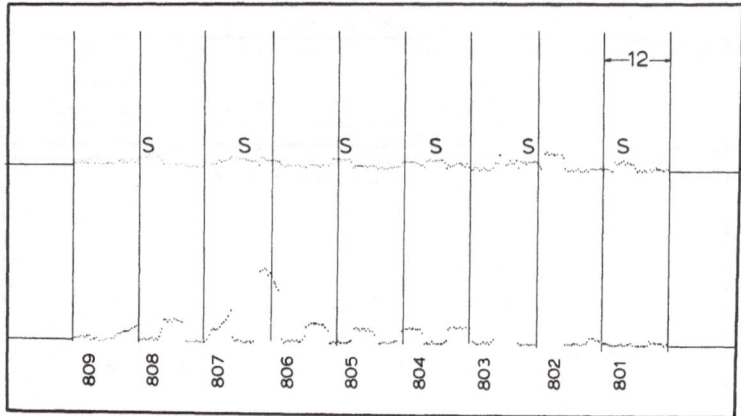

Fig. 6.   The same as in Figure 5. The example of telemetry is at the bottom and the reduced readings
are at the top. By 'Earth' is denoted those intervals of time when the
photometer was directed to the Earth.

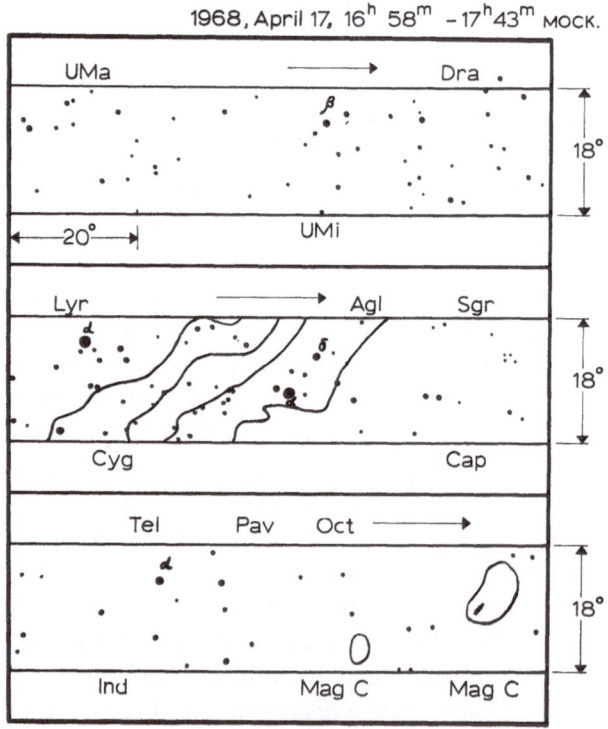

Fig. 7.   The region on the sky crossed by the view-field of the photometer (in the direction of arrow)
for the second experiment (K 213).

As the telemetry gives us the deflection against the radioactive standard, the brightness of the night sky in the visual region can be evaluated. (The phosphor was calibrated in the number of 10th magnitude stars per square degree at the Earth, by

comparing the brightness of two different regions of the sky at equal zenith distances with the brightness of the phosphor.) We can see from Figures 5 and 6 that at some places the brightness of the sky is very low (even on the day-side of the Earth). The minimum brightness (near Draco) in the first experiment was estimated to be

$$I(\text{visual}) = 85 \text{ 10th magn. stars/square degree.}$$

If we reduce this for zodiacal light (about 50–70 stars at least) we have the very low value of about 15–35 10th magnitude stars.

In the second experiment the same value is estimated at 100, but here we had a loss of sensitivity in the V-channel and this value is more uncertain.

The calculation of expected (theoretical) ratio of fluxes U/V was made by the actual counts of stars of different spectral classes and different magnitudes in the H.D. catalogue and their corresponding contributions to the flux in the regions of U- and V-sensitivity of our photometers according to the formula

$$\frac{U}{V} = \frac{b_u}{b_v} \frac{\displaystyle\sum_{m<0}^{8,0} \sum_k v_k(m) \, 10^{-0.4\,m} \int_0^{\lambda_0} \varphi_1 F_k(\lambda) \, d\lambda + U_z}{\displaystyle\sum_{m<0}^{8,0} \sum_k v_k(m) \, 10^{-0.4\,m} \int_{\lambda_0}^{\infty} \varphi_2 F_k(\lambda) \, d\lambda + V_z}, \tag{1}$$

where summation is over all magnitudes m (up to 8th) and spectral classes k with the weighting function $v_k(m)$, characterizing the relative abundance (percentage) of stars of different spectral classes (determined in [2]), and U_z and V_z are the contributions due to zodiacal light. The energy distributions for different spectral classes were taken according to [3], where the scale of effective temperatures was adopted as given in Table II.

TABLE II

	T_e (K) (adopted)	T_e (K) (Morton-Adams)
B0	34000	30900
B5	18000	15600
A0	10000	9600
A5	7000	
F0	6000	
F5	5000	

Denoting the sum by U^* and V^* we have in the region with large U/V the following approximate expression:

$$\frac{U}{V} \cong \frac{b_u}{b_v} \frac{U^*}{V^*} \frac{1}{1 + \dfrac{V_z}{V^*}}, \tag{2}$$

TABLE III

Comparison between calculated and observed values U/V

Region	$(U/V)^*$	$(1 + (V_2/V^*))^{-1}$	$(U/V)_c$	$(U/V)_0$	$(O - C)$	Remarks
Cyg-Aql, M.W. $\alpha = 19^h25^m, \delta = +20°$	0.67	0.82	K 51 0.55	0.64	+0.16	C is not corrected for reddening
Cass. M.W. $\alpha = 0^h30^m, \delta = +57°$	0.60	0.69	0.41	0.47	+0.15	C is not corrected for reddening
α C Ma	0.67	0.80	0.54 K 213	0.40	-0.32	C is not corrected for reddening
Lyr-Cyg-Aql I, $\alpha = 19^h25^m, \delta = +20°$	3.42	0.73	2.50			
II, $\alpha = 19^h45^m, \delta = 0°$	3.40	0.48	1.64			C is not corrected for reddening
		mean	2.07	1.98	-0.05	

because the contribution of zodiacal light in U-flux is negligible (type G) as compared to that of the stars. The relation (2) permits us approximately to exclude the influence of zodiacal light by taking V^* (in numbers of 10th magnitude stars per square degree) and V_z (in the same units) from [4] and [5] correspondingly.

The comparison of such calculations with the observations is shown in Table III. The calculated U/V are not reduced for interstellar reddening, because according to [6] there is no essential interstellar absorption for stars brighter than 10th magnitude in the regions considered here.

We see from Table III the satisfactory agreement between O and C in the limits of $\Delta m = \pm 0.15$ Å for the Lyra/Cygnus-area, but there is some ultraviolet deficiency for Sirius and the surrounding stars, which is larger than the probable errors of observations ($\Delta m \simeq \pm 0.15$) (compare with [7]). The most gratifying result is that the data from both experiments in Kosmos 51 and Kosmos 213 are in good agreement.

If, however, we take into account that the Morton-Adams scale of T_e is more adequate for U region data (and thus that $(U/V)_c$ should be diminished), the conclusion about the existence of some ultraviolet excess in the Milky Way seems to be inevitable. (The influence of faint stars and the Milky Way itself on C-values can lead only to the decrease of the ratio U/V, because the mean contribution of the Milky Way to U is much smaller than to V if calculated according to the mean spectral type of our Galaxy.) Further investigations should probably explain the nature and the source of such U-excess.

References

[1] Dimov, N. and Severny, A.: 1965, I LIL Symp., Athens. 16 Sept.; 111, 1966; XI COSPAR, Japan, May 1968 (in press).
[2] Nort, H.: 1950, *Bull. Astron. Inst. Netherl.* **22**, 181.
[3] Saper, A. and Kuuzik, I.: 1963, *Soobszen. Astron. Obs. Tartu* N7.
[4] Roach, F. E. and Megill, L. R.: 1961, *Astrophys. J.* **133**, 228.
[5] Dumont, R.: 1965, *Publ. Obs. Haute-Provence* VII. No. 42.
[6] Ichsanov, R. N.: 1959, *Izv. Krimsk. Astrof. Obs.* **21**, 257.
[7] Bless, R. C., Code, A. D., and Houck, T. E.: 1968, *Astrophys. J.* **153**, 561.

Discussion

Malaise: You mentioned sodium salycilate for calibrating your UV channel; this is a laboratory transfer standard. What did you use as the primary standard for calibration in the UV?

Severny: We used a hydrogen lamp calibrated with the aid of a black-body and also a tungsten band lamp calibrated in the same way. The calibration of the radioactive ascintillation was made by comparing two areas on the sky at the same zenith distances and knowing the difference in brightness of the out-of-atmosphere components.

Humphries: I would like to report a measurement obtained by Sudbury with the spectrophotometer which he described yesterday. The background levels of approximately 2000 counts/sec give a measured upper limit for the brightness of a night sky in a broad band from 1700 Å to 2500 Å. The value near the galactic poles was equivalent to that from a 6.5 visual magnitude, early B-type star in a 200 Å band in this spectral region; this is equivalent to about five 10th magnitude B-type stars per square degree. This may include zodiacal light and stray light contributions but the nature of the variations around the sky seems to indicate a largely galactic origin.

Severny: I wish to emphazise that the influence of zodiacal light can be very significant even in the far ultraviolet because sometimes we have appreciable enhancement of UV-emission of the Sun due to flare activity and zodiacal light can probably reflect these variations.

Campbell: (a) Could you repeat the fields of view of your photometers? (b) Is it possible to define the effective passband of the 2700 Å photometers? (c) What is the absolute sensitivity of your photometers?

Severny: (a) the photometric half-width of the field of view of our photometers was large, $\sim 18°$. (b) The effective passband of our UV photometer was about 600–700 Å. (c) I think we are not prepared at the present to specify the absolute values of flux in ergs cm^{-2} s^{-1}. (We have made the estimates but we should check them.) But we have made estimates in terms of the number of stars of 10^m0 per square degree (on the scale of visual magnitudes) and I have presented in my talk the minimal brightness of the stellar component in the vicinity of the ecliptic pole as 15 stars of 10^m0 per square degree.

MARINER 5 MEASUREMENTS OF ULTRAVIOLET EMISSION
FROM THE GALAXY

CHARLES A. BARTH

Dept. of Astro-Geophysics, and Laboratory for Atmospheric and Space Physics,
University of Colorado, Boulder, Colo., U.S.A.

Abstract. The Mariner 5 ultraviolet measurements obtained while the spacecraft was in interplanetary flight are interpreted as Lyman-α radiation. This radiation may arise from (1) the scattering of solar Lyman-α radiation by interplanetary atomic hydrogen, (2) the scattering of solar Lyman-α by atomic hydrogen that is present at the edge of the solar system, or (3) diffusely scattered Lyman-α radiation from the Galaxy. The Mariner 5 measurements show a symmetry with respect to the galactic equator which suggests that the major source of the observed emission is the diffuse galactic radiation. In the Münch model for the diffuse Lyman-α radiation from an H I region, the sources of the Lyman-α photons are stellar chromospheres, the photons are scattered a large number of times by interstellar atomic hydrogen, and are absorbed by interstellar dust. Some of the Mariner 5 measurements may be attributed to Lyman-α emission from an H II region since the field of observation scanned the edge of the Gum nebula.

1. Introduction

Mariner 5, a planetary spacecraft, performed a number of ultraviolet observations while in interplanetary flight to Venus and beyond. In the several weeks following the encounter with Venus on October 19, 1967, and before the cessation of radio communications with the Earth on December 1, 1967, the spacecraft executed a series of manoeuvres designed to measure Lyman-α radiation emanating from the Galaxy. The ultraviolet photometer experiment which was designed to measure the distribution of atomic hydrogen in the outer atmosphere of Venus was used in these experiments (Barth *et al.*, 1967). The observations were unique in that they were conducted in interplanetary space far from the hydrogen coronas of Venus and the Earth, and from a spacecraft that was rolling in such a way that the photometer field of view made a nearly perpendicular crossing of the galactic plane.

2. Experimental

The Mariner 5 ultraviolet photometer consisted of three separate photomultiplier tubes each with a caesium iodide photocathode and a lithium fluoride window. Two of the channels had their short wavelength responses limited by calcium fluoride and barium fluoride filters to provide wavelength bands of 1250–2200 Å and 1350–2200 Å, respectively. The channel without a filter responded in the 1050–2200 Å wavelength interval. The experimental technique, which was designed for planetary observations, used the difference in signals between the lithium fluoride and calcium fluoride channels as a measure of hydrogen Lyman-α 1216 Å radiation. The difference in signals between the calcium fluoride and barium fluoride channels was a measure of the atomic oxygen 1304 Å line. In astronomical observations all three

Houziaux and Butler (eds.), Ultraviolet Stellar Spectra and Ground-Based Observations, 334–340.
All Rights Reserved. Copyright © 1970 by the IAU.

channels respond to radiation from early type stars. It is, however, still possible to interpret the difference in signals between the lithium fluoride and calcium fluoride channels as Lyman-α radiation. All three channels had field of view limiters but no optics. The full field of view for the lithium fluoride channel as it responded to a point source was 6° and for the other two channels was 2.4°.

The photometer was mounted on the spacecraft so that its view direction made an angle of 90° with the spacecraft–Sun line and an angle of 95° with the plane containing the Sun, the spacecraft, and the star Canopus. Since the spacecraft was usually stabilized on the Sun and Canopus, the ultraviolet photometer viewed a region near the ecliptic which slowly changed in ecliptic longitude as the spacecraft traversed its orbit about the Sun. On several occasions the spacecraft was released from stabilization on Canopus and rolled about the spacecraft–Sun line. During these manoeuvres, the ultraviolet photometer field of view swept along lines of ecliptic longitude. Two such series of rolls, one on November 7, 1967, and the other on November 19, 1967, are the basis of the observations reported in this paper. At these times the spacecraft was approxi. mately 1×10^8 km from the Sun and over 5×10^6 km from Venus.

3. Observations

The regions of the celestial sphere that were observed during the rolling manoeuvres are shown in Figure 1, which is a chart of ultraviolet stars plotted in galactic coordinates. The relative size of the dots represents the magnitudes of the stars as they would be measured in the 1350–1600 Å wavelength region where the three ultraviolet photometer channels had their maximum response. The track of the November 7 roll began in the Southern galactic hemisphere and moved initially to decreasing longitude

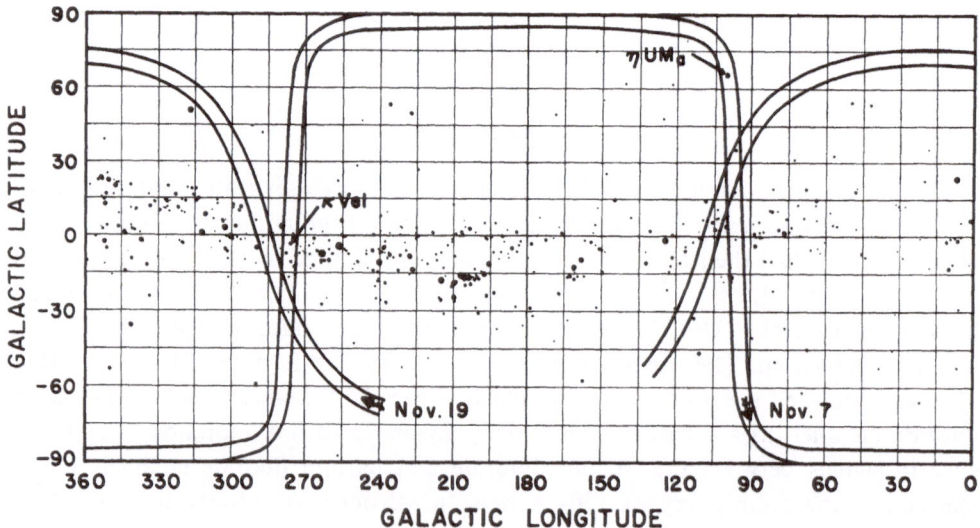

Fig. 1.   Ultraviolet star chart which shows regions of observation during November 7 and 19 rolls.

and then northward in latitude. The observation track crossed the plane of the Galaxy in the constellation Vela, sweeping to Northern galactic latitudes and then southward crossing the galactic plane again, this time in the constellation Cygnus. The width of track shown in the figure is 6° corresponding to the full field of view of the lithium fluoride channel.

Data obtained on the November 7 roll are shown in Figure 2. The intensity of radiation measured by the lithium fluoride channel is shown as a linear function of the position in the roll. In addition, galactic latitude and ecliptic latitude coordinates are plotted along the abscissa. Figure 2 displays the data from three sequential rolls of the spacecraft, each approximately of 1 hour's duration. This lithium fluoride channel data shows responses to discrete ultraviolet objects such as the Large Magellanic Cloud, κ Vel, and η UMa. When the lithium fluoride channel responds to these objects, the calcium fluoride and barium fluoride channels respond as well. However, the lithium fluoride channel measures a signal throughout the roll which the other two do not. This signal is interpreted as Lyman-α radiation. The striking thing about this Lyman-α signal is that it is not uniform in all directions. To the South of the galactic plane, the flux is 5×10^{-4} ergs cm^{-2} s^{-1} ster^{-1}, while to the North it is 7×10^{-4} ergs cm^{-2} s^{-1} ster^{-1}. When the field of view crossed the galactic plane in Vela, an enhancement of Lyman-α flux was observed over a region 20° wide. When the

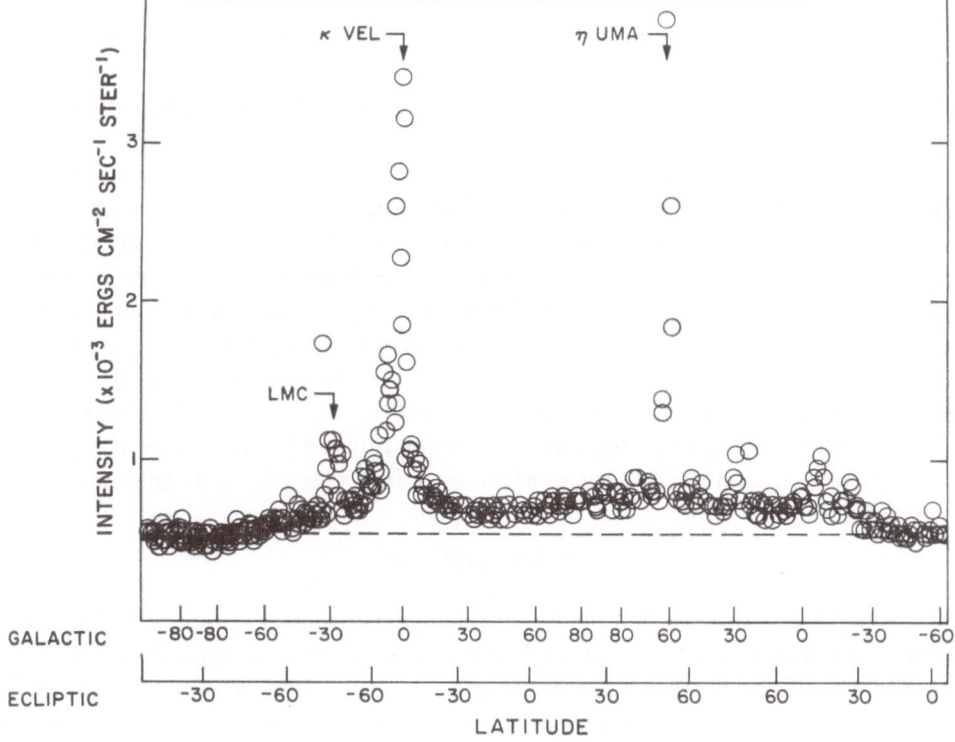

Fig. 2.   Radiation measured by lithium fluoride channel during November 7 roll.

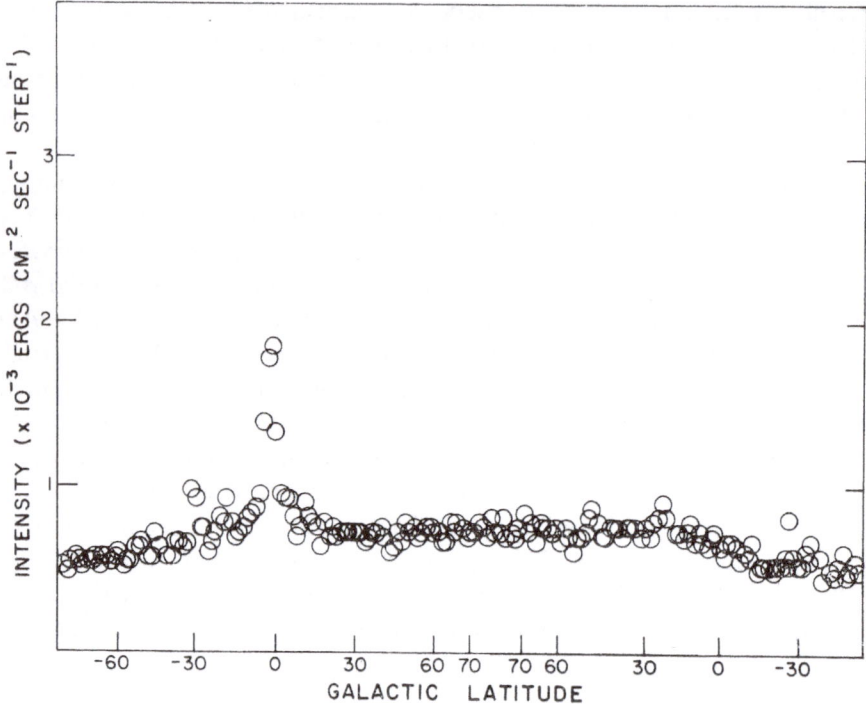

Fig. 3.   Radiation measured by lithium fluoride channel during November 19 roll.

field of view crossed the galactic plane in Cygnus, however, a comparable enhancement was not observed.

The observation track of the November 19, 1967, roll is also shown in Figure 1. The observations also began in the Southern galactic hemisphere, crossed the galactic plane in Vela, scanned through the Northern galactic hemisphere crossing the plane again in Cygnus, stopping in the Southern hemisphere. This was a single roll of approximately 315°. The data from the lithium fluoride channel for this roll are shown in Figure 3. All of the general characteristics of the November 7 rolls are present. All three channels respond to hot stars and the Large Magellanic Cloud. The lithium fluoride channel alone shows a background signal throughout the roll. The Northern galactic hemisphere shows a larger value than the Southern hemisphere. The galactic plane crossing in Vela shows an enhancement in Lyman-α, while the crossing in Cygnus does not.

4. Interpretation

The diffuse Lyman-α glow observed by Mariner 5 may originate from atomic hydrogen located (1) within the solar system, (2) at the edge of the solar system, or (3) in the interstellar medium. If resonance reradiation of solar Lyman-α by interplanetary atomic hydrogen is the explanation of the Mariner results, a density of the order of 10^{-2} atoms cm^{-3} is required if the atomic hydrogen is distributed uniformly throughout the solar system (Barth *et al.*, 1968). For observations made perpendicular to the spacecraft–Sun line, as was the case with Mariner 5, the emission rate is equal to that produced by the local volume density of atoms times an equivalent path length of $\pi/2$ times the distance of the instrument from the Sun. An interplanetary density of 1×10^{-2} atoms cm^{-3} would produce an optical depth of approximately $\tau = 0.1$, a value that is not in conflict with any spectroscopic observational results.

Even if the interplanetary medium is devoid of atomic hydrogen, there may be a buildup of atoms at the edge of the solar system where the solar wind runs into the interstellar medium (Patterson *et al.*, 1963). For a shell of hydrogen atoms at 10 AU illuminated by solar Lyman-α, a density of 10^{-1} atoms cm^{-3} is required to explain the observed Mariner 5 intensities.

These two mechanisms in which the hydrogen atoms scatter Lyman-α radiation which originates in the Sun predict that the diffuse Lyman-α glow be distributed uniformly as viewed from the Mariner spacecraft or at most with a symmetry that is oriented with respect to an ecliptic coordinate system. The observations that are plotted in Figure 2 show that the asymmetric intensity distribution is oriented with respect to the galactic coordinate system and not the ecliptic. This observational result suggests that at least some, if not all, of the Lyman-α radiation must come from the Galaxy.

A model of diffuse galactic Lyman-α radiation has been given by Münch (1962). In this model, the source of the Lyman-α photons is the chromospheres of late type stars. The stellar Lyman-α photons are scattered by interstellar atomic hydrogen until they are finally absorbed by interstellar dust. Münch (1962) has calculated that the mean distance that a Lyman-α photon travels before being absorbed is 50 pc. In traversing that distance some 10^9 scatterings occur. Münch also calculated that the expected surface brightness of the diffuse Lyman-α radiation should be 6.4×10^{-4} erg cm^{-2} s^{-1} ster^{-1}. That this number agrees so well with the Mariner observations is surprising but perhaps fortuitous.

To explain the different intensities observed by Mariner 5 in the Northern and Southern galactic hemispheres using the Münch model, one may assume that the ratio of interstellar hydrogen to dust is different above the galactic plane than it is below the plane.

Münch (1962) has also pointed out that large local sources of Lyman-α radiation may be expected from H II regions, in particular, the extensive Gum H II region that is excited by γ Vel and ζ Pup (Gum, 1952). In an H II region, the ultimate source of photons is stellar flux shortward of the Lyman limit from early type stars. Lyman-α photons are produced during the radiative recombination of ionized hydrogen. The Lyman-α photons are lost in the H II region either by two-photon decay or ionization of the excited atom. Lyman-α photons escaping from the H II region eventually are lost by absorption on interstellar dust.

The increase in the diffuse Lyman-α glow that occurred when the Mariner 5 photometer swept through Vela may be an observation of the edge of the Gum H II region. Unfortunately, it was not possible to obtain observations closer to the center of this region from Mariner 5. If the identification of the brightening of the Lyman-α glow with the Gum region can be verified by future observations, it may be possible to determine how close this H II region extends to the solar system.

Observations of Lyman-α radiation were also made in interplanetary space from Venera 4 (Kurt and Dostovalov, 1968). The intensities observed from this spacecraft were about a factor of 3 lower than those observed from Mariner 5. It has not been possible to compare results from the two spacecraft as a function of galactic coordinates.

5. Summary

The diffuse Lyman-α glow observed by Mariner 5 is attributed at least in part to radiation scattered by hydrogen atoms in the interstellar medium. In addition, Lyman-α radiation from an H II region may have been observed in Vela.

Acknowledgement

The Mariner 5 project is to be commended for conducting the operations necessary for these astrophysical observations after the completion of the planetary mission. This research has been supported by the National Aeronautics and Space Administration under grant number NGL 06-003-052.

References

Barth, C. A., Pearce, J. B., Kelly, K. K., Wallace, L., and Fastie, W. G. 1967, *Science* **158**, 1675.
Barth, C. A., Wallace, L., and Pearce, J. B.: 1968, *J. Geophys. Res.* **73**, 2541.
Gum, C. S.: 1952, *Observatory* **72**, 151.
Kurt, V. G. and Dostovalov, S. B.: 1968, *Nature* **218**, 258.
Münch, G.: 1962, in *Space Age Astronomy* (ed. by A. J. Deutsch and W. B. Klemperer), Academic Press, New York, ch. 10, p. 219.
Patterson, T. N. L., Johnson, F. S., and Hanson, W. B.: 1963, *Planet. Space Sci.* **11**, 767.

Discussion

Severny: I have two questions. First: How do you explain the high increase in Lyman-α intensity in high galactic latitudes? Second: do you observe the increase of intensity outside Lyman-α when crossing the Milky Way?

Barth: If the model put forth by Münch (1962) is the correct explanation of the galactic Lyman-α intensity, then the difference in intensity between the Northern and Southern galactic hemispheres may be explained by a difference in the ratio of neutral hydrogen to dust in these two regions. The photometer channels that measured outside of the Lyman-α region did not measure an increase in signal when crossing the Milky Way other than that produced by early stars.

Carruthers: What density of interstellar hydrogen was assumed in deriving the 50 pc mean free path for Lyman-α scattering?

Barth: In the calculations that were done by Münch (1962) for Lyman-α scattering in an H I region, the density of interstellar hydrogen that was used was 1 atom cm^{-3}.

Carruthers: If the density is 0.1 cm^{-3} instead of 1.0 cm^{-3}, then the scattering path would be comparable with the thickness of the galactic disk, and hence one would more likely expect an increase in the observed emission toward the galactic plane.

Underhill: What constellation did you cross through when going from the N side of the galactic plane to the S side?

Barth: Cygnus.

Henize: Can you give any estimate of the densities required if the Vela radiation originates in the Gum H II region?

Barth: There does not seem to be much doubt that H II regions will produce Lyman-α radiation. The key question in the case of the Gum H II region is whether there is an intervening H I region of sufficient density to obscure the Lyman-α. This problem can be solved if additional observations are made in the Vela-Puppis region.

Henize: I note a large deflection due to the Large Magellanic Cloud. Is it possible to use this data to estimate the ability of the Lyman-α radiation to penetrate the interstellar medium?

Barth: The observation of galaxies in Lyman-α radiation is potentially a powerful technique to learn something about the density of intergalactic hydrogen. A preliminary examination of the data in all three channels from the Mariner 5 photometer has not revealed any excess Lyman-α. That may be attributed to the Large Magellanic Cloud. The signal that you see in the figure may be explained as coming from a large number of early type stars.

Herbig: If the mean free path of a Lyman-α photon is so short in the interstellar medium, would it not be more significant to correlate these data not with galactic latitude, but with latitude measured with respect to the plane of Gould's Belt?

Barth: That is an excellent suggestion.

Sunayev: (a) The stars of the Galaxy must give emission both in the Lyman-α line and in the continuous spectrum. In this case the response of 'Venus' counters would be commensurate, whereas their ratio [(1050–1340 Å)/(1222–1340 Å)] is 100.

(b) Gaseous nebulae and the aggregate of H II regions surrounding hot stars also cannot ensure the observed intensity of Lyman-α emission since their emission is the result of the redistribution of stellar radiation beyond the Lyman limit, which cannot significantly (100–1000 times) exceed the total emission in the 1225–1340 Å band (Kurt's measurements).

(c) Subcosmic ray ionization losses in the interstellar medium are not sufficient because their energy density is not high enough.

Probably all Lyman-α radiation originates from the interaction of the stellar wind due to a mass loss with the interstellar medium. In this case the width of the line is very large and shifted Lyman-α quanta reached the observer without scattering in H I regions. The other possibility is that the origin of this Lyman-α radiation is in the inter planetary medium. It may be that the density of the solar wind is higher in the galactic plane because of the influence of the Galactic magnetic field on the solar wind.

THE ULTRAVIOLET BACKGROUND (INTERGALACTIC GAS, THE GALAXY, AND SUBCOSMIC RAYS)

V. G. KURT and R. A. SUNYAEV

Academy of Sciences, U.S.S.R. and Sternberg Astronomical Institute

Abstract. (1) *Observations*. – A survey is made of observations of the background radiation at UV wavelengths from above the atmosphere. Sources of the background radiation and ways of determining the extragalactic component of the background are discussed. Future observations are also discussed.

(2) *Cosmology*. – Limits to the properties of the intergalactic gas follow from observations of the UV background. The problem of detecting galaxies at early stages in their evolution is considered.

(3) *The Galaxy*. – Observations and theoretical estimates are given for the integrated brightness of the Galaxy at UV wavelengths beyond the Lyman-α line. Also discussed are the nature of the Lyman-α emission from the Milky Way and the principal ways of constructing the luminosity function of stars in the Galaxy from observations of the integrated UV spectrum.

(4) *Subcosmic Rays in the Interstellar Medium*. – Fast neutral excited hydrogen atoms can be formed from charge-exchange interactions between subcosmic-ray protons and neutral interstellar gas. Upper limits are given to the energy density of subcosmic rays having $E \sim 100$ keV.

(5) *Limits on the Background Radiation in the range* <912 $\mathring{A}$. – The distribution of neutral hydrogen in the peripheries of galaxies allows limits to be obtained for the flux of metagalactic ionising radiation. The heating and ionisation of the interstellar medium by X-rays is considered.

1. Introduction

It is now relatively easy to make observations in the near ultraviolet region of the spectrum (1000–3000 Å) from space vehicles above the atmosphere. Particularly great progress has been made in the study of the spectra of stars of early spectral type with high spectral resolution ($\Delta\lambda \sim 1$ Å). This is the work of Carruthers, Morton, Smith, Stecher, and others.

Below we discuss observations of the ultraviolet background radiation with lower spectral resolution (~ 100 Å) and the conclusions which can be drawn from the existing observational data concerning

(a) the density and temperature of the intergalactic gas which make it possible to determine the mean density of matter in the Universe (Kurt and Sunyaev, 1967a, b).

(b) the integrated ultraviolet radiation of our Galaxy (Kurt and Sunyaev, 1967c; Kurt and Dostovalov, 1968).

(c) the energy density of subcosmic rays in our Galaxy (Kurt and Sunyaev, 1968).

In addition we discuss indirect methods of determining the flux of background UV radiation in the wavelength range shorter than 912 Å from the distribution of neutral hydrogen in the peripheries of galaxies (Sunyaev, 1969a). It is clearly impossible to observe this radiation from the solar system but it has an important influence on the properties of the interstellar medium, such as its heating and its state of ionisation (Sunyaev, 1969b).

In this survey we propose a series of new experiments which will give important information about the properties of the intergalactic gas, the overall properties of the stellar population and the interstellar medium.

Houziaux and Butler (eds.), Ultraviolet Stellar Spectra and Ground-Based Observations, 341–348.

2. The Observations

In the observations in the UV region, photon counters and photomultipliers with a
field of view 1–10° have been used. In these observations the measured signal must be
separated from background noise due to cosmic rays and the dark current of the
photoelectric apparatus. The main contribution to the observed signal in near UV and
in the visible region of the spectrum results from the integral effect of stars in our
Galaxy even in the direction of the galactic pole and from zodiacal light. Therefore
in future observations, we must use apparatus with a smaller field of view to obtain a
smaller contribution from stars and electronic systems to decrease the contribution
from charged particles. In Figure 1 we present the observational data in the range
5000–1000 Å. In the range < 2000 Å, there seems to be a possibility of obtaining data
without the influence of galactic stars at high galactic latitudes. In this spectral region,
the principle contribution comes from early-type stars the number of which, per
square degree drops rapidly with increasing temperature. This question has been
discussed in detail by us (Kurt and Sunyaev, 1967b) for the spectral range 1300 Å
where the principal role is played by A-B stars. If the field of view of the apparatus is
~ 0.1 square degree, such stars result in a background of less than 10^{-23}–10^{-22} erg
cm^{-2} ster^{-1} Hz^{-1} at high galactic latitudes. According to our calculations, this value
is of the same order as the intensity expected from interplanetary space and the
extragalactic background from galaxies. For similar observations in the visible region
of the spectrum, the background results from the sum of stars of spectral class similar
to the Sun and zodiacal light, the colour temperature of which is similar to that of the
Sun, and the separation of the non-stellar component is extremely complicated.

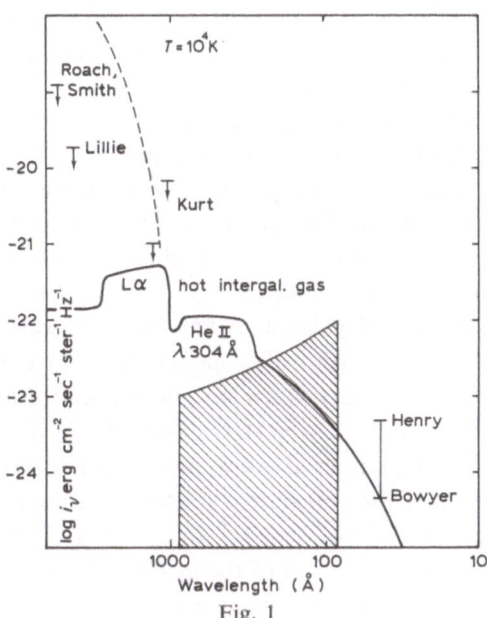

Fig. 1

According to the new data of Lillie (1969), the density of radiation near the Sun is 5×10^{-17} erg cm^{-3} Å^{-1} in the range 2000–5600 Å which corresponds to 1.5×10^{-7} erg cm^{-2} s^{-1} ster^{-1} Å^{-1} or 10^{-18} erg cm^{-2} s^{-1} ster^{-1} Hz^{-1}. The extragalactic contribution of this radiation at 4100 Å is less than 10^{-20} erg cm^{-2} s^{-1} ster^{-1} Hz^{-1}. This result is one order of magnitude lower than that of Roach and Smith (1968) and equal to that expected from the integrated effect of galaxies. This implies that any evolution of the brightness of galaxies with cosmological epoch must be small.

We have also observed the short wavelength range 1050–1180 Å with a field of view $\sim 20°$. The observed value of the background intensity, 7×10^{-21} erg cm^{-2} s^{-1} ster^{-1} Hz^{-1}, clearly gives an upper limit to the metagalactic radiation. It is possible that the main contribution to the measured signal, which is 5–10 times greater than the background due to cosmic rays, is the result of stars with maxima in the range 2000 Å. In this range, the quantum efficiency is less than or equal to 10^{-4}, which is not sufficient to reduce the background due to stars in the range 2000 Å.

3. The Background Radiation from Intergalactic Space for Wavelengths Shorter than 912 Å

Direct observations of the background radiation from intergalactic space in the range 912–100 Å is in principle impossible because it is completely absorbed by neutral hydrogen in our Galaxy. At present we have only observations in the soft X-ray region 44–70 Å (Bowyer *et al.*, 1968; Henry *et al.*, 1968) and the region close to Lyman α (1225–1340 Å) which has been discussed above. The observed soft X-ray intensity is $0.5 \times 10^{-24} < I_v < 5 \times 10^{-24}$ erg cm^{-2} s^{-1} ster^{-1} Hz^{-1} about 44 Å; the exact value of this flux is not known because we do not know how strongly the radiation is absorbed in our Galaxy. In the range 1225–1340 Å, the extragalactic background is not greater than 10^{-21} erg cm^{-2} s^{-1} ster^{-1} Hz^{-1}. At the present day, there is only one way of observing the background radiation for wavelengths less than 912 Å which is from observations of the peripheries of galaxies in the 21-cm line of neutral hydrogen. It is clear that as a result of interactions between the background ionising radiation (<912 Å) and neutral hydrogen in the peripheries of galaxies, there must exist ionisation zones analogous to Strömgren spheres around hot stars. Observational techniques for observing the 21-cm line are now highly developed and the distribution of hydrogen in the Andromeda Nebula (M31), the Magellanic clouds and at least 10 other galaxies and systems of galaxies is known. Observations have shown that the hydrogen halo extends far beyond the optical image of the galaxy (at least 1.5 times) and that neighbouring galaxies which have no connection at optical wavelengths are joined by bridges of neutral hydrogen (NGC 4631–4656). The dimensions of these bridges are very large but the number of atoms along the line of sight is much smaller than in the central regions of the galaxies; it follows that the density of hydrogen is very small and must be of the order of 3×10^{-3} cm^{-3} or less. An ionisation front forms at the periphery of the galaxy as a result of the incident flux of ionising radiation and moves into the galaxy with velocity $V = I/n_H$, where I is the flux of radiation in quanta

$cm^{-2} s^{-1}$ and n_H is the density of hydrogen in front of the ionisation front. The ionisation front continues to move into the galaxy with this velocity as long as the optical depth of the gas behind the front is not too large, i.e.

$$\int_0^\infty n_e^2 \alpha dl < I,$$

i.e. the number of recombinations for unit time in a column of gas behind the front must not exceed the number of incident quanta (α is the recombination coefficient and n_e the electron density behind the front).

If the total number of neutral and ionised particles along the line of sight does not increase with distance from the centre of the galaxy, then from the observations of the Andromeda Nebula if $I > 10^4$ quanta $cm^{-2} s^{-1}$, it is highly unlikely that there should exist a stationary Strömgren zone surrounding the neutral hydrogen halo with $n_H \sim 2 \times 10^{-3}$ cm^{-3} and $l \sim 3$ kpc. That is, if $I > 10^4$ quanta $cm^{-2} s^{-1}$, which corresponds to $I_\nu > 10^{-23}$ erg $cm^{-2} s^{-1} ster^{-1} Hz^{-1}$ in the range 912 Å $< \lambda < 304$ Å, the time scale for ionizing the halo be less than 10^{15} sec, which is less than the time scale of the galaxy. Similar calculations may be made for other galaxies. It is clear that even if we have a small flux of ionising radiation, a stationary or moving ionisation front must exist somewhere in the distant periphery of the galaxy and must result in a sharp boundary to the distribution of neutral hydrogen. We also note that if the ionising flux with $\lambda < 912$ Å were greater than $I_\nu \sim 4 \times 10^{-21}$ erg $cm^{-2} s^{-1} ster^{-1} Hz^{-1}$, the neutral hydrogen in the plane of our Galaxy in the region of the Sun could be completely ionised $(n_H l \sim 2.5 \times 10^{20}$ $cm^{-2};$ $n_H \sim 0.1$ $cm^{-3})$ (Sunyaev, 1969a, b). Therefore it appears that neutral hydrogen in the peripheries of galaxies is a more sensitive method of detecting the ultraviolet background radiation than existing experiments in the range $\lambda > 912$ Å.

4. The Heating and Ionisation of the Interstellar Gas by UV and soft X-Ray Radiation

The photons of the soft X-ray background with wavelengths around 40 Å which have been detected (Bowyer et al., 1968; Henry et al., 1968), heat and partially ionise the interstellar gas (Sunyaev, 1969b), influence the physical conditions in the interstellar gas and, in particular, hinder the formation of clouds of neutral hydrogen in the peripheries of galaxies and in the regions between spiral arms – wherever $n_H \sim 10^{-2}$ cm^{-3}. This is an example, similar to the relict radiation, of how the background radiation can influence the physical conditions in the Galaxy.

Pikel'ner (1967) has considered the heating and partial ionisation of the interstellar gas by subcosmic rays. It is considered that the subcosmic rays are injected from supernova explosions. Calculations on the explosion of supernovae carried out by V.S. Imshennyk and his co-workers have shown that a significant part of the energy is emitted as ionising radiation in the region 200–30 Å. This burst of radiation heats and ionises the interstellar gas to a distance of hundreds of parsecs from the super-novae; the relaxation time is very large. If we suppose that in the interstellar medium $n_e \sim 3 \times 10^{-3}$ cm^{-3} and $T_e \sim 10^2$ K, then the recombination time turns out to be of the order of $t \sim 1/\alpha n_e \sim 10^{14}$ sec. Comparison of this time with the frequency of super-nova outbursts in the Galaxy ($t \sim 10^9$–10^{10} sec) shows that the UV ionising radiation must have a significant influence on the properties of the interstellar medium. This conclusion is confirmed by the existence of a large number of X-ray sources with spectra which rise sharply in the ultraviolet, additional to which normal stars may be powerful sources at about 100 Å because of the existence of coronae similar to the Sun's.

5. Cosmological Aspects of Measurements of UV Radiation

A. INTERGALACTIC GAS

It is well known that if the mean density of matter in the Universe exceeds its critical value ($\varrho_{cr} = (3H_0^2)/(8\pi g) = 2 \times 10^{-29}$ g cm^{-3}), the Universe is closed and the expansion observed at the present day must eventually reverse to a contracting phase. If $\varrho < \varrho_{cr}$, the Universe is open. The matter contained in galaxies (i.e. visible matter) and radiation imply $\varrho \ll \varrho_{cr}$ and is insufficient to close the Universe. In principle, it is possible that there exists a diffuse intergalactic gas with density $\varrho \sim \varrho_{cr}$ which determines the dynam-ics of the Universe. The present observational data at radio ($\lambda 21$ cm) and soft X-ray wavelengths (44–60 Å) set limits to the permissible range of temperature of the inter-galactic gas of $10^4 - 3 \times 10^6$ K if $\varrho \gtrsim \varrho_{cr}$. If the gas has such a density and its temperature lies in this range, it radiates principally in the ultraviolet. Therefore measurements of the intergalactic ultraviolet background can give information about the density and temperature of the intergalactic gas and the past history of the Universe. Measure-ments in the range 1225–1340 Å have enabled upper limits to be set to the flux in the intergalactic redshifted Lyman α line. It was shown (Kurt and Sunyaev, 1967a, b) that these measurements set an upper limit of $\varrho < 5\varrho_{cr}$ for any temperature of the inter-galactic gas and $\varrho < \varrho_{cr}$ for the narrower temperature interval $3 \times 10^4 > T > 10^4$ K. The indirect method of determining the UV background described above ($\lambda < 912$ Å) im-plies that for any temperature $\varrho < 0.3\varrho_{cr}$. It is clear, however, that we must continue these direct observations in the UV region of the spectrum with higher sensitivity. Measurements in the spectral region $\lambda < 1216$ Å will enable the separation of the con-tributions from Lyman α lines, the continuous hydrogen spectrum and the HeII 304Å line, all of which are displaced due to the effect of the cosmological expansion.

B. OBSERVATIONS OF THE SPECTRUM OF 3C273

Above we discussed the possibility of observing radiation from the hot ionised gas. A much more sensitive method of measuring the density of a neutral intergalactic gas than observations of the 21-cm line is observation of quasars in the ultraviolet region of the spectrum with $\lambda > 1216$ Å. This has been done for 3C9 by Gunn and Peterson (1965) who have given an upper limit $n_H > 6 \times 10^{-11}$ cm^{-3}. The upper limit to the density of neutral hydrogen which may be evaluated from absorption in the optical spectra of quasars may be expressed as the optical depth

$$\tau = \frac{5 \times 10^{10}(1+z)^2\, n_{\mathrm{H}}}{\sqrt{1 + (\varrho/\varrho_{cz})\, z}}.$$

For $z \ll 1$, if $\tau \gtrsim 0.3$ (which could be observed), $n_{\mathrm{H}} \gtrsim 6 \times 10^{-12}$ cm^{-3}. The most convenient quasar to use is 3C273 with $z = 0.16$, $m = 13^{\mathrm{m}}$. It would be extremely interesting to obtain spectra of this quasar in the range 2000–912 Å and also the spectra of quasars with $z \gtrsim 2$ which would allow absorption due to intergalactic neutral helium He I 584 Å to be measured. Such spectra would also give much information about the quasars themselves. The spectrum of 3C273 in the region 1000–2000 Å is probably of the form $I_\nu \sim \nu^{-\alpha}$ with $\alpha \approx 0.7$ as is found in quasars with $z \sim 2$. In this case, in the UV region of the spectrum, the quasar would have intensity $\sim 10^{-25}$ erg cm^{-2} s^{-1} Hz^{-1} at 1300 Å. With a bandwidth $\Delta\lambda \sim 10$ Å such a flux could be observed with a telescope of diameter of about 20 cm. There remains of course the problem of orientating the telescope on to an object which is very faint at optical wavelengths but, at the same time, anomalously bright in the UV region. 3C9 would be 20 times weaker and therefore requires a 1-m telescope with the same spectral resolution. Both of these objects are excellent candidates for observations with the OAO.

C. YOUNG GALAXIES

Partridge and Peebles (1967) have considered the possibility of observing the radiation from young galaxies. In their model, galaxies were born very early at a redshift of about 20–30. In this case, most of their radiation falls in the infrared region where observations are extremely difficult because of the very high background and because of the low sensitivity of the detectors.

If galaxy formation continues to redshifts $z \lesssim 1$, the radiation of such objects falls principally in the ultraviolet region of the spectrum, where observing conditions are much more favourable.

Let us consider a galaxy of the type of our Galaxy in the process of collapse before star formation takes place. In this case the release of gravitational energy is 10^{59} ergs and the principal part of this energy will be emitted as Lyman α quanta over a time-scale $\sim 10^{14}$–10^{15} sec. Such an object will have luminosity $\sim 10^{44}$–10^{45} erg s^{-1} which will be emitted in a very narrow line of band width about 1 Å. Such an object can even be observed at cosmological distances. In this calculation we have not considered absorption by dust which is formed at a later stage and also assumed that all the radiation escapes from the galaxy. If we include nuclear sources of energy and suppose that in the beginning stars of early spectral type are formed, then the energy output increases by a very large factor.

6. Ultraviolet Radiation from the Galaxy

UV measurements from the Venus space-probes determine the radiation from the

Milky Way in the Lyman α line and the continuum in the range 1225–1340 Å (Kurt and Dostovalov, 1968).

A. THE RADIATION IN LYMAN α

In the range 1050–1340 Å which includes the Lyman-α line, the intensity at galactic longitude $l^{II} = 100°$ is approximately 2.5×10^{-4} erg cm^{-2} s^{-1} ster^{-1} and at $l^{II} = 300°$ roughly 10 times less. A similar effect has been observed in the region $l^{II} = 80°$ and $l^{II} = 120°$, where the intensities are very similar and equal to $5–10 \times 10^{-5}$ erg cm^{-2} sec^{-1} ster^{-1}. The characteristic property of this emission is that it strongly depends upon galactic longitude.

B. CONTINUOUS EMISSION

At the same time in the channel sensitive to the band 1225–1340 Å (i.e. without Lyman α) the intensity of the Milky Way was equal to only 10^{-6} erg cm^{-2} ster^{-1}. This radiation is the result of the population of hot stars.

We now have estimates of the diffuse UV radiation from stars both from studies of experimental spectra and from theoretical model stellar atmospheres. In the spectral region 1000–1500 Å, the calculated values of the intensity exceed the observed values by a factor of about 10. Such a discrepancy could be explained in the following ways:

(1) An abrupt 'cut-off' in the spectra of stars in the range $\lambda < 1500$ Å as has been observed in stars of earlier spectral type than A0. One could suppose that the effective temperature of A and B type stars in the ultraviolet is 2000–3000° less than T_{eff} in the visual region, which would result in a decrease in intensity by a factor of 10.

(2) Blanketing of absorption lines in the range 1225–1340 Å. The equivalent width grows rapidly with decreasing stellar temperature. The radiation from the Milky Way in this spectral region consists principally of stars of type B5-A5, where the strongest absorption lines are C II, C III, Si III, S II (Morton, and his co-workers) and also Lyman α. Calculations for O5, B0 and B4 stars and also spectra obtained in the U.S.A. from rockets have enabled the background spectrum due to stars to be estimated on different assumptions about their luminosity function and interstellar absorption of UV radiation. At the present day, we know the standard luminosity function for nearby stars within a distance of 10 pc. For observations in the range 1000–2000 Å absorption by interstellar dust makes it possible to observe stars within a sphere of about 300 pc. If the bandwidth of the spectrum considered is about 100 Å wide, the contribution at different wavelengths is determined by stars of a very narrow class of spectral types. This makes it possible to improve the luminosity function for O–F stars.

7. Subcosmic Rays

Non-resonant charge-exchange interactions of the proton component of subcosmic rays

$$P + H \rightarrow H(2p) + P$$

and excitations due to fast neutral atoms with neutral interstellar hydrogen

$$H + H \to H(2p) + H$$

result in the formation of excited atoms having large velocities. These atoms radiate Lyman α quanta with wavelength shifted because of the Doppler effect. In the region of interest to us (1225–1340 Å) radiation comes from particles with energy between 26 keV and several MeV. The main contribution results from particles with energy less than 100 keV. This is because the cross-section for resonance and charge-exchange interactions decreases rapidly for energies greater than 100 keV. If we know an upper limit to the intensity of ultraviolet radiation, it is easy to give an upper limit to the density and energy density of subcosmic rays. These calculations were carried out by us (Kurt and Sunyaev, 1968) and we found

$$n(E) < \frac{10^{-9}}{n_H} \text{ cm}^{-3} \text{ keV}^{-1}$$

$$W < 5 \times 10^{-3} \text{ eV cm}^{-3}.$$

The above rough estimates can be improved by narrowing the spectral interval because $E = 100$ keV corresponds to a maximum bandwidth of 17 Å. In the existing measurements the bandwidth was 10 times broader. Observations with small angular resolution perpendicular to the Galactic plane can exclude the background due to stars. It would also be very interesting to study the spectra of the background for $\lambda < 1216$ Å, where the radiation of stars is very weak.

Acknowledgement

We are very grateful to M. Longair for his considerable help.

References

Bowyer, C. S., Field, G. B., and Mack, J. E.: 1968, *Nature* **217**, 32.
Gunn, J. E. and Peterson, B. A.: 1965, *Astrophys. J.* **142**, 1633.
Henry, R. C., Fritz, G., Meekins, J. E., Friedman, H., and Byram, E. T.: 1968, *Astrophys. J.* **153**, L11.
Kurt, V. G. and Dostovalov, S. B.: 1968, *Nature* **217**, 219.
Kurt, V. G. and Sunyaev, R. A.: 1967a, *JETP Letters* **5**, 299.
Kurt, V. G. and Sunyaev, R. A.: 1967b, *Cosmic Research* [Kosmicheskie issledovanija] **5**, 496.
Kurt, V. G. and Sunyaev, R. A.: 1967c, *Astron. J. USSR* **44**, 1157.
Kurt, V. G. and Sunyaev, R. A.: 1968, *JETP Letters* **7**, 215.
Lillie, C. F.: 1969, *Bull. Am. Astron. Soc.* **1**, 132.
Partridge, R. B. and Peebles, P. J. E.: 1967, *Astrophys. J.* **177**, 868.
Pikel'ner, S. B.: 1967, *Astron. J. USSR* **44**, 1915.
Roach, F. F. and Smith, L. L.: 1968, *Geophys. J.* **15**, 227.
Sunyaev, R. A.: 1969a, *Astrophys. Letters* **3**, 33.
Sunyaev, R. A.: 1969b, *Astron. J. USSR* **46**, 929.

LYMAN-α RADIATION FROM NEBULAR OBJECTS

N. PANAGIA and M. FULCHIGNONI

Laboratorio di Astrofisica, Frascati, Italy

Abstract. For a simplified model of a gaseous nebula, that is spherical, isothermal, homogeneous and composed of pure hydrogen, the amount of Lyman-α radiation emitted from the surface of the nebula is evaluated for various physical conditions. The problem of observability of the Lyman-α emission line is examined, taking into account absorption by dust and by interstellar neutral hydrogen.

In this note preliminary results of a theoretical study of the Lyman-α radiation from nebular objects are presented.

The great difficulty in observing the Lyman-α line is that the flux is much reduced by interstellar absorption, due not only to interstellar dust but also to neutral hydrogen.

For the present analysis we have assumed a simple model nebula, namely one that is spherical, isothermal, homogeneous and composed of pure hydrogen. This model corresponds to that of Gerola and Panagia (1969), so that the calculations of level populations were performed using the same procedure.

The Lyman-α line intensity radiated from the nebula is obtained by integrating the radiative transfer equation in the direction of the line of sight throughout the whole sphere. Assuming complete redistribution in frequencies, the transfer equation along a line can be written as:

$$\frac{dI_\nu}{ds} = j_\nu - k_\nu I_\nu,$$
(1)

where

$$j_\nu = j_0 H(a, x); \qquad k_\nu = k_0 H(a, x).$$
(2)

In (2) one has:

$$x = \frac{\nu - \nu_0}{\Delta\nu_{Doppler}}; \qquad a = \frac{A(2p, 1s)}{4\pi\Delta\nu_{Doppler}},$$
(3)

where ν_0 indicates the central frequency of the line and $A(2p, 1s)$ is the Einstein coefficient of spontaneous emission for the transition $2p - 1s$ which generates the Lyman-α line; $H(a, x)$ represents the well known Voigt profile (Hummer, 1965). Clearly for the adopted model, j_0 and k_0 do not depend on the position into the nebula.

Integrating (1) throughout the sphere, along the line of sight, one obtains:

$$I(\nu)\, d\nu = 2\pi R^2 \frac{j_0}{k_0}\, p(x)\, d\nu$$
(4)

$$p(x) = \frac{1}{2} + \frac{\exp[-2\tau(x)]\,[2\tau(x) + 1] - 1}{[2\tau(x)]^2}$$
(5)

Houziaux and Butler (eds.), Ultraviolet Stellar Spectra and Ground-Based Observations, 349–354.

$$\tau(x) = \tau_0 H(a, x); \qquad \tau_0 = k_0 R,$$ (6)

where R is the radius of the nebula.

The profile of the emergent line is then determined by the function $p(x)$. In Figure 1 are plotted some profiles (dashed lines) as functions of x, corresponding to various values of τ_0 and for $a = 4.72 \times 10^{-4}$. The lines have a flat and broad core with constant value $\frac{1}{2}$, which extends approximately until $\tau(x)$ becomes less than 4; then for a given value of a, the width of the core is determined by the value of τ_0.

In these calculations the continuous radiation was not included explicitly; however, some evaluations of this factor were made. The results were that, in nebular conditions, the contribution of the continuous emission is always negligible in a large region of the spectrum centred on the Lyman-α line.

To take into account interstellar absorption, it is still possible to use Equation (1), referred to interstellar conditions, putting:

$$j_v = 0$$ (7)
$$k_v = k_{0i} H(a_i, x_i) + k_{\text{dust}},$$ (8)

where the subscript 'i' denotes quantities referred to interstellar conditions. The quantity x_i is simply related to x as follows from the definition (3a):

$$x_i = \left(\frac{T_{\text{neb}}}{T_i}\right)^{1/2} \cdot x.$$ (9)

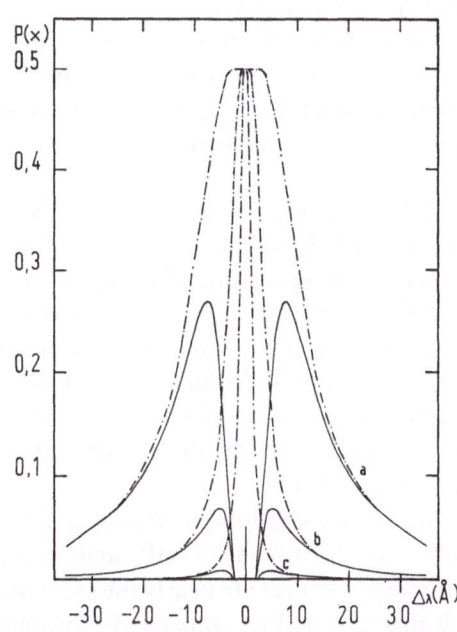

Fig. 1.   Solid lines show profile of the Lyman-α line after absorption by a column of neutral hydrogen of 5.4×10^{20} atoms cm^{-2} at a kinetic temperature of 110 K. The curves refer to nebulae with $T_e = 10^4$ K and (a) $\tau_0 = 10^8$; (b) $\tau_0 = 10^7$; (c) $\tau_0 = 10^6$ respectively. For comparison the emergent profiles (broken lines) are shown for the same cases.

Finally the flux per unit wavelength that is received outside the atmosphere is given by:

$$F(\lambda)\,d\lambda = \frac{c}{\lambda^2}\frac{I(\nu)}{4\pi\,d^2}\exp\left[-\tau_i(x) - 0.921\,A_\lambda\,d_{kpc}\right]d\lambda$$

$$= 4.16 \times 10^{18}\left(\frac{R}{d}\right)^2\frac{N_{2p}}{N_{1s}}\,p(x)\exp\left[-\tau_i(x) - 0.921\,A_\lambda\,d_{kpc}\right]d\lambda,$$

$$\tag{10}$$

where the term $0.921\,A\,d_{kpc}$ in the exponential factor accounts for the absorption due to dust, A being monochromatic extinction in magnitudes per pc and d_{pc} the distance in pc. The factor $(4\pi d^2)^{-1}$ represents the geometrical dilution of the radiation.

In Figure 1 some profiles of the received flux of Lyman-α radiation are shown (solid lines): they correspond to the fluxes after absorption by a column of neutral hydrogen of 5.4×10^{20} atoms cm^{-2} with a kinetic temperature of 110K. It is evident that the greater τ_0, the broader the emergent line from the nebula and then the total flux will be less affected by interstellar hydrogen absorption.

Of course, the amount of radiation received, for given conditions of interstellar medium, depends not only upon the received profile, which is determined by τ_0, but also essentially on the ratio of the populations of the levels $2p$ and $1s$ of hydrogen. These are determined not only by self-absorption or Lyman-α radiation within the nebula, but also by other physical parameters, particularly the electron temperature and density and the mean degree of ionization.

Examining the equilibrium equation for the $2p$ level and the transfer problem through the interstellar medium, one finds that in the range of conditions typical of planetary nebulae, for given values of R, N_e, T_e, that is for a nearly constant behaviour of the optical spectrum, the lower the degree of ionization of hydrogen the greater the flux of Lyman-α radiation.

In Figure 2 the fluxes of the Lyman-α line are plotted corresponding to a nebula of four sq sec of area on the sky with $T_e = 10^4$ K, $N_e = 10^4$ cm^{-3}, $R = 10^{17}$ cm and which is 1 kpc from the Sun. A kinetic temperature of 110K and a mean density of interstellar neutral hydrogen of 0.2 atoms cm^{-3} are assumed. A mean extinction by dust of 3.75 magnitudes per kpc was adopted; this value is consistent with a reasonable extrapolation of Stecher's data (1965), having assumed 3 for the ratio of visual absorption to color excess and an extinction of 1 magnitude par kpc in the visual.

The abscissa shows wavelengths in Å and the ordinate gives the logarithm of the flux, in units of 10^{-11} ergs cm^{-2} s^{-1} Å^{-1}.

The curves correspond, from top to bottom, to degrees of ionization from 5.0×10^{-1} to 5.0×10^{-4} (corresponding to τ_0 between 2.9×10^7 and 2.9×10^4). One can see that the separations between successive curves are nearly the same on the logarithmic scale: this means that the flux is proportional to some nearly constant power of the degree of ionization which in this case is about the square root. Indeed these profiles all correspond to cases in which ionization and excitation of nebular hydrogen are due essentially to radiation from the central star.

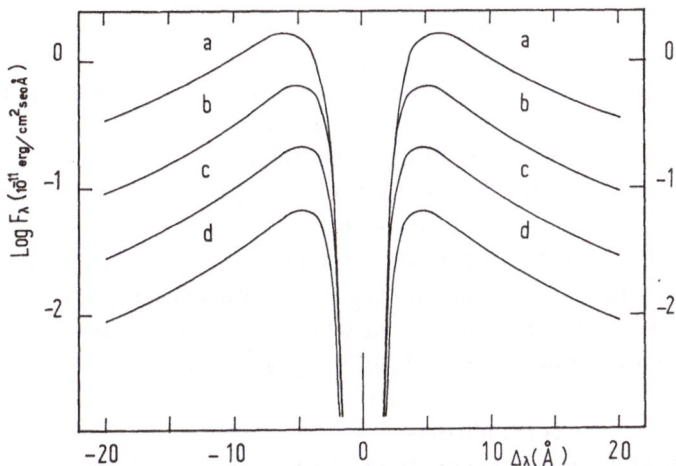

Fig. 2. Fluxes of Lyman-α radiation received from an area of a nebula subtending four square second of arc with $T_e = 10^4$ K, $N_e = 10^4$ cm^{-3}, $R = 10^{17}$ cm which is 1 kpc from the Sun. Mean density of neutral hydrogen of 0.2 atoms cm^{-3} with kinetic temperature of 110K, and mean dust extinction of 3.75 magnitudes per kpc are assumed. The abscissa shows wavelengths in Å and the ordinate the logarithm of the flux in units of 10^{-11} ergs cm^{-2} s^{-1} Å^{-1}. The curves correspond to:

(a) $N_1/N_p = 5.0 \times 10^{-1}$, $\tau_0 = 2.9 \times 10^7$;
(b) $N_1/N_p = 5.0 \times 10^{-2}$, $\tau_0 = 2.9 \times 10^6$;
(c) $N_1/N_p = 5.0 \times 10^{-3}$, $\tau_0 = 2.9 \times 10^5$;
(d) $N_1/N_p = 5.0 \times 10^{-4}$, $\tau_0 = 2.9 \times 10^4$.

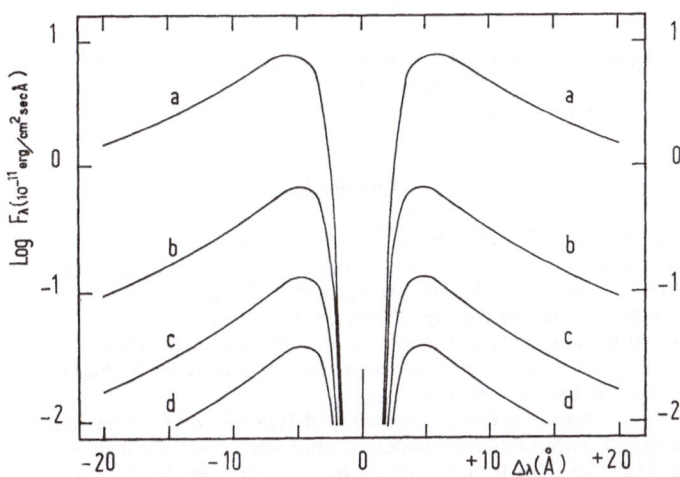

Fig. 3. Fluxes of Lyman-α radiation received from an area of a nebula subtending four square second of arc with $T_e = 1.4 \times 10^4$ K; the other parameters and the units of the plots are the same as for Figure 2. The same assumptions on the interstellar matter are also made. The curves correspond to:

(a) $N_1/N_p = 3.6 \times 10^{-1}$, $\tau_0 = 1.8 \times 10^7$;
(b) $N_1/N_p = 3.6 \times 10^{-2}$, $\tau_0 = 1.8 \times 10^6$;
(c) $N_1/N_p = 3.6 \times 10^{-3}$, $\tau_0 = 1.8 \times 10^5$;
(d) $N_1/N_p = 3.6 \times 10^{-4}$, $\tau_0 = 1.8 \times 10^4$.

Incidentally the two lower curves are representative of the physical conditions of planetary nebulae.

In Figure 3 the fluxes of the Lyman-α line received are shown for the case of a nebula with $T_e = 1.4 \times 10^4$ K and with the other parameters the same. Identical conditions of interstellar matter have also been considered.

The curves correspond, from top to bottom, to degrees of ionization from 3.6×10^{-1} to 3.6×10^{-4}, corresponding to τ_0 from 1.8×10^7 to 1.8×10^4.

It is to be noted that the separation of the upper curve from the one immediately beneath is greater than the separation between the other successive curves. In fact, the upper curve corresponds to a nebula in which collisions are responsible for about 10% of the total ionization, whereas the other profiles correspond to a nearly radiative situation.

Summarizing, we can conclude that:

(1) In suitable but not severe conditions the Lyman-α emission line can be detected, although it suffers considerable interstellar absorption;

(2) In the range of physical conditions typical of planetary nebulae, the intensity is very sensitive to the degree of ionization; therefore measurements of Lyman-α radiation can give much more accurate determinations of the true degree of ionization of hydrogen than those which can be obtained from the ground-based optical spectrum;

(3) A larger Lyman-α emission by low excitation nebulae with respect to other nebulae is to be expected.

References

Gerola, H. and Panagia, N.: 1969, Rapp. Int. 69/17, Laboratorio di Astrofisica, Frascati.
Hummer, D. G.: 1965, *Mem. Roy. Astron. Soc.* **70**, 1.
Stecher, T. P.: 1965, *Astrophys. J.* **142**, 1683.

Discussion

Sunyaev: Have you the estimation of Hα emission from nebulae?

Panagia: I have no detailed results corresponding to the same conditions for which I have shown Lyman-α line profiles. These calculations are right now being carried out in collaboration with H. Gerola of Buenos-Aires University. However I can say that, for given values of radius electronic temperature and density of the nebula, in collisional cases Hα emission is greater than in radiative cases by a factor twenty, more or less (see for instance R. A. R. Parker, 1964, *Astrophys. J.* **138**, 208), just as well as Lyman-α radiation.

Sunyaev: The observations of profile and intensity of Hα emission from nebulae and the Milky Way can give us the choice between different models of observed α-emission from the Milky Way. I think Courtès (France) and Tscheglor from Sternberg Astronomical Institute can find the answer to this question.

Courtès: The width of Hα in the classical galactic nebulae is of the order of 0.4 Å. There exists in the inner part of the galaxy (Sagittarius clouds) a faint extended general emission which is broadened by differential rotation but this emission is not very broad, 1 Å approximately.

We find the same order of width in the central disks of M33 and the Large Magellanic Cloud; some other local group galaxies show a similar Hα emission.

The emission suggested by Sunyaev is perhaps observable in the periphery of the Milky Way if we try to look in the directions of the largest geometrical depth.

Of course with poor resolution nebulae spectrographs it is difficult to select all these emissions from the Hα geocoronal emission, but with the use of the Pérot-Fabry interferometer it is easy to separate the general galactic emission from the geocoronal Hα emission owing to the difference of their profile.

Hekela: This is interesting theoretical work but has a lack. It is not possible to compare this procedure with observation and check it. I suggest you to solve it in a little bit difficult manner together with equations of ionization structure.

Panagia: You are right; however I have presented here only preliminary results, using a very simplified model, to give some evaluations of the expected flux of the Lyman-α line. On the other hand it is easy to see that the values I obtained are lower limits for the true flux: in fact, if ionization structure is taken into account, the emergent line has not longer flat core and monotonically decreasing wings, but it has in the center a relative minimum and presents some symmetric prominent feature before to decrease (C. M. Walmsley and W. C. Mathews 1969, *Astrophys. J.* **155**, 57). In this case clearly, being the same as the total energy emitted in the line, the received flux, after absorption of interstellar neutral hydrogen, is greater than that I have shown here.

SPACE AND GROUND-BASED STELLAR
SPECTROPHOTOMETRY; A SUMMARY

C. DE JAGER

The Astronomical Institute, Utrecht, The Netherlands

1. Introduction

One of the most important results of the present Symposium is the obvious necessity of close cooperation between ground-based and UV stellar spectral observers. Either of the two observational techniques yields only part of the total information about the stars and the interstellar medium; together they enable the scientist to construct the imposing image of modern stellar spectroscopy that we have seen growing during this Symposium.

2. Calibration

It is the purpose of all objective spectrophotometry to express the instrumental output as photon fluxes, either on a relative or an absolute scale. Absolute photometry can only be done through a thorough calibration of the instrument, i.e. the telescope, the spectrograph and the detector. For the calibration one needs a laboratory source or an in-flight calibration source; the latter is often a secondary or tertiary standard, calibrated by means of a well tested laboratory source.

As stated by Boldt, the fundamental problem in the field of absolute intensity calibration is the development of primary standards. Such a standard should be really absolute, constant over long periods, and, if possible, easy to manipulate. Up to now a few primary standards have been developed, which should be able to guarantee an accuracy of the order of $\pm 10\%$. Therefore, the development of other primary standards does not seem of high interest. However, it is highly important to inter-compare the available primary standards with each other in order to find out whether they really agree within the above error limit of 10%.

Boldt described three calibration methods based on laboratory sources: the branching ratio method, the synchrotron radiation method, and the blackbody radiation method. Bless described a synchrotron source with electrons of energies up to 240 MeV in a vacuum of 10^{-9} Torr; the electrons can be distinguished individually and have lifetimes of hours, and thus provide a stable source of known flux.

For relative in-flight calibration a Čerenkov source seems to be the best; theoretically the output should be constant with time and known in absolute units. As another secondary standard sodium salicylate was mentioned; it transforms photons from energies below 3000 Å to photons of energies above 4000 Å with an efficiency which is nearly independent of wavelength of the ultraviolet radiation below 3000 Å.

Of certain detectors, like ionization chambers and sodium salicylate, the efficiency should be known *a priori* and be constant in time. Carruthers mentioned that unity

Houziaux and Butler (eds.), Ultraviolet Stellar Spectra and Ground-Based Observations, 355–361.

gain ion chambers, particularly the windowless free-flow variety satisfy this demand fairly well. However, it is well-known and was stressed again by Davis, that the efficiency of photon counters actually may change rapidly and decrease after days, which necessitates regular in-flight calibration.

A remarkable kind of in-flight calibration was mentioned by Severny: occasionally one may use the Sun through its reflection against a satellite with a known UV albedo, at a known distance.

3. Interstellar Extinction

The amount of interstellar extinction can only be determined by comparing the intensities in spectra of two stars of the same spectral type, one being nearby and the other far away. The main problem is how certain one may be that two stars which are seemingly similar in the visual spectral region are still similar in the ultraviolet. A few estimates of the interstellar extinction have been made by comparing a nearby main sequence star with a distant supergiant; this seems a dangerous procedure, but OAO results seem to indicate that early-type stars with the same visual colors and spectral types are indeed similar in the 3000–1000 Å region.

The importance of the new UV stellar spectral observations is, of course, that the region over which extinction is determined extends no longer only from $\lambda^{-1} = 1$ to 3 μ^{-1}, but now up to 9 μ^{-1}. This proves to be important for the determination of properties of the grains.

Stecher showed a fine extinction curve for ζ Per, while Bless and Savage showed extinction curves for ten pairs of stars. There is variation from one star to the other but the general trend is that a hump exists near 2200 Å. The character of the extinction curve for very short wavelengths, around 1100 Å, is not yet certain; most people assume a continued rise for great λ^{-1}, after the peak at $\lambda^{-1} \approx 5 \mu^{-1}$, but Carruthers showed that in special cases, as in the Orion Nebula region, the extinction law can be drastically different from that found generally applicable, which has also been found true at the longer wavelengths accessible from the ground.

With regard to the interpretation of these results it should be mentioned that the peak in the extinction curve was predicted earlier by theorists, but the wavelength was expected to be longer (near $\lambda^{-1} \approx 3$ or 4 μ^{-1}). This previous broad hump (at $\lambda^{-1} \approx 3.5$) was interpreted by assuming the grains to be large enough that extinction saturation was reached at that wavelength. The narrow new peak near $\lambda^{-1} \approx 5.5 \mu^{-1}$ would appear to be due rather to absorption characteristics of the grain material – either an absorption band or the beginning of an absorption edge. This hump would be obscured if the particles were too large. However, the principal observed indication that the particles are smaller than assumed previously is that the extinction curve is continuing its rise beyond $\lambda^{-1} \approx 6 \mu^{-1}$, implying that the *saturation hump* occurs very far out (assuming of course that the extinction at the short wavelengths is still due to the grains).

The composition of the grains is not yet certain. Various models were discussed and seem to be able to explain the observed extinction curves equally well: the graphite core ice mantle (Greenberg and Shah), pure ice or silicate grains, and silicate core-ice mantle particles (Greenberg), a mixture of graphite and silicate particles with or without mantles of ice or solid H_2 (Wickramasinghe). It was suggested that the observed peak in the extinction curve would correspond to a transition to the conduction band of the π-electron in graphite; it seems too narrow for ice (Stecher, Wickramasinghe).

Solomon discussed the possibility of the occurrence of interstellar hydrogen molecules, which might form through the intermediacy of interstellar grains as a catalyser; the resulting expected molecular densities were depressingly low (see also Section 8). The suggestion that solid H_2 could be expected on the grains was questioned by Van de Hulst and Greenberg; they wondered whether the grain temperatures could be low enough to maintain a solid H_2 mantle.

4. Absolute Intensities

For a few stars absolute intensities have been determined; Gingerich and Latham, and Morton presented model-atmosphere calculations of stellar spectral intensity distributions which were compared with Stecher's observations of Sirius and ζ Puppis. The determination of the stellar diameters by Hanbury Brown and his colleagues then enables the effective temperature of the stars to be determined directly.

Generally the observed intensities in the far ultraviolet (around 1200 Å and shorter) are somewhat lower than those predicted by the models; for the main sequence early-type stars the discrepancy is small and may be due to the agglomeration of unresolved Fraunhofer lines in the spectrum, not accounted for in the model calculations. The influence of carbon absorption competes with Lyman α in determining the shape of the stellar intensity curve near 1100 Å.

In the case of bright giant, and supergiant A and F stars the computed ultraviolet intensities are systematically larger than the observed values; in these cases expanding shells of gas may be involved.

There are observational as well as theoretical problems with regard to the absolute intensity in the UV. With regard to the reality of the UV deficiencies various authors stressed that one should be extremely careful in ascertaining that absolute calibration has been done correctly. It is only necessary to recall that the *solar* continuous spectrum in the visual spectral range has been determined now by many authors in long, quiet periods of observations in ideal observing conditions and that even now differences of 20% between one author and the other still occur. Therefore, one should not be surprised about differences of factors of 2 between one author and another in the absolute calibration of UV stellar spectra.

From the theoretical point of view Gingerich noted that the introduction of convective models (should these exist in the range of hot stars!) would not help substantially in changing the spectral continuum intensities. Miss Underhill mentioned that many of the Fraunhofer lines are formed in outer stellar shells with excitation conditions that may strongly deviate from those in the stellar photosphere.

5. Observations of Stellar Spectra

Davis showed TV pictures of star fields obtained by the Celescope experiment in the Orbiting Astronomical Observatory. The observations of line spectra were reviewed by Wilson; the following groups have now obtained line spectra of stars in the UV: Jenkins-Morton (Princeton); Carruthers (Naval Research Laboratory); Stecher (Goddard) and Smith (Goddard); the Wisconsin group with the OAO.

Stecher showed spectrophotometer runs of several stellar spectra between 1150 and 3200 Å with a resolution of 10 Å.

Bless had OAO spectra of similar resolution between 1100 and 1800 Å and showed a slide giving the variation of the strong spectral features with spectral type for main sequence B stars.

Smith showed a photographic spectrum of ζ Puppis extending from 921 to 1360 Å with a resolution of 0.8 Å. There are many weak lines which exhibit a range of ionization (like N III, N IV, N V) and excitation (0–76 eV) essentially in agreement with visual spectral observations of similar type stars. One may assume these lines to arise in a stationary part of the stellar atmosphere. In addition strong P Cygni type profiles were observed for resonance transitions in O VI, N III and S VI. These lines show mean radial velocities between 1000 and 2000 km s^{-1}, and may originate in a circumstellar envelope.

Also other observers stressed the important point that many giant stars show strong lines with P Cygni profiles in the rocket UV. These lines indicate outstreaming motion, at velocities up to several thousand km s^{-1}.

6. Outstreaming Motions

Observational results on outstreaming motions were discussed by Feast and Hutchings. Mass loss is not generally detectable in stars but is evident in very luminous early-type stars in the visible as well as in the UV spectrum; in actual fact for stars with absolute visual magnitudes brighter than -6 mass loss is a normal feature. The P Cygni profiles of far UV spectral lines in giant and supergiant stars were mentioned in Section 5. Many UV velocities are larger than velocities determined in the visual spectral range (Hutchings), which is, of course, only a question of having the proper kind of lines to observe the moving layers; the lines with P Cygni profiles occur mainly in the middle UV. In cool red giants there is clear evidence for chromospheric activity (Deutsch), as is shown by observations of the Balmer lines, the Ca II H and K lines, and the NaD lines. The time scale in these changes is of the order of a few weeks or months. For the K line in α Tauri (K5 III) a quasi period has been found.

The theory of these lines is difficult since existing stellar photospheric models are not appropriate in the range of heights where the strong lines are formed (Underhill); in some cases the opacity in strong lines may be 10^6 to 10^8 times the values for the continuum; obviously these lines should be interpreted by non-LTE physics. Also fluorescence phenomena generated by UV lines in molecules might play a part (Swings and Swings).

The explanation of the outstreaming motion is perhaps due to radiation pressure exerted by the many strong ultraviolet lines occurring between 900 and 3000 Å. A detailed discussion of this problem was given by Solomon and Lucy, who discussed the hydrodynamic problem of flow and gave theoretical results; a mass loss of a few times 10^{-8} solar masses per year was predicted for O and B stars of high luminosity. A semi-empirical model of an expanding envelope based on Morton's UV spectra in Orion was basically in agreement with this picture (Hutchings).

7. The Sun

Pottasch showed that observations of the solar continuous spectrum in the ultraviolet spectral range may yield a model of the photospheric and chromospheric regions; this is obviously so because the continuous absorption coefficient increases strongly towards shorter wavelengths. This method is even more powerful than the use of center-to-limb observations of the continuous spectrum since the average depth of emission of ultraviolet observations varies over a large range between the visual and the far UV wavelengths. It seems appropriate to apply this method to stellar spectra also. However, one may doubt whether the method would work for hot stars where the continuous extinction is for a great part due to Thompson scattering by free electrons and is virtually constant as a function of the wavelength. The behavior of the Al I discontinuity at 2080 Å was discussed by Bonnet and Gingerich; according to the former the photo-ionization of aluminum must undoubtedly cause an effect on the solar continuous spectrum but quantitatively the discontinuity can only be explained with a fairly large Al abundance, by the introduction of non-LTE physics with regard to the ground level population of Al I, and by introducing inhomogeneities in the transition zone photosphere-chromosphere.

On the other hand Gingerich showed that the new Harvard-Smithsonian Reference model of the photosphere and low chromosphere is able to account for the aluminum discontinuity at the centre of the disc with an LTE model and a 'normal' Al abundance. However, the limb behavior cannot be explained by that model.

Vial and Lemaire tried to use existing chromospheric models to explain the observed limb-behavior of the Lyman α line and the Mg II lines and found it hard to obtain agreement with observations. Splendid spectral observations of the Mg II lines at 2800 Å were shown by Lemaire and by Wilson; the latter was the first who obtained interferometric spectra in the rocket UV. The flare-associated behavior of various groups of lines was studied by Prokofiev and co-workers who found a great diversity in their behavior.

8. Interstellar Absorption Lines

The first interstellar Lyman-α absorption data obtained by OAO were presented and compared with rocket data. The OAO data taken with a resolution of about 10 Å are blended with N v at 1230 Å and Si III at 1200 Å, so that it is difficult to find the real equivalent widths of the Lyman α line. The OAO data yield equivalent widths, about two to three times those obtained by rocket spectrographs, where the resolution was generally better than with OAO. This large factor seems not to be due entirely to the above mentioned blending, but may be due to the choice of the height of the continuum level. The problem of where to draw the continuum is not yet solved satisfactorily.

A comparison of the interstellar neutral hydrogen density derived from 21 cm data and from Lyman α is extremely interesting. If one uses the equivalent widths estimated from high resolution spectra, a discrepancy seems to occur between these two data, which is particularly apparent in the Orion complex, but which is also evident for some other parts of the Milky Way. In the Scorpius region, however, the Lyman α data appear in general agreement with the 21 cm data.

Two suggestions were put forward for explaining the apparent weakness of the Lyman α interstellar absorption lines in the Orion regions. Jenkins considered the effect of neutral hydrogen gas clouds with diameters smaller than 0.1 parsec, a mass of $5\ M_\odot$, and temperatures between 100 and 1000 K; these clouds would have to be distributed in such a way that up to now all stellar observations have missed the clouds, while the 21 cm observations, taken with a broader detecting angle should integrate over the clouds and the region in between.

The other suggestion is that neutral hydrogen was blown away around the Orion stars, which would certainly reduce the circumstellar component of Lyman α interstellar absorption. This hypothesis may be supported by the observation that there exists a reasonable correlation between equivalent widths of interstellar Lyman α and NaD lines.

Other possibly observable interstellar lines, not yet within the reach of present equipment, but perhaps observable when at a later time higher resolution spectrographs exist, are lines of C, N, Al, Si, P and S; Herbig listed 23 lines of these elements. A problem is the theoretical prediction of the line strengths since several of these lines should be due to highly ionized atoms; the computation of the degree of ionization in interstellar space is difficult. It is remarkable that the Ca/Na abundance ratio in interstellar space seems about 10^3 times smaller than the normal value for most stars. Various suggestions to explain this difference were listed by Herbig.

9. Interstellar Emission Lines

Observations of the interstellar Lyman α emission were reported by Sunyaev (Venera flights) and by Barth (Mariner flights). The spacecraft performed several rolls when on their way to Venus and could thus define the distribution of Lyman α emission along various tracks at right angles to the Milky Way. There are several contributions: the interplanetary component, the galactic component, perhaps mainly due to a circumsolar region with a radius of about 50 pc, and partly also the Lyman α component from the H I regions; theoretical calculations of that latter aspect were shown by Panagia and Fulchignoni. However, it is far from clear how the above three components contribute to the observed Lyman α distribution.

Severny reported Cosmos satellite observations of the emission of the Milky Way in the near UV (about 2700 Å). The field of view was large, and it was not possible to separate the effect of the many faint stars from that of a true diffuse background.

Further, hypothetical, contributions to the UV emission flux were summarized by Sunyaev: low energy cosmic rays could perhaps contribute. Furthermore, there could also be an inter- and an extragalactic component, due to quasars and to young galaxies. All this is highly fascinating and makes one look forward to further observations in this spectral range.

Acknowledgements

A draft of this paper was sent to a number of participants of the Symposium, who by their stimulating criticism and helpful remarks greatly helped to improve its quality. Their contributions are very thankfully mentioned.